T0331626

OPTIMAL MASS TRANSPORT ON EUCLIDEAN SPACES

Over the past three decades, optimal mass transport has emerged as an active field with wide-ranging connections to the calculus of variations, partial differential equations (PDEs), and geometric analysis. This graduate-level introduction covers the field's theoretical foundation and key ideas in applications. By focusing on optimal mass transport problems in a Euclidean setting, the book is able to introduce concepts in a gradual, accessible way with minimal prerequisites, while remaining technically and conceptually complete. Working in a familiar context will help readers build geometric intuition quickly and give them a strong foundation in the subject. This book explores the relation between the Monge and Kantorovich transport problems, solving the former for both the linear transport cost (which is important in geometric applications) and the quadratic transport cost (which is central in PDE applications), starting from the solution of the latter for arbitrary transport costs.

Francesco Maggi is Professor of Mathematics at the University of Texas at Austin. His research interests include the calculus of variations, partial differential equations, and optimal mass transport. He is the author of *Sets of Finite Perimeter and Geometric Variational Problems: An Introduction to Geometric Measure Theory,* published by Cambridge University Press.

Optimal Mass Transport on Euclidean Spaces

FRANCESCO MAGGI

University of Texas at Austin

Shaftesbury Road, Cambridge CB2 8EA, United Kingdom

One Liberty Plaza, 20th Floor, New York, NY 10006, USA

477 Williamstown Road, Port Melbourne, VIC 3207, Australia

314–321, 3rd Floor, Plot 3, Splendor Forum, Jasola District Centre, New Delhi – 110025, India

103 Penang Road, #05–06/07, Visioncrest Commercial, Singapore 238467

Cambridge University Press is part of Cambridge University Press & Assessment, a department of the University of Cambridge.

We share the University's mission to contribute to society through the pursuit of education, learning and research at the highest international levels of excellence.

www.cambridge.org
Information on this title: www.cambridge.org/9781009179706

DOI: 10.1017/9781009179713

First published 2023

A catalogue record for this publication is available from the British Library

Library of Congress Cataloging-in-Publication Data
Names: Maggi, Francesco, 1978– author.
Title: Optimal mass transport on Euclidean spaces / Francesco Maggi.
Description: Cambridge ; New York, NY : Cambridge University Press, 2023. |
Series: Cambridge studies in advanced mathematics ; 207 | Includes
bibliographical references and index.
Identifiers: LCCN 2023016425 | ISBN 9781009179706 (hardback) | ISBN
9781009179713 (ebook)
Subjects: LCSH: Transport theory – Mathematical models. | Mass transfer. |
Generalized spaces.
Classification: LCC QC175.2 .M34 2023 | DDC 530.13/8–dc23/eng/20230817
LC record available at https://lccn.loc.gov/2023016425

ISBN 978-1-009-17970-6 Hardback

"Francesco Maggi's book is a detailed and extremely well written explanation of the fascinating theory of Monge–Kantorovich optimal mass transfer. I especially recommend Part IV's discussion of the 'linear' cost problem and its subtle mathematical resolution."

– Lawrence C. Evans, UC Berkeley

"Over the last three decades, optimal transport has revolutionized the mathematical analysis of inequalities, differential equations, dynamical systems, and their applications to physics, economics, and computer science. By exposing the interplay between the discrete and Euclidean settings, Maggi's book makes this development uniquely accessible to advanced undergraduates and mathematical researchers with a minimum of prerequisites. It includes the first textbook accounts of the localization technique known as needle decomposition and its solution to Monge's centuries old cutting and filling problem (1781). This book will be an indispensable tool for advanced undergraduates and mathematical researchers alike."

– Robert McCann, University of Toronto

to Vicki

Contents

Preface

In a hypothetical hierarchy of mathematical theories, the theory of optimal mass transport (OMT hereafter) lies at quite a fundamental level, yielding a formidable descriptive power in very general settings. The most striking example in this direction is the theory of curvature-dimension conditions, which exploits OMT to construct fine analytic and geometric tools in ambient spaces as general as metric spaces endowed with a measure. The Bourbakist aesthetics would thus demand OMT to be presented in the greatest possible generality from the onset, narrowing the scope of the theory only when strictly necessary. In contrast, this book stems from the pedagogically more pragmatic viewpoint that many key features of OMT (and of its applications) already appear in full focus when working in the simplest ambient space, the Euclidean space $\mathbb{R}^n$, and with the simplest transport costs per unit mass, namely, the "linear" transport cost $c(x, y) = |x - y|$ and the quadratic transport cost $c(x, y) = |x - y|^2$. Readers of this book, who are assumed to be graduate students with an interest in Analysis, should find in these pages sufficient background to start working on research problems involving OMT – especially those involving partial differential equations (PDEs); at the same time, having mastered the basics of the theory in its most intuitive and grounded setting, they should be in an excellent position to study more complete and general accounts on OMT, like those contained in the monographs [Vil03, Vil09, San15, AGS08]. For other introductory treatments that could serve well to the same purpose, see, for example [ABS21, FG21].

The story of OMT began in 1781, that is, in the midst of a founding period for the fields of Analysis and PDE, with the formulation of a *transport problem* by Monge. Examples of famous problems formulated roughly at the same time include the wave equation (1746), the Euler equations (1757), Plateau's problem[1] and the minimal surface equation (1760), and the heat equation and the Navier–Stokes equations (1822). An interesting common trait of all these problems is that the frameworks in which they had been originally formulated have all proved inadequate for their satisfactory solutions. For example, the study of heat and wave equations has stimulated the investigation of Fourier's series, with the corresponding development of functional and harmonic analysis, and of the notion of *distributional* solution. Similarly, the study of the minimal surface equation and of the Plateau's problem has inspired a profound revision of the notion of surface itself, leading to the development of geometric measure theory. The Monge problem has a similar history:[2] an original formulation (essentially intractable), and a modern reformulation, proposed by Kantorovich in the 1940s, which leads to a broader class of problems and to many new questions. The main theme of this book is exploring the relation between the Monge and Kantorovich transport problems, solving the former both for the linear transport cost (the one originally considered by Monge, which is of great importance in geometric applications) and for the quadratic transport cost (which is central in applications to PDE), starting from the solution of the latter for arbitrary transport costs.

The book is divided in four parts, requiring increasing levels of mathematical maturity and technical proficiency at the reader's end. Besides a prerequisite[3] familiarity with the basic theory of Radon measures in $\mathbb{R}^n$, the book is essentially self-contained.

Part I opens with an introduction to the original minimization problem formulated by Monge in terms of **transport maps** and includes a discussion about the intractability of the Monge problem by a direct approach, as well as some basic examples of (sometimes optimal) transport maps (Chapter 1). It then moves to the solution of the **discrete** OMT problem with a generic transport cost $c(x, y)$ (Chapter 2), which serves to introduce in a natural way three key ideas behind Kantorovich's approach to OMT problems: the notions of **transport plan**, ***c*-cyclical monotonicity**, and **Kantorovich duality**. Kantorovich's theory is then presented in Chapter 3, leading to existence and characterization results for optimal transport plans with respect to generic transport costs.

---

[1] The minimization of surface area under a prescribed boundary condition (together with the related minimal surface equation) has been studied by mathematicians at least since the work of Lagrange (1760). The modern consolidated terminology calls this minimization problem "Plateau's problem," although Plateau's contribution was dated almost a century later (1849) and consisted in extensive experimental work on soap films.

[2] For a complete and accurate account on the history of the Monge problem and on the development of OMT, see the bibliographical notes of Villani's treatise [Vil09].

[3] All the relevant terminologies, notations, and required results are summarized in Appendix A, which is for the most part a synopsis of [Mag12, Part I].

The optimal transport *plans* constructed in Kantorovich's theory are more general and flexible objects than the optimal transport *maps* sought by Monge, which explains why solving the Kantorovich problem is way easier than solving the Monge problem. Moreover, a transport map canonically induces a transport plan with the same transport cost, thus leading to the fundamental question: *When are optimal transport plans induced by optimal transport maps?* Parts II and IV provide answers to these questions in the cases of the quadratic and linear transport costs, respectively.

Part II opens with the **Brenier theorem** (Chapter 4), which asserts the existence of an optimal transport map in the Monge problem with quadratic cost under the assumptions that the origin mass distribution is absolutely continuous with respect to the Lebesgue measure and that both the origin and final mass distributions have finite second order moments; moreover, this optimal transport map comes in the form of the *gradient of a convex function*, and is *uniquely* determined, and therefore called *the* **Brenier map** from the origin to the final mass distribution. In Chapter 5, we establish some sharp results on the first order differentiability of convex functions, which we then use in Chapter 6 to prove McCann's remarkable extension of the Brenier theorem – in which the absolute continuity assumption on the origin mass distribution is sharply weakened, and the finiteness of second order moments is entirely dropped. In both the Brenier theorem and the Brenier–McCann theorem, the transport condition is expressed in a measure-theoretic form (see (1.6) in Chapter 1) which is weaker than the "infinitesimal transport condition" originally envisioned by Monge (see (1.1) in Chapter 1). The former implies the latter for transport maps that are Lipschitz continuous and injective, but, unfortunately, both properties are not generally valid for gradients of convex functions. To address this point, in Chapter 7, we provide a detailed analysis of the second order differentiability properties of convex functions, which we then exploit in Chapter 8 to prove the validity of the Monge–Ampère equation for Brenier maps between absolutely continuous distributions of mass. In turn, the latter result is of key technical importance for the applications of quadratic OMT problems to PDE and geometric/functional inequalities.

Part III has two main themes: the first one describes some celebrated applications of Brenier maps to mathematical models of physical interests; the second one introduces the geometric structure of the **Wasserstein space**. That is the space of finite second-moment probability measures $\mathcal{P}_2(\mathbb{R}^n)$ endowed with the distance $\mathbf{W}_2$ defined by taking the square root of the minimum value in the Kantorovich problem with quadratic cost. The close relation between the purely geometrical properties of the Wasserstein space and the inner workings of many mathematical models of basic physical importance is one of the most charming and inspiring traits of OMT theory and definitely the reason why OMT is so relevant for mathematicians with such largely different backgrounds.

We begin the exposition of these ideas in Chapter 9, where we present OMT proofs of two inequalities of paramount geometric and physical importance, namely, the Euclidean isoperimetric inequality and the Sobolev inequality. We then continue in Chapter 10 with the analysis of a model for self-interacting gases at equilibrium. While studying the uniqueness problem for minimizers in this model, we naturally introduce an OMT-based notion of convex combination between probability measures, known as **displacement interpolation**, together with a corresponding class of **displacement convex** "internal energies." The latter include an example of paramount physical importance, namely, the (negative) entropy S of a gas. As a further application of displacement convexity (beyond the uniqueness of equilibria of self-interacting gases), we close Chapter 10 with an OMT proof of another key geometric inequality, the Brunn–Minkowski inequality.

In Chapter 11, we introduce the Wasserstein space $(\mathcal{P}_2(\mathbb{R}^n), W_2)$, build up some geometric intuition on it by a series of examples, prove that it is a complete metric space, and explain how to interpret displacement convexity as *geodesic interpolation* in $(\mathcal{P}_2(\mathbb{R}^n), W_2)$. We then move, in the two subsequent chapters, to illustrate how to interpret many parabolic PDEs as *gradient flows* of displacement convex energies in the Wasserstein space. In Chapter 12, we introduce the notion of gradient flow, discuss why interpreting an evolution equation as a gradient flow is useful, how it is possible that the same evolution equation may be seen as the gradient flow of different energies, and how to construct gradient flows through the **minimizing movements scheme**. Then, in Chapter 13, we exploit the minimizing movements scheme framework to prove that the Fokker–Planck equation (describing the motion of a particle under the action of a chemical potential and of white noise forces due to molecular collisions) can be characterized (when the chemical potential is convex) as the gradient flow of a displacement convex functional on the Wasserstein space, and, as a further application, we derive quantitative rates for convergence to equilibrium in the Fokker–Planck equation.

In Chapter 14, we obtain additional insights about the geometry of the Wasserstein space by looking at the **Euler equations** for the motion of an incompressible fluid. The Euler equations describe the motion of an incompressible fluid in the absence of friction/viscosity and can be characterized as geodesics equations in the (infinite-dimensional) "manifold" M of volume-preserving transformations of a domain. At the same time, geodesics (on a manifold embedded in some Euclidean space) can be characterized by a limiting procedure involving an increasing number of "mid-point projections" in the ambient space: there lies the connection with OMT, since the Brenier theorem allows us to characterize L^2-projections over $\mathcal{M}$ as compositions with Brenier maps. The analysis of the Euler equations serves also to introduce two crucial objects: the **action functional** of an incompressible fluid (time integral of the total kinetic energy) and the **continuity equation** (describing how the Eulerian velocity of the fluid transports its mass density). In Chapter 15, we refer to these objects when formally introducing the concepts of **(Eulerian) velocity of a curve of measures** in $\mathcal{P}_2(\mathbb{R}^n)$ and characterize the Wasserstein distance between the end points of such a curve in terms of minimization of a corresponding action functional. This is the celebrated **Benamou–Brenier formula**, which provides the entry point to understand the "Riemannian" (or "infinitesimally Hilbertian") structure of the Wasserstein space. We only briefly explore the latter direction, by quickly reviewing the notion of gradient induced by such Riemannian structure (Otto's calculus).

Part IV begins with Chapter 16, where a sharp result for the existence of optimal transport maps in dimension one is presented. In Chapter 17, we first introduce the fundamental *disintegration theorem* and then exploit it to give a useful geometric characterization of transport plans induced by transport maps and to prove the equivalence of infima for the Monge and Kantorovich problems when the origin mass distribution is atomless. We then move, in Chapter 18, to construct optimal transport maps for the Monge problem with linear transport cost. We do so by implementing a celebrated argument due to Sudakov, which exploits disintegration theory to reduce the construction of an optimal transport map to the solution of a family of one-dimensional transport problems. The generalization of Sudakov's argument to more general ambient spaces (like Riemannian manifolds or even metric measure spaces) lies at the heart of a powerful method for proving geometric and functional inequalities, known as the "needle decomposition method," and usually formalized in the literature as a "localization theorem." In Chapter 19, we present these ideas in the Euclidean setting. Although this restricted setting does not allow us to present the most interesting applications of the technique itself, its discussion seems, however, sufficient to illustrate several key aspects of the method, thus putting readers in an ideal position to undertake further reading on this important subject.

Having put the focus on the clarity of the mathematical exposition above anything else, the main body of this book contains very few bibliographical references and almost no bibliographical digressions. A set of bibliographical notes has been included in the appendix with the main intent of acknowledging the original papers and references used in the preparation of the book and of pointing students to a few of the many possible further readings.

For comprehensive bibliographies and historical notes on OMT, we refer to [AGS08, Vil09, San15].

This book originates from a short course on OMT I taught at the Universidad Autónoma de Madrid in 2015 at the invitation of Matteo Bonforte, Daniel Faraco, and Juan Luis Vázquez. An expanded version of the lecture notes of that short course formed the initial core of a graduate course I taught at the University of Texas at Austin during the fall of 2020, and whose contents roughly correspond to the first 14 chapters of this book. The remaining chapters have been written in a more advanced style and without the precious feedback generated from class teaching. For this reason, I am very grateful to Fabio Cavalletti, Nicola Gigli, Carlo Nitsch, and Aldo Pratelli who, in reading those final four chapters, have provided me with very insightful comments, spotted subtle problems, and suggested possible solutions that led to some major revisions (and, I think, eventually, to a very nice presentation of some deep and beautiful results!). I would also like to thank Lorenzo Brasco, Kenneth DeMason, Luigi De Pascale, Andrea Mondino, Robin Neumayer, Daniel Restrepo, Filippo Santambrogio, and Daniele Semola for providing me with additional useful comments that improved the correctness and clarity of the text. Finally, I thank Alessio Figalli for his initial encouragement in turning my lecture notes into a book.

With my gratitude to Luigi Ambrosio and Cedric Villani, from whom I have first learned OMT about 20 years ago during courses at the Scuola Normale Superiore di Pisa and at the Mathematisches Forschungsinstitut Oberwolfach, and with my sincere admiration for the many colleagues who have contributed to the discovery of the incredibly beautiful Mathematics contained in this book, I wish readers to find here plenty of motivations, insights, and enjoyment while learning about OMT and preparing themselves for contributing to this theory with their future discoveries!

Notation

a.e., almost everywhere

s.t., such that

w.r.t., with respect to

$\mathbb{N}, \mathbb{Z}, \mathbb{Q}, \mathbb{R}$, natural, integer, rational, and real numbers

$\mathbb{R}^n$, the n-dimensional Euclidean space

$B_r(x)$, open ball in $\mathbb{R}^n$ with center x and radius r (Euclidean metric)

$\mathrm{Int}(E)$, interior of a set $E \subset \mathbb{R}^n$ (Euclidean topology)

$\mathrm{Cl}(E)$, closure of a set $E \subset \mathbb{R}^n$ (Euclidean topology)

∂E, boundary of a set $E \subset \mathbb{R}^n$ (Euclidean topology)

$\mathbb{S}^n$, the n-dimensional sphere in $\mathbb{R}^{n+1}$

ν_E, the outer unit normal $\nu_E : \partial E \to \mathbb{S}^{n-1}$ of a set $E \subset \mathbb{R}^n$ with C^1-boundary

$\mathbb{R}^{n \times m}$, matrices with n-rows and m-columns

A^*, the transpose of a matrix/linear operator

$\mathbb{R}^{n \times n}_{\mathrm{sym}}$, matrices $A \in \mathbb{R}^{n \times n}$, with $A = A^*$

$w \otimes v$ ($v \in \mathbb{R}^n$, $w \in \mathbb{R}^m$), the linear map from $\mathbb{R}^n$ to $\mathbb{R}^m$
  defined by $(w \otimes v)[e] = (v \cdot e) w$

$\mathcal{B}(\mathbb{R}^n)$, the Borel subsets of $\mathbb{R}^n$

$\mathcal{L}^n$, the Lebesgue measure on $\mathbb{R}^n$

$\mathcal{H}^k$, the k-dimensional Hausdorff measure on $\mathbb{R}^n$

$\mathcal{P}(\mathbb{R}^n)$, probability measures on $\mathbb{R}^n$

$\mathcal{P}_{\mathrm{ac}}(\mathbb{R}^n)$, measures in $\mathcal{P}(\mathbb{R}^n)$ that are absolutely continuous w.r.t. $\mathcal{L}^n$

$\mathcal{P}_p(\mathbb{R}^n)$, measures in $\mathcal{P}(\mathbb{R}^n)$ with finite p-moment ($1 \leq p < \infty$)

$\mathcal{P}_{p,\mathrm{ac}}(\mathbb{R}^n) = \mathcal{P}_{\mathrm{ac}}(\mathbb{R}^n) \cap \mathcal{P}_p(\mathbb{R}^n)$

$\mu \ll \nu$, μ is absolutely continuous w.r.t. ν (μ, ν measures on $\mathbb{R}^n$)

$L^p(\mathbb{R}^n)$, p-summable functions w.r.t. $\mathcal{L}^n$ ($1 \leq p \leq \infty$)

$L^p(\mu)$, p-summable functions w.r.t. a Borel measure μ ($1 \leq p \leq \infty$)

$C^0_c(\mathbb{R}^n)$, functions on $\mathbb{R}^n$ that are continuous with compact support

$C_b^0(\mathbb{R}^n)$, functions on $\mathbb{R}^n$ that are continuous and bounded

$\overset{*}{\rightharpoonup}$, weak-star convergence of Radon measures (C_c^0-test functions)

$\overset{n}{\rightharpoonup}$, narrow convergence of Radon measures (C_b^0-test functions)

$C_c^0(\mathbb{R}^n) \otimes C_c^0(\mathbb{R}^m)$, functions $f(x,y) = g(x)\,h(y)$, $(x,y) \in \mathbb{R}^n \times \mathbb{R}^m$,
   with $g \in C_c^0(\mathbb{R}^n)$, $h \in C_c^0(\mathbb{R}^m)$

$C^{0,\alpha}(\mathbb{R}^n)$, α-Hölder continuous functions ($\alpha \in (0,1]$, $C^{0,1} = \mathrm{Lip}$)

$C^{k,\alpha}(\mathbb{R}^n)$, k-times differentiable functions whose kth gradient is in $C^{0,\alpha}$

Df, distributional derivative of $f \in L_{\mathrm{loc}}^1(\mathbb{R}^n)$

∇f, pointwise gradient of f or density of Df w.r.t. $\mathcal{L}^n$

$f\,d\mu$, or $f(x)\,d\mu(x)$, the measure defined
   by the integral of $f \in L_{\mathrm{loc}}^1(\mu)$ w.r.t. μ

$\int_X f(x,y)\,d\mu(x)$, integration w.r.t. μ occurs in the x-variable of $f(x,y)$

$f|_E$, the restriction of f to $E \subset F$ when f is a function defined on F

$\mathbf{p} : X \times Y \to X$, projection on the first factor of $X \times Y$, i.e., $\mathbf{p}(x,y) = x$

$\mathbf{q} : X \times Y \to Y$, projection on the second factor of $X \times Y$, i.e., $\mathbf{q}(x,y) = y$

$\mathbf{id}_X$, the identity map on the set X (X omitted if clear from the context)

$C(a,b,\ldots)$, a generic constant depending only $a,b,\ldots$
   whose value may increase at each subsequent appearance

Disambiguation: The terms "formal" and "formally" are used in this book to indicate the quality of being endowed with full mathematical rigor. This may create confusion since, in the Analysis literature, the term "formal" is sometimes (if not often) used to express the quality of "being presented without a full justification": e.g., expressions like "by a formal integration by parts, taken without discussing the negligibility of boundary terms" or "by a formal argument that does not take into account measurability issues" are quite common in the OMT literature. However, synonyms of "formal" are "official," "legal," "validated," and "authoritative," which definitely point to the quality of possessing full mathematical rigor; hence, the use of "formal" in this book.

PART I

The Kantorovich Problem

1

An Introduction to the Monge Problem

We begin this chapter by introducing two formulations of the Monge problem. The first one is closer to the original one proposed by Monge himself, and is based on a hardly manageable pointwise transport condition (Section 1.1). The second one exploits concepts of modern measure theory to formulate a more flexible transport condition and is the one adopted in the rest of the book (Section 1.2). We then close the chapter by building some intuition on the notions of the transport map and of the *optimal* transport map. This is achieved by looking into a simple "duality-based optimality criterion" (Section 1.3) and by exploiting monotonicity in the construction of transport maps, first in dimension 1 (Section 1.4) and then in higher dimensions (Section 1.5).

1.1 The Original Monge Problem

A modern-language proxy for Monge's original formulation of his eponymous transport problem can be introduced as follows. Given two smooth, nonnegative functions $\rho, \sigma : \mathbb{R}^n \to [0, \infty)$ with the dimensions of mass per unit volume, and assuming that the mass distributions $\rho(x)\, dx$ and $\sigma(y)\, dy$ have the same (unit) total mass, i.e.,

$$\int_{\mathbb{R}^n} \rho(x)\, dx = \int_{\mathbb{R}^n} \sigma(y)\, dy = 1,$$

we consider smooth, injective maps $T : \mathbb{R}^n \to \mathbb{R}^n$ that *transport* $\rho(x)\, dx$ to $\sigma(y)\, dy$, in the sense that the total infinitesimal volume of the origin mass distribution $\rho(x)\, dx$ at x is required to be equal to the total infinitesimal volume of the final mass distribution $\sigma(y)\, dy$ at $y = T(x)$. Since, by the change of variables formulae for smooth injective maps, we have $dy|_{y=T(x)} = |\det \nabla T(x)|\, dx$, the transport constraint takes the form

$$|\det \nabla T(x)|\, \sigma(T(x)) = \rho(x), \qquad \forall x \in \{\rho > 0\}. \tag{1.1}$$

3

We call (1.1) the **pointwise transport condition** from $\rho(x)\,dx$ to $\sigma(y)\,dy$. Taking $|y - x|$ as the transport cost[1] to move a unit mass from x to y, the **original Monge problem** (from $\rho(x)\,dx$ to $\sigma(y)\,dy$) is the minimization problem

$$M = \inf \left\{ \int_{\mathbb{R}^n} |T(x) - x|\,\rho(x)\,dx : T \text{ is smooth, injective and (1.1) holds} \right\}.$$
(1.2)

Since work has the dimensions of force times length, if λ denotes the amount of force per unit mass at our disposal to implement the "instructions" of transport maps, then $\lambda\,M$ is the **minimal amount of work** needed to transport $\rho(x)\,dx$ into $\sigma(y)\,dy$. Since work is a form of (mechanical) energy, (1.2) is, in precise physical terms, an "energy minimization problem."

From a mathematical viewpoint – even from a modern mathematical viewpoint that takes advantage of all sorts of compactness and closure theorems discovered since Monge's time – (1.2) is a *very* challenging minimization problem. Let us consider, for example, the problem of showing the mere existence of a minimizer. The baseline, modern strategy to approach this kind of question, the so-called *Direct Method of the Calculus of Variations*, works as follows. Consider an abstract minimization problem, $m = \inf\{f(x) : x \in X\}$, defined by a function $f : X \to \mathbb{R}$ such that $m \in \mathbb{R}$. By definition of infimum of a set of real numbers, we can consider a *minimizing sequence*[2] for m, that is, a sequence $\{x_j\}_j$ in X such that $f(x_j) \to m$ as $j \to \infty$. Assuming that: (i) there is a notion of convergence in X such that "$\{f(x_j)\}_j$ bounded in $\mathbb{R}$ implies, up to subsequences, that $x_j \to x \in X$," and (ii) "$f(x) \leq \liminf_j f(x_j)$ whenever $x_j \to x$," we conclude that any subsequential limit x of $\{x_j\}_j$ is a minimizer of m, since, using in the order, $x \in X$, properties (i) and (ii), and the minimizing sequence property, we find

$$m \leq f(x) \leq \liminf_j f(x_j) = m.$$

With this method in mind, and back to the original Monge problem (1.2), we assume that M is finite (i.e., we assume the existence of at least one transport map with finite transport cost) and consider a minimizing sequence $\{T_j\}_j$ for (1.9). Thus, $\{T_j\}_j$ is a sequence of smooth and injective maps with

$$\sigma(T_j(x))\,|\det \nabla T_j(x)| = \rho(x),$$
(1.3)

for all $x \in \{\rho > 0\}$ and $j \in \mathbb{N}$, and such that

$$\lim_{j \to \infty} \int_{\mathbb{R}^n} |T_j - x|\,\rho = M < \infty.$$
(1.4)

---

[1] The transport cost $|x - y|$ is commonly named the "linear cost," although evidently $(x, y) \mapsto |x - y|$ is not linear.

[2] Notice that a subsequence of a minimizing sequence is still a minimizing sequence.

Trying to check assumption (i) of the Direct Method, we ask if (1.4) implies the compactness of $\{T_j\}_j$, say, in the sense of pointwise (a.e.) convergence. Compactness criteria enforcing this kind of convergence, like the Ascoli–Arzelà criterion, or the compactness theorem of Sobolev spaces, would require some form of uniform control on the gradients (or on some sort of incremental ratio) of the maps T_j. It is, however, clear that no control of that sort is contained in (1.4). It is natural to think about pointwise convergence here, because should the maps T_j converge pointwise to some limit T, then by Fatou's lemma, we would find

$$\int_{\mathbb{R}^n} |T - x| \, \rho \le \lim_{j \to \infty} \int_{\mathbb{R}^n} |T_j - x| \, \rho = M,$$

thus verifying assumption (ii) of the Direct Method. Finally, even if pointwise convergence could somehow be obtained, we would still face the issue of showing that the limit map T belongs to the competition class (i.e., T is smooth and injective, and it satisfies the transport constraint (1.1)) in order to infer $\int_{\mathbb{R}^n} |T - x| \, \rho \ge M$ and close the Direct Method argument. Deducing all these properties on T definitely requires some form of convergence of ∇T_j toward ∇T (as is evident from the problem of passing to the limit the nonlinear constraint (1.3)) – a task that is even more out of reach than proving the pointwise convergence of T_j in the first place! Thus, even from a modern perspective, establishing the mere existence of minimizers in the original Monge problem is a formidable task.

1.2 A Modern Formulation of the Monge Problem

We now introduce the modern formulation of the Monge problem that will be used in the rest of this book. The first difference with respect to Monge's original formulation is that we extend the class of distributions of mass to be transported to the whole family $\mathcal{P}(\mathbb{R}^n)$ of probability measures on $\mathbb{R}^n$. We usually denote by μ the origin distribution of mass, and by ν the final one, thus going back to Monge's original formulation by setting $\mu = \rho \, d\mathcal{L}^n$ and $\nu = \sigma \, d\mathcal{L}^n$, where $\mathcal{L}^n$ is the Lebesgue measure on $\mathbb{R}^n$. This first change demands a second one, namely, we need to reformulate the pointwise transport condition (1.1) in a way that makes sense even when μ and ν are not absolutely continuous with respect to $\mathcal{L}^n$. This is done by resorting to the notion of push-forward (or direct image) of a measure through a map, which we now recall (see also Appendix A.4).

We say that T **transports** μ if there exists a Borel set $F \subset \mathbb{R}^n$ such that

$$T : F \to \mathbb{R}^n \text{ is a Borel map,} \tag{1.5}$$

and μ is concentrated on F (i.e., $\mu(\mathbb{R}^n \setminus F) = 0$).

Whenever $T : F \to \mathbb{R}^n$ transports μ, we can define a Borel measure $T_\# \mu$ (the push-forward of μ through T) by setting, for every Borel set $E \subset \mathbb{R}^n$,

$$(T_\# \mu)(E) = \mu(T^{-1}(E)),$$
$$\text{where } T^{-1}(E) = \{x \in F : T(x) \in E\}. \tag{1.6}$$

Notice that, according to this definition $(T_\# \mu)(\mathbb{R}^n) = \mu(F)$; therefore, the requirement that μ is concentrated on F in (1.5) is necessary to ensure that $T_\# \mu \in \mathcal{P}(\mathbb{R}^n)$ if $\mu \in \mathcal{P}(\mathbb{R}^n)$. Finally, we say that T **is a transport map from** μ **to** ν if

$$T_\# \mu = \nu.$$

Clearly, the transport condition (1.6) does not require T to be differentiable, nor injective; moreover, it boils down to the pointwise transport condition (1.1) whenever the latter makes sense, as illustrated in the following proposition.

Proposition 1.1 *Let $\mu = \rho \, d\mathcal{L}^n$ and $\nu = \sigma \, d\mathcal{L}^n$ belong to $\mathcal{P}(\mathbb{R}^n)$, μ be concentrated on a Borel set F, and $T : F \to \mathbb{R}^n$ be an injective Lipschitz map. Then, $T_\# \mu = \nu$ if and only if*

$$\sigma(T(x)) \, |\det \nabla T(x)| = \rho(x), \qquad \text{for } \mathcal{L}^n\text{-a.e. } x \in F. \tag{1.7}$$

Proof By the injectivity and Lipschitz continuity of T, the area formula,

$$\int_{T(F)} \varphi(y) \, \sigma(y) \, dy = \int_F \varphi(T(x)) \, |\det \nabla T(x)| \, \sigma(T(x)) \, dx, \tag{1.8}$$

holds for every Borel function $\varphi : T(F) \to [0, \infty]$ (see Appendix A.10). Since T is injective, for every Borel set $G \subset F$, we have $G = T^{-1}(T(G))$. Therefore, by definition of $T_\# \mu$ and by (1.8) with $\varphi = 1_{T(G)}$, we find

$$(T_\# \mu)(T(G)) = \mu(T^{-1}(T(G))) = \int_G \rho,$$

$$\nu(T(G)) = \int_{T(G)} \sigma(y) \, dy = \int_G |\det \nabla T(x)| \, \sigma(T(x)) \, dx.$$

By arbitrariness of $G \subset F$, we find that $T_\# \mu = \nu$ if and only if (1.7) holds. □

Based on these considerations, given $\mu, \nu \in \mathcal{P}(\mathbb{R}^n)$, we formally introduce the **Monge problem** from μ to ν by letting

$$\mathbf{M}_1(\mu, \nu) = \inf \left\{ \int_{\mathbb{R}^n} |T(x) - x| \, d\mu(x) : T_\# \mu = \nu \right\}. \tag{1.9}$$

Problem (1.9) is, in principle, more tractable than (1.2). Transport maps are no longer required to be smooth and injective, as reflected in the new transport condition (1.6). It is still unclear, however, if, given two arbitrary $\mu, \nu \in \mathcal{P}(\mathbb{R}^n)$, there always exists at least one transport map from μ to ν, and if such transport map can be found with finite transport cost; whenever this is not the case, we have $\mathbf{M}_1(\mu, \nu) = +\infty$, and the Monge problem is ill posed. A more fundamental issue is that, even in a situation where we know from the onset that $\mathbf{M}_1(\mu, \nu) < \infty$, it is still very much unclear how to verify assumption (i) in the Direct Method: what notion of subsequential convergence (for minimizing sequences $\{T_j\}_j$) is needed for passing the transport condition $(T_j)_\# \mu = \nu$ to a limit map T? This difficulty will eventually be solved by working with the Kantorovich formulation of the transport condition, which requires extending competition classes for transport problems from the family of transport *maps* to that of transport *plans*. From this viewpoint, the modern formulation of the Monge problem, as much as the original one, is still somehow untractable by a direct approach.

We can, of course, formulate the Monge problem with respect to a general[3] **transport cost** $c : \mathbb{R}^n \times \mathbb{R}^n \to \mathbb{R}$. Interpreting $c(x, y)$ as the cost needed to transport a unit mass from x to y (notice that c does not need to be symmetric in (x, y)!), we define the **Monge problem with transport cost** c by setting

$$\mathbf{M}_c(\mu, \nu) = \inf \left\{ \int_{\mathbb{R}^n} c(x, T(x)) \, d\mu(x) : T_\# \mu = \nu \right\}. \tag{1.10}$$

Our focus will be largely (but not completely) specific to the cases of the linear cost $c(x, y) = |x - y|$ and of the quadratic cost $c(x, y) = |x - y|^2$. In the following, when talking about "the Monge problem," we shall either assume that the transport cost under consideration is evident from the context or otherwise add the specification "with general cost," "with linear cost," or "with quadratic cost." From the historical viewpoint, of course, only the Monge problem with linear cost should be called "the Monge problem."

1.3 Optimality via Duality and Transport Rays

We now anticipate an observation that we will formally reintroduce later on[4] in our study of the Kantorovich duality theory and that provides a simple and effective criterion to check the optimality of a transport map in the Monge problem. The remark is that, if $f : \mathbb{R}^n \to \mathbb{R}$ is a Lipschitz function with $\mathrm{Lip}(f) \le 1$ (briefly, a 1-Lipschitz function), and if T is a transport map from μ to ν, then

$$\int_{\mathbb{R}^n} f \, d\nu - \int_{\mathbb{R}^n} f \, d\mu = \int_{\mathbb{R}^n} [f(T(x)) - f(x)] \, d\mu(x) \le \int_{\mathbb{R}^n} |T(x) - x| \, d\mu(x),$$

---

[3] In practice, we shall work with transport costs that are at least lower semicontinuous, thus guaranteeing the Borel measurability of $x \mapsto c(x, T(x))$.

[4] See Section 3.7.

so that one always has

$$\sup_{\mathrm{Lip}(f)\leq 1} \int_{\mathbb{R}^n} f \, dv - \int_{\mathbb{R}^n} f \, d\mu \leq \inf_{T_\#\mu=v} \int_{\mathbb{R}^n} |T(x) - x| \, d\mu(x). \qquad (1.11)$$

In particular, if for a given transport map T from μ and v, we can find a 1-Lipschitz function f such that

$$f(T(x)) - f(x) = |T(x) - x|, \qquad \text{for } \mu\text{-a.e. } x \in \mathbb{R}^n, \qquad (1.12)$$

then, integrating (1.12) with respect to $d\mu$ and exploiting $T_\#\mu = v$, we find

$$\int_{\mathbb{R}^n} f \, dv - \int_{\mathbb{R}^n} f \, d\mu = \int_{\mathbb{R}^n} |T(x) - x| \, d\mu(x),$$

thus deducing from (1.11) that T is a minimizer in the Monge problem $\mathbf{M}_1(\mu, v)$ (and, symmetrically, that f is a maximizer in the "dual" maximization problem appearing on the left-hand side of (1.11)). We illustrate this idea with the so-called "book-shifting example." Given $N \geq 2$, let us consider the Monge problem from μ to v with

$$\mu = \frac{1_{[0,N]}}{N} \, d\mathcal{L}^1, \qquad v = \frac{1_{[1,N+1]}}{N} \, d\mathcal{L}^1.$$

We can think of μ as a collection of N books of mass $1/N$ that we want to shift to the right (not necessarily in their original order) by a unit length. The map $T(t) = t + 1$ for $t \in \mathbb{R}$ (corresponding to shifting each book to the right by a unit length) is a minimizer in $\mathbf{M}_1(\mu, v)$ since it satisfies (1.12) with $f(t) = t$. By computing the transport cost of T, we see that $\mathbf{M}_1(\mu, v) = 1$. We easily check that transport map S defined by $S(t) = t$, for $t \in [1, N]$, and $S(t) = t + N$, for $t \in [0, 1)$, which corresponds to moving *only* the left-most book to the right by a length equal to N, has also a unit transport cost and thus is also optimal in $\mathbf{M}_1(\mu, v)$. This shows, in particular, that the Monge problem can admit multiple minimizers.

It is interesting to notice that the connection between optimal transport maps and 1-Lipschitz "potential functions" expressed in (1.12) was also clear to Monge, who rather focused on the more expressive identity

$$\nabla f(x) = \frac{T(x) - x}{|T(x) - x|}. \qquad (1.13)$$

The relation between (1.13) and (1.12) is clarified by noticing that $\mathrm{Lip}(f) \leq 1$ and (1.12) imply that f is affine with unit slope along the oriented segment from x to $T(x)$, that is,

$$f(x + t(T(x) - x)) = f(x) + t|T(x) - x|, \qquad \forall t \in [0, 1], \qquad (1.14)$$

from which (1.13) follows[5] if f is differentiable at x. Such oriented segments are called *transport rays*, and their study plays a central in the solution to the Monge problem (presented in Part IV). Notice that, by (1.14), the graph of f above the union of such segments is a *developable surface*; this connection seems to be the reason why Monge started (independently from Euler) the systematic study of developable surfaces.

1.4 Monotone Transport Maps

In dimension $n = 1$, it is particularly easy to construct *optimal* transport maps by looking at *monotone* transport maps. Here we just informally discuss this important idea, which will be addressed rigorously in Chapter 16.

It is quite intuitive that, in dimension 1, moving mass by monotone increasing maps must be a good transport strategy (the book-shifting example from Section 1.3 confirming that intuition). Considering the case when $\mu = \rho\, d\mathcal{L}^1$ and $\nu = \sigma\, d\mathcal{L}^1$, for an increasing map to be a transport map, we only need to check that the "rate of mass transfer," i.e., the derivative of the transport map, is compatible with the transport condition (1.1), and this can be achieved quite easily by defining $T(x)$ through the formula

$$\int_{-\infty}^{x} \rho = \int_{-\infty}^{T(x)} \sigma, \qquad \forall x \in \mathbb{R}. \tag{1.15}$$

In more geometric terms, we are prescribing that the mass stored by μ to the left of x corresponds to the mass stored by ν to the left of $T(x)$, i.e., we are setting $\mu((-\infty, x)) = \nu((-\infty, T(x)))$; see Figure 1.1. Indeed, an informal differentiation in x of (1.15) gives

$$\rho(x) = T'(x)\, \sigma(T(x)),$$

which (thanks to $T' \geq 0$) is the pointwise transport condition (1.1). The map T defined in (1.15) is called the **monotone rearrangement** of μ into ν and provides a minimizer in the Monge problem. We present here a simple argument in support of this assertion, which works under the assumption that $\{T \geq \mathbf{id}\} = \{x : T(x) \geq x\}$ and $\{T < \mathbf{id}\} = \{x : T(x) < x\}$ are equal, respectively, to complementary half-lines $[a, \infty)$ and $(-\infty, a)$ for some $a \in \mathbb{R}$: indeed, in this case, we can consider a 1-Lipschitz function $f : \mathbb{R} \to \mathbb{R}$ with

$$f'(x) = 1_{\{T \geq \mathbf{id}\}}(x) - 1_{\{T < \mathbf{id}\}}(x)$$

and see that if $x \in \{T \geq \mathbf{id}\}$, and thus $(x, T(x)) \subset \{T \geq \mathbf{id}\} = \{f' = 1\}$, we have

---

[5] See Proposition 18.6 for a formal discussion.

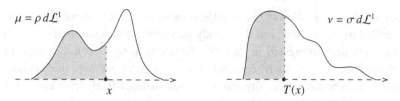

Figure 1.1 The amount of mass stored by $\mu = \rho\, d\mathcal{L}^1$ to the left of x corresponds to the amount of mass stored by $\nu = \sigma\, d\mathcal{L}^1$ to the left of $T(x)$; see (1.15).

$$f(T(x)) - f(x) = \int_x^{T(x)} f' = T(x) - x = |T(x) - x|\,;$$

while if $x \in \{T < \mathbf{id}\}$, and thus $(T(x), x) \subset \{T < \mathbf{id}\} = \{f' = -1\}$, we have

$$f(T(x)) - f(x) = \int_x^{T(x)} f' = x - T(x) = |T(x) - x|.$$

Hence (1.12) holds, and T is optimal in the Monge problem.

1.5 Knothe Maps

Monotone rearrangements can be used to define transport maps in higher dimensions. Given $\mu, \nu \in \mathcal{P}(\mathbb{R}^n)$, with $\mu, \nu \ll \mathcal{L}^n$, and an orthonormal basis $\tau = \{\tau_i\}_{i=1}^n$ of $\mathbb{R}^n$ (with coordinates $x^i = x \cdot \tau_i$), one can define **the Knothe map** T from μ to ν **(relative to the orthonormal basis** τ**)** by the following procedure (which, for the sake of simplicity, is discussed only informally here and in the case $n = 2$; an expert reader should have little difficulty in formalizing and extending to higher dimensions the following sketch). Writing $\mu = \rho\, d\mathcal{L}^2$ and $\nu = \sigma\, d\mathcal{L}^2$, we define the first component of T by a monotone rearrangement depending on the coordinate x^1 only, i.e., we set $T^1(x) = T^1(x^1)$ with

$$\int_{-\infty}^{x^1} ds \int_{\mathbb{R}} \rho(s,t)\, dt = \int_{-\infty}^{T^1(x^1)} ds \int_{\mathbb{R}} \sigma(s,t)\, dt, \qquad \forall x \in \mathbb{R}^2. \quad (1.16)$$

In this way, the total mass stored by μ inside the half-plane $\{z^1 < x^1\} = \{z \in \mathbb{R}^2 : z^1 < x^1\}$ is set to be equal to the total mass stored by ν inside the half-plane $\{z^1 < T^1(x^1)\}$. This choice of T^1 implies that points in the vertical line $\{z^1 = x^1\}$ are to be mapped by T inside the vertical line $\{z^1 = T^1(x^1)\}$. Thus, it is just natural to do this by a monotone rearrangement of $\rho(x^1, t)\, dt$ into $\sigma(T^1(x_1), t)\, dt$. Since these two measures have not the same total mass, we first normalize them into probability measures, and then we define $T^2(x) = T^2(x^1, x^2)$ by setting

$$\frac{\int_{-\infty}^{x^2} \rho(x^1,t)\,dt}{\int_{\mathbb{R}} \rho(x^1,t)\,dt} = \frac{\int_{-\infty}^{T^2(x^1,x^2)} \sigma(T^1(x^1),t)\,dt,}{\int_{\mathbb{R}} \sigma(T^1(x^1),t)\,dt} \qquad \forall x \in \mathbb{R}^2. \qquad (1.17)$$

Informal differentiations of (1.16) in x^1 and (1.17) in x^2 give

$$\int_{\mathbb{R}} \rho(x^1,t)\,dt = \frac{\partial T^1}{\partial x^1}(x) \int_{\mathbb{R}} \sigma(T^1(x^1),t)\,dt$$

$$\frac{\rho(x)}{\int_{\mathbb{R}} \rho(x^1,t)\,dt} = \frac{\partial T^2}{\partial x^2}(x) \frac{\sigma(T(x))}{\int_{\mathbb{R}} \sigma(T^1(x^1),t)\,dt},$$

while, evidently, $\partial T^1 / \partial x^2 = 0$; therefore,

$$\det \nabla T(x) = \frac{\partial T^1}{\partial x^1}(x) \frac{\partial T^2}{\partial x^2}(x) = \frac{\rho(x)}{\sigma(T(x))}.$$

Therefore (1.1) holds (notice that $\det \nabla T(x) \geq 0$), and T transports μ into ν.

The (formal) construction of Knothe maps proves the important point that there are always transport maps between two $\mathcal{L}^n$-absolutely continuous probability measures. Moreover, because of their componentwise monotonicity, Knothe maps can be used in place of optimal transport maps in certain arguments. For example, the proofs of the sharp Euclidean isoperimetric and Sobolev inequalities presented in Chapter 9 can be rigorously carried over using Knothe maps rather than (as done in that chapter) optimal transport maps in the Monge problem with quadratic transport cost (known, more briefly, as Brenier maps).

This said, when $n \geq 2$, we do not expect Knothe maps to be *optimal* transport maps for the linear and quadratic transport costs, as explained (only informally) in the following remarks.

Remark 1.2 (Knothe maps, in general, fail the noncrossing condition) In general, we do not expect Knothe maps to be optimal in the Monge problem with linear cost. To explain this point, we informally notice that for a transport map T from μ to ν to be optimal in $\mathbf{M}_1$, necessary condition is:[6] *for every* $x_1, x_2 \in \operatorname{spt} \mu$, *it holds*

$$|T(x_1) - x_1| + |T(x_2) - x_2| \leq |T(x_1) - x_2| + |T(x_2) - x_1|. \qquad (1.18)$$

Indeed, should (1.18) fail at $x_1 \neq x_2$, then one should be able to define a new transport map by sending small neighborhoods of x_1 and x_2, respectively, to

---

[6] In the terminology and notation to be introduced in the next two chapters condition (1.18) can be seen as a particular case of c-cyclical monotonicity condition (with respect to the linear cost $c(x,y) = |x - y|$) applied to $\operatorname{spt} \gamma_T$, $\gamma_T = (\mathbf{id} \times T)_{\#}\mu$; see Remark 3.10.

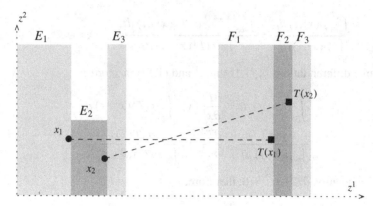

Figure 1.2 In this example $\mu = 1_E \, d\mathcal{L}^2$ and $\nu = 1_F \, d\mathcal{L}^2$, where E and F are two unit area regions, with each obtained as the union of a family of three rectangles, denoted respectively by $\{E_i\}_{i=1}^3$ and $\{F_i\}_{i=1}^3$. There are horizontal vectors v_1, v_2, and v_3 such that, if A denotes the diagonal matrix with entries $(1/2, 2)$, then $F_1 = E_1 + v_1$, $F_2 = A[E_2] + v_2$, and $F_3 = E_3 + v_3$. The Knothe map T from μ to ν (in the coordinate system (z^1, z^2)) is such that $T(x) = x + v_1$ if $x \in E_1$, $T(x) = A[x] + v_2$ if $x \in E_2$, and $T(x) = x + v_3$ if $x \in E_3$. (Notice that T is discontinuous on the segments separating E_1 from E_2 and E_2 from E_3.) In the figure, we have selected points $x_1, x_2 \in E$ such that the corresponding segments $[x_1, T(x_1)]$ and $[x_2, T(x_2)]$ intersect. Correspondingly, condition (1.18) does not hold, and T is not optimal in the Monge problem with linear cost.

small neighborhoods of $T(x_2)$ and $T(x_1)$, thus lowering the total transport cost. This said, it is easily seen that (1.18) can be violated by a Knothe map; see, for example, Figure 1.2.

Remark 1.3 (Knothe maps (in general) are not Brenier maps)   In general, we do not expect Knothe maps to be optimal in the Monge problem with quadratic cost. Indeed, looking at Theorem 4.2 in Chapter 4 and keeping in mind the informal character of this remark, this would mean that $\nabla T = \nabla^2 f$ for a convex function $f : \mathbb{R}^n \to \mathbb{R}$. In particular, ∇T would be symmetric, whereas gradients of Knothe maps are usually represented by triangular matrices with nontrivial off-diagonal entries. Incidentally, this method for excluding optimality in the Monge problem with quadratic cost does not apply to the example of the Knothe map depicted in Figure 1.2, since, in that case, ∇T takes only two symmetric values,

$$A_1 = \begin{pmatrix} 1 & 0 \\ 0 & 1 \end{pmatrix}, \qquad A_2 = \begin{pmatrix} 1/2 & 0 \\ 0 & 2 \end{pmatrix}.$$

However, should $\nabla^2 f = A_i$ hold on E_i for $i = 1, 2$, we could then find $a, b \in \mathbb{R}^2$ and $c \in \mathbb{R}$ such that $f(z) = (|z|^2/2) + a \cdot z$ for $z \in E_1$ and $f(z) = (z^1)^2/4 + (z^2)^2 + b \cdot z + c$ for $z \in E_2$. Notice, however, that there is no way to adjust a, b, and c so that f is continuous on the vertical segment at the interface between E_1 and E_2 (f being convex on $\mathbb{R}^n$, it must be continuous on E); indeed, the $(z^2)^2$-coefficients of the two polynomials describing f on E_1 and E_2 are different.

2

Discrete Transport Problems

In this chapter we introduce the basic ideas of Kantorovich's approach to transport problems by working in the discrete setting. From a formal viewpoint, this is just a particular case of the theory developed in Chapter 3, so its discussion in a separate chapter is mainly motivated by pedagogical reasons. The notions of transport plan (Section 2.1) and c-cyclical monotonicity (Section 2.2) arise effortlessly in this context and without any real technical burden. We then move to consider c-cyclical monotonicity in the model case of the quadratic transport cost $c(x,y) = |x - y|^2$, thus establishing its link with convexity and the corresponding Kantorovich duality (Sections 2.3 and 2.4). Finally, in Section 2.5, we consider the discrete Monge problem.

2.1 The Discrete Kantorovich Problem

Discrete transport problems involve origin and final mass distributions μ and ν that are concentrated at finitely many points; that is to say, we consider

$$\mu = \sum_{i=1}^{N} \mu_i \, \delta_{x_i}, \qquad \nu = \sum_{j=1}^{M} \nu_j \, \delta_{y_j}, \tag{2.1}$$

where $X = \{x_i\}_{i=1}^{N}$ and $Y = \{y_j\}_{j=1}^{M}$ are collections of *distinct*[1] points in $\mathbb{R}^n$, and with μ_i and ν_j *positive numbers* such that

$$\mu(\mathbb{R}^n) = \sum_{i=1}^{N} \mu_i = 1, \qquad \nu(\mathbb{R}^n) = \sum_{j=1}^{M} \nu_j = 1. \tag{2.2}$$

The corresponding Monge problem $\mathbf{M}_c(\mu, \nu)$ may be ill-posed (independently from the choice of a transport cost c) for the basic reason that there may be no transport maps from μ to ν.

---

[1] By this we mean that X is a family of N distinct points in $\mathbb{R}^n$ and that Y is a family of M distinct points in $\mathbb{R}^n$, although we are not requiring $X \cap Y$ to be empty.

Remark 2.1 (Nonexistence of transport maps)   Consider (2.1) and (2.2) with $N = 1$ (hence, $\mu_1 = 1$) and $M = 2$. Since μ is concentrated at x_1, every $\mathbb{R}^n$-valued map T defined at x_1 transports μ with $T_{\#}\mu = \delta_{T(x_1)}$. Thus, $T_{\#}\mu = \nu = \nu_1 \delta_{y_1} + \nu_2 \delta_{y_2}$ cannot hold, as both ν_1 and ν_2 are positive and $y_1 \neq y_2$. Hence, $\mathbf{M}_c(\mu,\nu) = +\infty$, with empty competition class.

In the situation of Remark 2.1, one would like to transport a ν_1-amount of the mass sitting at x_1 to y_1 and a $1 - \nu_1 = \nu_2$-amount to y_2: the only problem is that these simple "mass splitting instructions" cannot be described by maps. However, they can be efficiently and naturally described by using matrices,

$$\gamma = \{\gamma_{ij}\} \in \mathbb{R}^{N \times M},$$

so that $\gamma_{ij} \in [0,1]$ is the amount of mass sitting at x_i to be transported to y_j. Since the partial sum $\sum_{i=1}^{N} \gamma_{ij}$ represents the total mass received at the site y_j, and the partial sum $\sum_{j=1}^{M} \gamma_{ij}$ represents the total mass shipped from the site x_i, the conditions for γ to represent transport instructions from μ to ν are that

$$\mu_i = \sum_{j=1}^{M} \gamma_{ij}, \qquad \nu_j = \sum_{i=1}^{N} \gamma_{ij}. \tag{2.3}$$

Any $\gamma \in \mathbb{R}^{N \times M}$ with nonnegative entries and satisfying (2.3) for μ and ν as in (2.1) and (2.2) is called a **discrete transport plan** from μ to ν. The set of all transport plans from μ to ν is the convex[2] set $\Gamma \subset \mathbb{R}^{N \times M}$ defined by

$$\Gamma(\mu,\nu) = \left\{ \gamma \in \mathbb{R}^{N \times M} : \gamma_{ij} \geq 0, \ \mu_i = \sum_{k=1}^{M} \gamma_{ik}, \ \nu_j = \sum_{k=1}^{N} \gamma_j \text{ for every } i,j \right\}. \tag{2.4}$$

Given a cost function $c : \mathbb{R}^n \times \mathbb{R}^n \to \mathbb{R}$, the total cost associated to the discrete transport plan γ is then given by

$$\mathrm{Cost}(\gamma) = \sum_{i,j} c(x_i, y_j)\, \gamma_{ij}, \tag{2.5}$$

and we obtain **the discrete Kantorovich problem**,

$$\mathbf{K}_c(\mu,\nu) = \inf \left\{ \sum_{i,j} c(x_i, y_j)\, \gamma_{ij} : \gamma \in \Gamma(\mu,\nu) \right\}. \tag{2.6}$$

In sharp contrast with the case of the Monge problem (see Chapter 1 and Remark 2.1), establishing the existence of minimizers in (2.6) is actually *trivial*. A minimizer γ in $\mathbf{K}_c(\mu,\nu)$ is called an **optimal discrete transport plan**.

Theorem 2.2 (Optimal discrete transport plans)   *If $\mu, \nu \in \mathcal{P}(\mathbb{R}^n)$ are discrete (i.e., if (2.1) and (2.2) hold), then for every function $c : \mathbb{R}^n \times \mathbb{R}^n \to \mathbb{R}$ there is a*

---

[2]   A set $X \subset \mathbb{R}^n$ is convex if $t\,x + (1 - t)\,y \in X$ whenever $x, y \in X$ and $t \in (0, 1)$.

minimizer of the discrete Kantorovich problem $\mathbf{K}_c(\mu, \nu)$. *Moreover,* $\mathbf{M}_c(\mu, \nu) \geq \mathbf{K}_c(\mu, \nu)$.

Proof The function $\gamma \mapsto \text{Cost}(\gamma)$ is linear on $\mathbb{R}^{N \times M}$, while $\Gamma(\mu, \nu)$ is a non-empty ($\gamma_{ij} = \mu_i \nu_j$ always belongs to $\Gamma(\mu, \nu)$), convex, compact set in $\mathbb{R}^{N \times M}$ (the constraints defining $\Gamma(\mu, \nu)$ are clearly convex and closed, and $\Gamma(\mu, \nu)$ is bounded since $\gamma \in \Gamma(\mu, \nu)$ implies $0 \leq \gamma_{ij} \leq 1$ for every i, j). Therefore, the existence of a minimizer of $\mathbf{K}_c(\mu, \nu)$ is trivially established by the Direct Method.

The inequality $\mathbf{M}_c(\mu, \nu) \geq \mathbf{K}_c(\mu, \nu)$ is trivial if there are no transport maps. Now, if T transports μ into ν, then $T : \{x_i\}_{i=1}^N \to \mathbb{R}^n$ is such that $T_{\#}\mu = \sum_{i=1}^N \mu_i \, \delta_{T(x_i)}$ equals $\nu = \sum_{j=1}^M \nu_j \, \delta_{y_j}$. This means, necessarily, that $N \geq M$ and that there is $\sigma : \{1, \ldots, N\} \to \{1, \ldots, M\}$ surjective such that, for every i and j, respectively,

$$T(x_i) = y_{\sigma(i)}, \qquad \nu_j = \sum_{\{i : \sigma(i) = j\}} \mu_i.$$

Correspondingly, the plan[3]

$$\gamma_{ij} = \mu_i \, \delta_{\sigma(i), j}$$

is such that $\gamma \in \Gamma(\mu, \nu)$, with

$$\text{Cost}(\gamma) = \sum_{i=1}^N \sum_{j=1}^M c(x_i, y_j) \, \mu_i \, \delta_{\sigma(i), j} \tag{2.7}$$

$$= \sum_{i=1}^N c(x_i, T(x_i)) \, \mu_i = \int_{\mathbb{R}^n} c(x, T(x)) \, d\mu(x).$$

Hence, $\mathbf{K}_c(\mu, \nu) \leq \text{Cost}(\gamma) \leq \int_{\mathbb{R}^n} c(x, T(x)) \, d\mu(x)$ whenever T transports μ into ν, and $\mathbf{K}_c(\mu, \nu) \leq \mathbf{M}_c(\mu, \nu)$ follows by arbitrariness of T. $\qquad \square$

Two basic features of discrete Kantorovich problems are illustrated in the following remarks.

Remark 2.3 ($\Gamma(\mu, \nu)$ contains open segments if $N, M \geq 2$.) Let us notice, first of all, that $\Gamma(\mu, \nu)$ consists of a single element of $\mathbb{R}^{M \times N}$ unless $N \geq 2$ and $M \geq 2$. This is obvious if $N = M = 1$. If $N = 1$, $M \geq 1$, then (2.3) implies $\gamma_{1j} = \nu_j$ for every j, and similarly if $N \geq 2$, $M = 1$, then (2.3) gives $\gamma_{i1} = \mu_i$ for every i. This said, as soon as $N, M \geq 2$, for every $\gamma \in \Gamma(\mu, \nu)$ such that $\gamma_{ij} > 0$ for every i, j (one such element is always given by $\gamma_{ij} = \mu_i \nu_j$), there exists an open segment centered at γ which is entirely contained in $\Gamma(\mu, \nu)$. Indeed, if we define $\gamma^t = \{\gamma_{ij}^t\}$ by

---

[3] Here $\delta_{h,k}$ is the Kronecker symbol of h and k, not to be confused with δ_x, the Dirac mass concentrated at $x \in \mathbb{R}^n$.

$$\gamma_{11}^t = \gamma_{11} + t, \quad \gamma_{12}^t = \gamma_{12} - t, \qquad \gamma_{ij}^t = \gamma_{ij} \text{ if either } i \geq 3 \text{ or } j \geq 3,$$
$$\gamma_{21}^t = \gamma_{21} - t, \quad \gamma_{21}^t = \gamma_{21} + t,$$

then $\gamma^t \in \Gamma(\mu, \nu)$ for every sufficiently small value of $|t|$.

Remark 2.4 (Nonuniqueness of minimizers) $\mathbf{K}_c(\mu, \nu)$ may possess multiple minimizers. This is always the case, for example, if c, X and Y are such that,[4] for some $\lambda > 0$, $c(x_i, y_j) = \lambda$ for every i, j. Indeed, in that case,

$$\text{Cost}(\gamma) = \sum_{i,j} c(x_i, y_j)\, \gamma_{ij} = \lambda \sum_{i=1}^{N} \sum_{j=1}^{M} \gamma_{ij} = \lambda \sum_{i=1}^{N} \mu_i = \lambda,$$

i.e., cost is constant on $\Gamma(\mu, \nu)$, so every discrete transport plan is optimal.

2.2 c-Cyclical Monotonicity with Discrete Measures

We now further develop the remark made in Remark 2.3 to obtain a necessary optimality condition for minimizers γ in problem $\mathbf{K}_c(\mu, \nu)$. We start by noticing that we could well have $\gamma_{ij} = 0$ for some pair of indexes (i, j): when this happens, it means that the optimal plan γ has no convenience in sending any of the mass stored at x_i to the destination y_j. We thus look at those pairs (i, j) such that $\gamma_{ij} > 0$ and consider the set

$$S(\gamma) = \{(x_i, y_j) \in \mathbb{R}^n \times \mathbb{R}^n : \gamma_{ij} > 0\} \tag{2.8}$$

of those pairs of locations in the supports of μ and ν that are exchanging mass under the plan γ. We now formulate a necessary condition for the optimality of γ in terms of a geometric property of $S(\gamma)$.

Theorem 2.5 *If $\mu, \nu \in \mathcal{P}(\mathbb{R}^n)$ are discrete (i.e., if (2.1) and (2.2) hold) and γ is a minimizer in the discrete Kantorovich problem $\mathbf{K}_c(\mu, \nu)$, then for every finite subset $\{(z_\ell, w_\ell)\}_{\ell=1}^{L}$ of $S(\gamma)$ we have*

$$\sum_{\ell=1}^{L} c(z_\ell, w_\ell) \leq \sum_{\ell=1}^{L} c(z_{\ell+1}, w_\ell), \tag{2.9}$$

where $z_{L+1} = z_1$.

Remark 2.6 Based on Theorem 2.5, we introduce the following crucial notion, to be discussed at length in the sequel: a set $S \subset \mathbb{R}^n \times \mathbb{R}^n$ is c-**cyclically monotone** if every finite subset $\{(z_\ell, w_\ell)\}_{\ell=1}^{L}$ of S satisfies (2.9).

---

[4] This situation can of course be achieved in many different ways: for example, we could work in $\mathbb{R}^2$, with $x_1 = (0, 0)$, $x_2 = (1, 1)$, $y_1 = (1, 0)$, $y_2 = (0, 1)$, and with $c(x, y)$ being any nonnegative function of the Euclidean distance $|x - y|$; see Figure 2.5.

Proof of Theorem 2.5   By construction, $z_\ell = x_{i(\ell)}$ and $w_\ell = y_{j(\ell)}$ for suitable functions $i : \{1, \ldots, L\} \to \{1, \ldots, N\}$ and $j : \{1, \ldots, L\} \to \{1, \ldots, M\}$, and $\alpha_\ell = \gamma_{i(\ell)j(\ell)} > 0$ for every ℓ. Given $\varepsilon > 0$ with $\varepsilon < \min_\ell \alpha_\ell$, we construct a family of transport plans γ^ε by making, first of all, the following changes:

$$\gamma_{i(1)j(1)} \quad \to \quad \gamma^\varepsilon_{i(1)j(1)} = \gamma_{i(1)j(1)} - \varepsilon,$$

$$\gamma_{i(2)j(2)} \quad \to \quad \gamma^\varepsilon_{i(2)j(2)} = \gamma_{i(2)j(2)} - \varepsilon,$$

$$\cdots$$

$$\gamma_{i(L)j(L)} \quad \to \quad \gamma^\varepsilon_{i(L)j(L)} = \gamma_{i(L)j(L)} - \varepsilon.$$

i.e., we decrease by ε the amount of mass sent by γ from $z_\ell = x_{i(\ell)}$ to $w_\ell = y_{j(\ell)}$. Without further changes, the resulting plan γ^ε is not admissible: indeed, we have left unused an ε of mass at each origin site z_ℓ, while each of the destination sites w_ℓ is missing an ε of mass. To fix things, we transport the excess mass ε sitting at $z_{\ell+1}$ to w_ℓ and thus prescribe the following changes:

$$\gamma_{i(2)j(1)} \quad \to \quad \gamma^\varepsilon_{i(2)j(1)} = \gamma_{i(2)j(1)} + \varepsilon,$$

$$\gamma_{i(3)j(2)} \quad \to \quad \gamma^\varepsilon_{i(3)j(2)} = \gamma_{i(3)j(2)} + \varepsilon,$$

$$\cdots$$

$$\gamma_{i(L+1)j(L)} \quad \to \quad \gamma^\varepsilon_{i(L+1)j(L)} = \gamma_{i(1)j(L)} + \varepsilon,$$

where $i(L + 1) = 1$. Notice that, since $\varepsilon > 0$ by assumption, we do not need $\gamma_{i(\ell+1)j(\ell)}$ to be positive to prescribe the second round of changes. Finally, by setting $\gamma^\varepsilon_{ij} = \gamma_{ij}$ for every $(i, j) \notin \{(i(\ell), j(\ell)) : 1 \le \ell \le L\}$, we find that $\gamma^\varepsilon \in \Gamma(\mu, \nu)$ and therefore that

$$0 \le \text{Cost}(\gamma^\varepsilon) - \text{Cost}(\gamma) = \sum_{\ell=1}^{L} -\varepsilon\, c(z_\ell, w_\ell) + \varepsilon\, c(z_{\ell+1}, w_\ell).$$

Given that $\varepsilon > 0$, we deduce the validity of (2.9).   □

   When minimizing (as done in $\mathbf{K}_c(\mu, \nu)$) a linear function f on a compact convex set K, if f is nonconstant on K, then any minimum point x_0 will necessarily lie on ∂K. In particular, given a unit vector τ with $x_0 + t\tau \in K$ for every sufficiently small and positive t, by differentiating in t the inequality $f(x_0 + t\tau) \ge f(x_0)$, we find that $\nabla f(x_0) \cdot \tau \ge 0$. The family of inequalities $\nabla f(x_0) \cdot \tau \ge 0$ indexed over all the admissible directions τ is then a necessary and sufficient condition for x_0 to be a minimum point of f on K.

   From this viewpoint, in Theorem 2.5 we have identified a family of "directions τ" that can be used to take admissible one-sided variations of an optimal transport plan γ. Understanding if c-cyclical monotonicity is not only a necessary condition for minimality but also a *sufficient* one is tantamount to prove that such one-sided variations exhaust all the admissible ones. The Kantorovich duality theorem (Theorem 3.13) provides an elegant way to prove that this is indeed the case – i.e., that c-cyclical monotonicity fully characterizes optimality in $\mathbf{K}_c(\mu, \nu)$. The main difficulties related to the Kantorovich duality theorem are not of a technical character – actually the theorem is deduced by somehow elementary considerations – but rather conceptual – for example, if one insists (as it would seem natural when working with transport problems) in having a clear geometric understanding of things. Indeed, while the combinatorial nature of c-cyclical monotonicity is evident, its geometric content is definitely less immediate.

Luckily, in the case of the quadratic cost $c(x, y) = |x - y|^2$, c-cyclical monotonicity is immediately related to convexity. By examining this relation in detail, and by moving in analogy with it in the case of general costs, we will develop geometric and analytical ways to approach c-cyclical monotonicity, as well as develop the Kantorovich duality theory. To explain the relation with convexity, it is sufficient to look at (2.9), with $c(x, y) = |x - y|^2$, to expand the squares $|z_\ell - w_\ell|^2$ and $|z_{\ell+1} - w_\ell|^2$, to cancel out the sums over ℓ of $|z_\ell|^2$, $|w_\ell|^2$ and $|z_{\ell+1}|^2$ (as we can thanks to $z_{L+1} = z_1$), and, finally, to obtain the equivalent condition

$$\sum_{\ell=1}^{L} w_\ell \cdot (z_{\ell+1} - z_\ell) \leq 0, \qquad \text{for all } \{(z_\ell, w_\ell)\}_{\ell=1}^{L} \subset S. \tag{2.10}$$

This condition is (very well) known in convex geometry as **cyclical monotonicity** (of S). As proved in the next section, (2.10) is equivalent to require that S lies in the graph of the gradient of a convex function on $\mathbb{R}^n$. We can quickly anticipate this result by looking at the simple case when $n = 1$ and $L = 2$, and (2.10) just says

$$0 \leq w_2(z_2 - z_1) - w_1(z_2 - z_1) = (w_2 - w_1)(z_2 - z_1),$$

that is,

$$\begin{cases} (z_1, w_1), (z_2, w_2) \in S, \\ z_1 \leq z_2, \end{cases} \Rightarrow \qquad w_1 \leq w_2. \tag{2.11}$$

The geometric meaning of (2.11) is absolutely clear (see Figure 2.1): S must be contained in the extended graph of a monotone increasing function from $\mathbb{R}$ to $\mathbb{R}$ (where the term "extended" indicates that vertical segments corresponding to jump points are included in the graph). Since monotone functions are the gradients of convex functions, the connection between convexity and OMT problems with quadratic transport cost is drawn.

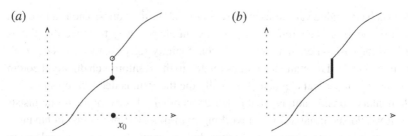

Figure 2.1 (a) The graph of an increasing function $f : \mathbb{R} \to \mathbb{R}$ with a discontinuity at a point x_0. The black dot indicates that the function takes the lowest possible value compatible with being increasing; (b) The extended graph of f is a subset of $\mathbb{R}^2$ which contains the graph of f and the whole vertical segment of values that f may take at x_0 without ceasing to be increasing. The property of being contained into the extended graph of an increasing function is easily seen to be equivalent to (2.11).

2.3 Basics about Convex Functions on $\mathbb{R}^n$

We now review some key concepts concerning convex functions on $\mathbb{R}^n$ that play a central role in our discussion. In OMT it is both natural and convenient to consider convex functions taking values in $\mathbb{R} \cup \{+\infty\}$. Since this setting may be unfamiliar to some readers, we offer here a review of the main results, including proofs of the less obvious ones.

Convex sets: A **convex set** in $\mathbb{R}^n$ is a set $K \subset \mathbb{R}^n$ such that $t x + (1 - t) y \in K$ whenever $t \in (0, 1)$ and $x, y \in K$. If $K \neq \emptyset$, the **(affine) dimension of K** is defined as the dimension of the smallest affine space containing K. The **relative interior** $\mathrm{Ri}(K)$ of a convex set K is its interior as a subset of the smallest affine space containing it; of course $\mathrm{Ri}(K) = \mathrm{Int}(K)$, where $\mathrm{Int}(K)$ is the set of interior points of K as a subset of $\mathbb{R}^n$, whenever K has dimension n. Given $E \subset \mathbb{R}^n$ we say that $z \in \mathbb{R}^n$ is a **convex combination in E** if $z = \sum_{i=1}^{N} t_i x_i$ for some coefficients $t_i \in [0, 1]$ such that $\sum_{i=1}^{N} t_i = 1$ and some $\{x_i\}_{i=1}^{N} \subset E$. The **convex envelope** $\mathrm{conv}(E)$ of $E \subset \mathbb{R}^n$ is the collection of all the convex combinations in E. The convex envelope can be characterized as the intersection of all the convex sets containing E. Of course, K is convex if and only if $K = \mathrm{conv}(K)$.

Convex functions: A function $f : \mathbb{R}^n \to \mathbb{R} \cup \{+\infty\}$ is a **convex function** if

$$f(t x + (1 - t) y) \leq t f(x) + (1 - t) f(y), \qquad \forall t \in [0, 1], x, y \in \mathbb{R}^n, \quad (2.12)$$

or, equivalently, if the **epigraph** of f, $\mathrm{Epi}(f) = \{(x, t) : t \geq f(x)\} \subset \mathbb{R}^{n+1}$, is a convex set in $\mathbb{R}^{n+1}$. The **domain of f**, $\mathrm{Dom}(f) = \{x \in \mathbb{R}^n : f(x) < \infty\} = \{f < \infty\}$, is a convex set in $\mathbb{R}^n$. Notice that, with this definition, whenever f

is not identically equal to $+\infty$, the dimension of $\mathrm{Dom}(f)$ could be any integer between 0 and n. Given a convex set K in $\mathbb{R}^n$ and a function $f : K \to \mathbb{R}$ satisfying (2.12) for $x, y \in K$, by extending $f = +\infty$ on $\mathbb{R}^n \setminus K$, we obtain a convex function with $K = \mathrm{Dom}(f)$; therefore, the point of view adopted here includes what is probably the more standard notion of "finite-valued, convex function defined on a convex set" that readers may be familiar with. The **indicator function** I_K of a convex set $K \subset \mathbb{R}^n$, defined by setting $I_K(x) = 0$ if $x \in K$ and $I_K(x) = +\infty$ if $x \notin K$, is a convex function. In particular, the basic optimization problem "minimize a finite-valued convex function g over a convex set K" can be simply recast as the minimization over $\mathbb{R}^n$ of the $\mathbb{R} \cup \{+\infty\}$-valued function $f = g + I_K$, that is

$$\inf_K g = \inf_{\mathbb{R}^n} \{ g + I_K \}. \tag{2.13}$$

Finally, a more practical reason for considering $\mathbb{R} \cup \{+\infty\}$-valued convex functions is that, as well shall see subsequently, many natural convex functions arise by taking suprema of families of affine functions, and thus they may very well take the value $+\infty$ outside of a convex set.

Lipschitz continuity and a.e. differentiability: We prove that *convex functions are always locally Lipschitz*[5] *in the relative interiors of their domains.* We give details in the case when $\mathrm{Dom}(f)$ has affine dimension n, since the general case is proved similarly. Let $\Omega = \mathrm{IntDom}(f)$, and let us first prove that f **is locally bounded in** Ω. To this end, given $B_r(x) \subset\subset \Omega$ we notice that, for every $z \in B_r(x)$, $y = 2x - z \in B_r(x)$ is such that $x = (y + z)/2$. Hence, $f(x) \le (f(y) + f(z))/2$, which gives

$$\inf_{B_r(x)} f \ge 2 f(x) - \sup_{B_r(x)} f,$$

i.e., f is locally bounded in Ω if it is locally bounded *from above* in Ω. To show boundedness from above, let us fix $n + 1$ unit vectors $\{v_i\}_{i=1}^{n+1}$ in $\mathbb{R}^n$ so that the simplex Σ with vertexes v_i – defined as the set of all the convex combinations $\sum_{i=1}^{n+1} t_i v_i$ corresponding to $0 < t_i < 1$ with $\sum_{i=1}^{n+1} t_i = 1$ – is an open set in $\mathbb{R}^n$ containing the origin in its interior. Now, for each $x \in \Omega$, we can find $r > 0$ such that $\Sigma_{x,r} = x + r \Sigma$ is contained in Ω, and since the convexity of f implies that

$$f\left(\sum_{i=1}^N t_i x_i \right) \le \sum_{i=1}^N t_i f(x_i), \qquad \begin{cases} \forall N \in \mathbb{N}, \forall \{t_i\}_{i=1}^N \in [0,1] \text{ s.t. } \sum_{i=1}^N t_i = 1, \\ \forall \{x_i\}_{i=1}^N \subset \mathbb{R}^n, \end{cases}$$
$$\tag{2.14}$$

---

[5] See Appendix A.10 for the basics on Lipschitz functions.

we conclude that

$$\sup_{\Sigma_{x,r}} f \leq \max_{1 \leq i \leq n+1} f(x + r\, v_i) < \infty.$$

By a covering argument, we conclude that f is locally bounded (from above and, thus, also from below) in Ω. We next exploit local boundedness to show that f **is locally Lipschitz in** Ω. Indeed, if B_{2r} is a ball of radius $2r$ compactly contained in Ω, and if B_r is concentric to B_{2r} with radius r, then

$$\mathrm{Lip}(f; B_r) \leq \frac{1}{r}\left(\sup_{B_{2r}} f - \inf_{B_{2r}} f\right).$$

To show this, pick $x, y \in B_r$ and write $y = t\,x + (1 - t)\,z$ for some $z \in \partial B_{2r}$. Then, $|x - y| = |1 - t|\,|x - z| \geq r\,|1 - t|$ so that

$$f(y) - f(x) \leq t\, f(x) + (1 - t)\, f(z) - f(x) \leq |1 - t|\,|f(x) - f(z)|$$

$$\leq \frac{|x - y|}{r}\left(\sup_{B_{2r}} f - \inf_{B_{2r}} f\right).$$

In particular, by Rademacher's theorem, f **is a.e. differentiable in** Ω, and a simple consequence of (2.12) shows that

$$f(y) \geq f(x) + \nabla f(x) \cdot (y - x), \qquad \forall y \in \mathbb{R}^n, \tag{2.15}$$

whenever f is differentiable at x with gradient $\nabla f(x)$. Condition (2.15) expresses the familiar property that convex functions lie above the tangent hyperplanes to their graphs whenever the latter are defined. In fact, inequality (2.15) points at a very fruitful way to think about convex functions, which we are now going to discuss.

Convex functions as suprema of affine functions: If $\mathcal{A}$ is any family of **affine functions** on $\mathbb{R}^n$ (i.e., if $\alpha \in \mathcal{A}$, then $\alpha(x) = a + y \cdot x$ for some $a \in \mathbb{R}$ and $y \in \mathbb{R}^n$), then it is trivial to check that

$$f = \sup_{\alpha \in \mathcal{A}} \alpha \tag{2.16}$$

defines a convex function on $\mathbb{R}^n$ with values in $\mathbb{R} \cup \{+\infty\}$. A convex function defined in this way is automatically lower semicontinuous on $\mathbb{R}^n$, as it is the supremum of continuous functions. Of course, not every convex function is going to be lower semicontinuous on $\mathbb{R}^n$ (e.g., $f(x) = I_{[0,1)}(x)$ is not lower semicontinuous at $x = 1$), so not every convex function will satisfy an identity like (2.16) on the whole $\mathbb{R}^n$. However, it is not hard to deduce from (2.15) that

$$f(z) = \sup\left\{f(x) + \nabla f(x) \cdot (z - x) : f \text{ is differentiable at } x\right\} \quad \forall z \in \mathrm{IntDom}(f), \tag{2.17}$$

so that (2.16) always holds on IntDom(f) if $\mathcal{A} = \{\alpha_x\}_x$ for x ranging among the points of differentiability of f, $\alpha_x(z) = a_x + y_x \cdot z$, $a_x = f(x) - \nabla f(x) \cdot x$, and $y_x = \nabla f(x)$. We now introduce the concepts of subdifferential and Fenchel–Legendre transform of a convex function. These concepts lead to a representation formula for convex functions similar to (but more robust than) (2.17).

Subdifferential at a point: Given a convex function f, a point $x \in \text{Dom}(f)$, and a hyperplane L in $\mathbb{R}^{n+1}$, we say that L is a **supporting hyperplane of** f **at** x if L is the graph of an affine function α on $\mathbb{R}^n$ such that $\alpha \leq f$ on $\mathbb{R}^n$ and $\alpha(x) = f(x)$. If $a \in \mathbb{R}$ and $y \in \mathbb{R}^n$ are such that $\alpha(z) = a + y \cdot z$ for all $z \in \mathbb{R}^n$, then y is called the **slope of** L. The **subdifferential** $\partial f(x)$ **of** f **at** x is defined as follows: if $x \in \text{Dom}(f)$, then we set

$$\partial f(x) = \Big\{ \text{slopes of all the supporting hyperplanes of } f \text{ at } x \Big\}$$

$$= \left\{ y \in \mathbb{R}^n : \exists\, a \in \mathbb{R} \text{ s.t.} \begin{array}{l} a + y \cdot z \leq f(z) \quad \forall z \in \mathbb{R}^n \\ a + y \cdot x = f(x). \end{array} \right\},$$

$$= \Big\{ y \in \mathbb{R}^n : f(z) \geq f(x) + y \cdot (z - x) \quad \forall z \in \mathbb{R}^n \Big\}; \qquad (2.18)$$

otherwise, i.e., if $f(x) = +\infty$, we set $\partial f(x) = \emptyset$. If f is differentiable at some $x \in \text{Dom}(f)$, then $x \in \text{IntDom}(f)$ (because f is finite in a neighborhood of x) and

$$\partial f(x) = \{\nabla f(x)\} \qquad (2.19)$$

(see Proposition 2.7 for the proof). At a generic point $x \in \text{Dom}(f)$, where f may not be differentiable, we always have that $\partial f(x)$ is a closed convex set in $\mathbb{R}^n$. For example, if $f(x) = |x|$, then $\partial f(0)$ is the closed unit ball in $\mathbb{R}^n$ centered at the origin (see Figure 2.2); if f is the maximum of finitely many affine functions α_i, with slope y_i, then $\partial f(x)$ is the convex envelope of those y_i such that x belongs to $\{f = \alpha_i\}$. In the following proposition we prove (2.19) together with a sort of continuity property of subdifferentials.

Proposition 2.7 (Continuity of subdifferentials) *If f is differentiable at x, then $\partial f(x) = \{\nabla f(x)\}$. Moreover, for every $\varepsilon > 0$ there exists $\delta > 0$ such that*

$$\partial f(B_\delta(x)) \subset B_\varepsilon(\nabla f(x)). \qquad (2.20)$$

Proof *Step one*: We prove (2.19). Given $y \in \partial f(x)$, set

$$F_{x,y}(z) = f(z) - f(x) - y \cdot (z - x), \qquad z \in \mathbb{R}^n.$$

Notice that $F_{x,y}$ has a minimum at x, since $F_{x,y}(z) \geq 0 = F_{x,y}(x)$ for every $z \in \mathbb{R}^n$. In particular, if f is differentiable at x, then $F_{x,y}$ is differentiable at x with $0 = \nabla F_{x,y}(x) = \nabla f(x) - y$ so that (2.19) is proved.

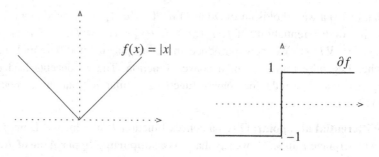

Figure 2.2 The subdifferential of $f(x) = |x|$ when $n = 1$.

Step two: If (2.20) fails, then there exist $\varepsilon > 0$ and $x_j \to x$ as $j \to \infty$ such that $|y_j - \nabla f(x)| \geq \varepsilon$ for every j and $y_j \in \partial f(x_j)$. It is easily seen that since f is bounded in a neighborhood of x, the sequence $\{y_j\}_j$ must be bounded in $\mathbb{R}^n$ and thus up to extracting subsequences, that $y_j \to y$ as $j \to \infty$. By taking limits as $j \to \infty$ in "$f(z) \geq f(x_j) + y_j \cdot (z - x_j)$ for every $z \in \mathbb{R}^n$," we deduce that $y \in \partial f(x) = \{\nabla f(x)\}$, in contradiction with $|y_j - \nabla f(x)| \geq \varepsilon$ for every j. $\square$

Fundamental theorem of (convex) Calculus: It is well-known that a smooth function $g : \mathbb{R} \to \mathbb{R}$ is the derivative f' of a smooth *convex* function $f : \mathbb{R} \to \mathbb{R}$ if and only if g is increasing on $\mathbb{R}$. The notion of subdifferential and a proper generalization of monotonicity to $\mathbb{R}^n$ allow to extend this theorem to $\mathbb{R} \cup \{+\infty\}$-valued convex functions on $\mathbb{R}^n$. First of all, let us introduce the notion of **(total) subdifferential** of f, defined as

$$\partial f = \bigcup_{x \in \mathbb{R}^n} \{x\} \times \partial f(x); \qquad (2.21)$$

see Figure 2.2. Notice that ∂f is a subset of $\mathbb{R}^n \times \mathbb{R}^n$, which is closed as soon as f is lower semicontinuous. Recalling that, as set in (2.10), $S \subset \mathbb{R}^n \times \mathbb{R}^n$ is **cyclically monotone** if for every finite set $\{(x_i, y_i)\}_{i=1}^N \subset S$ one has

$$\sum_{i=1}^{N} y_i \cdot (x_{i+1} - x_i) \leq 0, \qquad \text{where } x_{N+1} = x_1. \qquad (2.22)$$

Thus, we have the following theorem.

Theorem 2.8 (Rockafellar theorem) *Let $S \subset \mathbb{R}^n \times \mathbb{R}^n$ be a non-empty set: S is cyclically monotone if and only if there exists a convex and lower semicontinuous function $f : \mathbb{R}^n \to \mathbb{R} \cup \{+\infty\}$ such that*

$$S \subset \partial f. \qquad (2.23)$$

Remark 2.9 Notice that $S \neq \emptyset$ and $S \subset \partial f$ imply $\mathrm{Dom}(f) \neq \emptyset$.

Proof Proof that (2.23) *implies cyclical monotonicity*: Let us consider a finite subset $\{(x_i, y_i)\}_{i=1}^N$ of S. Then, $y_i \in \partial f(x_i)$ implies $\partial f(x_i) \neq \emptyset$, and thus $f(x_i) < \infty$. For every $i = 1, \ldots, N$, we know that

$$f(x) \geq f(x_i) + y_i \cdot (x - x_i), \qquad \forall x \in \mathbb{R}^n,$$

so that testing the i-th inequality at $x = x_{i+1}$ and summing up over $i = 1, \ldots, N$ gives

$$\sum_{i=1}^N f(x_{i+1}) \geq \sum_{i=1}^N f(x_i) + y_i \cdot (x_{i+1} - x_i).$$

We find (2.22) since $\sum_{i=1}^N f(x_{i+1}) = \sum_{i=1}^N f(x_i)$ by the convention $x_{N+1} = x_1$.

Proof that cyclical monotonicity implies (2.23): We need to define a convex function f which contains S in its subdifferential. To this end, we fix[6] $(x_0, y_0) \in S$ and define

$$f(z) = \sup \left\{ y_N \cdot (z - x_N) + \sum_{i=1}^{N-1} y_i \cdot (x_{i+1} - x_i) + y_0 \cdot (x_1 - x_0) : \{(x_i, y_i)\}_{i=1}^N \subset S \right\} \tag{2.24}$$

for $z \in \mathbb{R}^n$. Clearly, f is a convex and lower semicontinuous function on $\mathbb{R}^n$ with values in $\mathbb{R} \cup \{+\infty\}$. We also notice that $f(x_0) \in \mathbb{R}$. Indeed, by applying (2.22) to $\{(x_i, y_i)\}_{i=0}^N \subset S$ we find

$$y_N \cdot (x_0 - x_N) + \sum_{i=1}^{N-1} y_i \cdot (x_{i+1} - x_i) + y_0 \cdot (x_1 - x_0) \leq 0 \qquad \forall \{(x_i, y_i)\}_{i=1}^N \subset S, \tag{2.25}$$

so that $f(x_0) \leq 0$. (Actually, we can even say that $f(x_0) = 0$, since $f(x_0) \geq 0$ by testing (2.24) with $\{(x_1, y_1)\} = \{(x_0, y_0)\}$ at $z = x_0$.) Interestingly, the proof that $f(x_0) < \infty$ is the only point of this argument where cyclical monotonicity plays a role.

We now prove that $S \subset \partial f$. Indeed, let $(x_*, y_*) \in S$ and let $t \in \mathbb{R}$ be such that $t < f(x_*)$. By definition of $f(x_*)$, we can find $\{(x_i, y_i)\}_{i=1}^N \subset S$ such that

$$y_N \cdot (x_* - x_N) + \sum_{i=0}^{N-1} y_i \cdot (x_{i+1} - x_i) \geq t. \tag{2.26}$$

If we now define $\{(x_i, y_i)\}_{i=1}^{N+1} \subset S$ by setting $x_{N+1} = x_*$ and $y_{N+1} = y_*$, then, by testing the definition of f with $\{(x_i, y_i)\}_{i=1}^{N+1} \subset S$, we find that, for every $z \in \mathbb{R}^n$,

---

[6] The choice of (x_0, y_0) is analogous to the choice of an arbitrary additive constant in the classical fundamental theorem of Calculus.

$$f(z) \geq y_{N+1} \cdot (z - x_{N+1}) + \sum_{i=0}^{N} y_i \cdot (x_{i+1} - x_i)$$

$$= y_* \cdot (z - x_*) + y_N \cdot (x_* - x_N) + \sum_{i=0}^{N-1} y_i \cdot (x_{i+1} - x_i)$$

$$\geq t + y_* \cdot (z - x_*), \tag{2.27}$$

where in the last inequality we have used (2.26). Since $f(z)$ is finite at $z = x_0$, by letting $t \to f(x_*)^-$ in (2.27) first with $z = x_0$, we see that $x_* \in \text{Dom}(f)$, and then, by taking the same limit for an arbitrary z, we see that $y_* \in \partial f(x_*)$. $\quad\square$

Fenchel–Legendre transform: Given a convex function f, and the slope y of a supporting hyperplane to f, we know that there exists $a \in \mathbb{R}$ such that

$$a + y \cdot x \leq f(x), \qquad \forall x \in \mathbb{R}^n.$$

The largest value of $a \in \mathbb{R}$ such that this condition holds can be obviously characterized as $a = -f^*(y)$, where

$$f^*(y) = \sup \left\{ y \cdot x - f(x) : x \in \mathbb{R}^n \right\}. \tag{2.28}$$

The function f^* is called the **Fenchel–Legendre transform** of f. It is a convex function, and it is automatically lower semicontinuous on $\mathbb{R}^n$. Moreover, as it is easily seen, f^{**} is the lower semicontinuous envelope of f – i.e., the largest lower semicontinuous function lying below f: in particular, if f is convex and lower semicontinuous, then $f = f^{**}$, i.e.,

$$f(x) = \sup \left\{ x \cdot y - f^*(y) : y \in \mathbb{R}^n \right\} \qquad \forall x \in \mathbb{R}^n. \tag{2.29}$$

This is the "more robust" reformulation of (2.17). The last basic fact about convex functions that will be needed in the sequel is contained in the following two assertions:

$$f(x) + f^*(y) \geq x \cdot y, \qquad \forall x, y \in \mathbb{R}^n, \tag{2.30}$$

$$f(x) + f^*(y) = x \cdot y, \qquad \text{iff } x \in \text{Dom}(f) \text{ and } y \in \partial f(x). \tag{2.31}$$

Notice that (2.30) is immediate from the definition (2.28). If $f(x) + f^*(y) = x \cdot y$, then $x \cdot y - f(x) = f^*(y) \geq y \cdot z - f(z)$, i.e., $f(z) \geq f(x) + y \cdot (z - x)$ for every $z \in \mathbb{R}^n$, i.e., $y \in \partial f(x)$; and, vice versa, if $y \in \partial f(x)$, then $x \cdot y - f(x) \geq y \cdot z - f(z)$ for every $z \in \mathbb{R}^n$ so that $f^*(y) \leq x \cdot y - f(x)$ – which combined with (2.30) gives $f^*(y) = x \cdot y - f(x)$.

Many common inequalities in analysis can be interpreted as instances of the **Fenchel–Legendre inequality** (2.30): for example, if $1 < p < \infty$ and

$f(x) = |x|^p/p$, one computes that $f^*(y) = |y|^{p'}/p'$ for $p' = p/(p - 1)$, and thus finds[7] that (2.30) boils down to the classical **Young's inequality**.

Extremal points and the Choquet theorem:[8] Given a convex set K, we say that x_0 is an **extremal point** of K if $x_0 = (1 - t) y + t z$ with $t \in [0, 1]$ and $y, z \in K$ implies that either $t = 0$ or $t = 1$. We claim that

$$\begin{aligned} &\textit{if } K \subsetneq \mathbb{R}^n \textit{ is non-empty, closed, and convex,} \\ &\textit{then } K \textit{ has at least one extremal point.} \end{aligned} \qquad (2.32)$$

To this end, we argue by induction on n, with the case $n = 1$ being trivial. If $n \geq 2$, since K is not empty and not equal to $\mathbb{R}^n$, there is a closed half-space H such that $K \subset H$ and $\partial H \cap \partial K \neq \emptyset$. In particular, $J = K \cap \partial H$ is a convex set with affine dimension $(n - 1)$, and, by inductive hypothesis, there is an extremal point x_0 of J. We conclude by showing that x_0 is also an extremal point of K. Should this not be the case, we could find $t \in (0, 1)$ and $x, y \in K$ such that $x_0 = (1 - t) x + t y$. On the one hand, it must be $x, y \in \partial H$: otherwise, assuming, for example, that $x \in \text{Int}(H)$, by $x_0 \in \partial H$ and $t \in (0, 1)$ we would then find $y \notin H$, against $y \in K$; on the other hand, $x, y \in \partial H$ implies $x, y \in J$, and thus $x_0 = (1 - t) x + t y$ with $t \in (0, 1)$ would contradict the fact that x_0 is an extremal point of J. Having proved (2.32), we deduce from it the following statement (known as the **Choquet theorem**):

$$\begin{aligned} &\textit{if } K \subset \mathbb{R}^n \textit{ is convex and compact,} \\ &\textit{and } f : \mathbb{R}^n \to \mathbb{R} \cup \{+\infty\} \textit{ is convex and lower semicontinuous,} \\ &\textit{then there is an extremal point } x_0 \textit{ of } K \textit{ such that } f(x_0) = \inf_K f. \end{aligned} \qquad (2.33)$$

This is trivially true by (2.32) if $\text{Dom}(f) = \emptyset$. Otherwise, since f is lower semicontinuous and K is compact, we can apply the Direct Method to show that the set J of the minimum points of f over K is non-empty and compact. Since f is convex, J is also convex. Hence, by (2.32), J admits an extremal point, and (2.33) is proved.

2.4 The Discrete Kantorovich Problem with Quadratic Cost

We now use the fundamental theorem of Calculus for convex functions proved in Section 2.3 to give a complete discussion of the discrete transport problem

---

[7] Of course, we have not just discovered an incredibly short proof of Young's inequality: indeed, showing that $f^*(y) = |y|^{p'}/p'$ is equivalent to prove Young's inequality! From this viewpoint, the importance of the Fenchel's inequality is more conceptual than practical.

[8] These results are only used in Section 2.5 and can be omitted on a first reading.

with quadratic transport cost $c(x,y) = |x - y|^2$. In particular, we make our first encounter with the Kantorovich duality formula; see (2.36), which comes into play as our means for proving that cyclical monotonicity is a sufficient condition for minimality in the transport problem.

Theorem 2.10 *If $\mu, \nu \in \mathcal{P}(\mathbb{R}^n)$ are discrete (i.e., if (2.1) and (2.2) hold) and $c(x,y) = |x - y|^2$, then, for every discrete transport plan $\gamma \in \Gamma(\mu, \nu)$, the following three statements are equivalent: (i) γ is a minimizer of $\mathbf{K}_c(\mu, \nu)$; (ii) $S(\gamma) = \{(x_i, y_j) : \gamma_{ij} > 0\} \subset \mathbb{R}^n \times \mathbb{R}^n$ is cyclically monotone; (iii) there exists a convex function f such that $S(\gamma) \subset \partial f$. Moreover, denoting $\mathcal{H}$ as the family of pairs (α, β) such that $\alpha, \beta : \mathbb{R}^n \to \mathbb{R}$ satisfy*

$$\alpha(x) + \beta(y) \leq -x \cdot y \qquad \forall x, y \in \mathbb{R}^n \tag{2.34}$$

and defining $H : \mathcal{H} \to \mathbb{R}$ by setting

$$H(\alpha, \beta) = \sum_{i=1}^{N} \alpha(x_i)\, \mu_i + \sum_{j=1}^{M} \beta(y_j)\, \nu_j,$$

we have, for every $\gamma \in \Gamma(\mu, \nu)$ and $(\alpha, \beta) \in \mathcal{H}$,

$$\sum_{i,j} |x_i - y_j|^2 \gamma_{ij} \geq \sum_{i=1}^{N} |x_i|^2 \mu_i + \sum_{j=1}^{M} |y_j|^2 \nu_j + 2\, H(\alpha, \beta). \tag{2.35}$$

Finally, if γ satisfies (iii), then (2.35) holds as an identity with $(\alpha, \beta) = (-f, -f^*)$. In particular,

$$\mathbf{K}_c(\mu, \nu) = \sum_{i=1}^{N} |x_i|^2 \mu_i + \sum_{j=1}^{M} |y_j|^2 \nu_j + 2 \sup_{(\alpha, \beta) \in \mathcal{H}} H(\alpha, \beta). \tag{2.36}$$

Remark 2.11   Theorem 2.10 is, of course, a particular case of Theorem 3.20, in which the same assertions are proved without the discreteness assumption on μ and ν.

Proof of Theorem 2.10   Step one: We prove that (i) implies (ii) and that (ii) implies (iii). If (i) holds, then, by Theorem 2.5, we have

$$\sum_{\ell=1}^{L} |z_\ell - w_\ell|^2 \leq \sum_{\ell=1}^{L} |z_{\ell+1} - w_\ell|^2, \tag{2.37}$$

whenever $\{(z_\ell, w_\ell)\}_{\ell=1}^{L} \subset S(\gamma)$ and $z_{L+1} = z_1$. By expanding the squares in (2.37),

$$\sum_{\ell=1}^{L} w_\ell \cdot (z_{\ell+1} - z_\ell) \leq 0, \qquad \forall \{(z_\ell, w_\ell)\}_{\ell=1}^{L} \subset S(\gamma), \tag{2.38}$$

so that $S(\gamma)$ is cyclically monotone. In turn, if (ii) holds, then (iii) follows immediately by Rockafellar's theorem (Theorem 2.8).

Step two: For every $\gamma \in \Gamma(\mu, \nu)$ we have

$$\text{Cost}(\gamma) = \sum_{i,j} |x_i - y_j|^2 \gamma_{ij} = \sum_{i=1}^{N} |x_i|^2 \mu_i + \sum_{j=1}^{M} |y_j|^2 \nu_j + 2 \sum_{i,j} (-x_i \cdot y_j) \gamma_{ij},$$

so that (2.34) gives

$$\text{Cost}(\gamma) - \sum_{i=1}^{N} |x_i|^2 \mu_i - \sum_{j=1}^{M} |y_j|^2 \nu_j = -2 \sum_{i,j} x_i \cdot y_j \, \gamma_{ij} \geq 2 H(\alpha, \beta),$$

that is (2.35). Now, if γ satisfies (iii), then, by the Fenchel–Legendre inequality (2.30), we have $(-f, -f^*) \in \mathcal{H}$, while (2.31) and $S(\gamma) \subset \partial f$ give

$$f(x_i) + f^*(y_j) = x_i \cdot y_j, \qquad \text{if } \gamma_{ij} > 0, \qquad (2.39)$$

which in turn implies that (2.35) holds as an identity if we choose $(\alpha, \beta) = (-f, -f^*)$. This shows at once that (2.36) holds and that γ is a minimizer of $\mathbf{K}_c(\mu, \nu)$. □

The following three remarks concern the lack of uniqueness in the discrete Kantorovich problem.

Remark 2.12   We already know that uniqueness does not hold in problem $\mathbf{K}_c(\mu, \nu)$ for arbitrary data; recall Remark 2.4. However, the following statement (which will be proved in full generality in Theorem 3.15) provides a "uniqueness statement of sorts" for the quadratic transport cost: *If $S = \bigcup_\gamma S(\gamma)$, where γ ranges over all the optimal plans in the quadratic-cost transport problem defined by two discrete measures μ and ν, then S is cyclically monotone; in particular, there exists a convex function f such that $S(\gamma) \subset \partial f$ for every such optimal plan γ.* We interpret this as a uniqueness statement since the subdifferential ∂f appearing in it has the property of "bundling together" all the optimal transport plans of problem $\mathbf{K}_c(\mu, \nu)$. As done in Remark 2.13, this property can be indeed exploited to prove uniqueness in special situations. Notice that the cyclical monotonicity of $S = \bigcup_\gamma S(\gamma)$ is not obvious, since, in general, the union of cyclically monotone sets is not cyclically monotone; see Figure 2.3. The reason why $S = \bigcup_\gamma S(\gamma)$ is, nevertheless, cyclically monotone lies in the linearity of Cost combined with the convexity of $\Gamma(\mu, \nu)$. Together, these two properties imply that the set $\Gamma_{\text{opt}}(\mu, \nu)$ of all optimal transport plans for $\mathbf{K}_c(\mu, \nu)$ is convex: in particular, if γ^1 and γ^2 are optimal in $\mathbf{K}_c(\mu, \nu)$, then $(\gamma^1 + \gamma^2)/2$ is an optimal plan, and thus $S((\gamma^1 + \gamma^2)/2)$ is cyclically monotone, and since

$$S\left(\frac{\gamma^1 + \gamma^2}{2}\right) = S(\gamma^1) \cup S(\gamma^2).$$

we conclude that $S(\gamma^1) \cup S(\gamma^2)$ is also cyclically monotone.

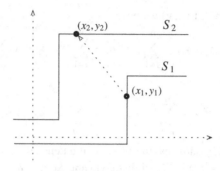

Figure 2.3 Two sets S_1 and S_2 that are cyclically monotone in $\mathbb{R} \times \mathbb{R}$, whose union $S_1 \cup S_2$ is not cyclically monotone. Indeed, by suitably picking points $(x_1, y_1) \in S_1$ and $(x_2, y_2) \in S_2$, we see that $(y_2 - y_1)(x_2 - x_1) < 0$, thus violating the cyclical monotonicity inequality on $\{(x_i, y_i)\}_{i=1}^{2}$.

Remark 2.13 (Uniqueness in dimension one)   When $n = 1$, the statement in Remark 2.12 can be used to prove the uniqueness of minimizers for the discrete Kantorovich problem with quadratic cost. We only discuss this result informally. First of all, we construct a **monotone discrete transport plan** γ^* from μ to ν. Assuming without loss of generality that $\{x_i\}_{i=1}^{N}$ and $\{y_j\}_{j=1}^{M}$ are indexed so that $x_i < x_{i+1}$ and $y_j < y_{j+1}$, we define γ_{ij}^* as follows: if $\mu_1 \leq \nu_1$, then set $\gamma_{11}^* = \mu_1$, and $\gamma_{1j}^* = 0$ for $j \geq 2$; otherwise, we let $j(1)$ be the largest index j such that $\mu_1 > \nu_1 + \ldots + \nu_j$ and set

$$
\gamma_{1j}^* = \begin{cases} \nu_j, & \text{if } 1 \leq j \leq j(1), \\ \mu_1 - (\nu_1 + \ldots + \nu_{j(1)}), & \text{if } j = j(1) + 1, \\ 0, & \text{if } j(1) + 1 < j \leq M. \end{cases}
$$

In this way we have allocated all the mass μ_1 sitting at x_1 among the first $j(1)+1$ receiving sites, with the first $j(1)$ receiving sites completely filled. Next, we start distributing the mass μ_2 at site x_2, start moving the largest possible fraction of it to $y_{j(1)+1}$ (which can now receive a $\nu_{j(1)+1} - [\mu_1 - (\nu_1 + \ldots + \nu_{j(1)})]$ amount of mass), and keep moving any excess mass to the subsequent sites $y_{j(1)+k}$, $k \geq 2$, if needed. Evidently, the resulting transport plan γ^* is such that $S(\gamma^*)$ is contained in the extended graph of an increasing function so that γ^* is indeed optimal in $\mathbf{K}_c(\mu, \nu)$. A heuristic explanation of why this is the unique optimal transport plan is given in Figure 2.4a. For a proof, see Theorem 16.1-(i,ii).

Remark 2.14   We review Remark 2.4 in light of the results of this chapter. Denoting with superscripts the coordinates of points, so that $p = (p^1, p^2)$ is the generic point of $\mathbb{R}^2$, we take

$$
x_1 = (0, 1), \qquad x_2 = (0, -1), \qquad y_1 = (-1, 0), \qquad y_2 = (1, 0).
$$

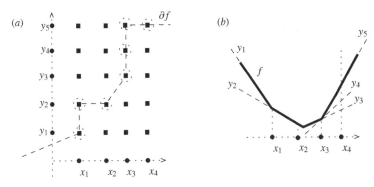

Figure 2.4 (a): A discrete transport problem with quadratic cost and $n = 1$. The black squares indicates all the possible interaction pairs (x_i, y_j). Weights μ_i and ν_j are such that $S(\gamma^*)$ consists of the circled black squares. If γ is another optimal transport plan, then, by the statement in Remark 2.12, $S(\gamma) \cup S(\gamma^*)$ is contained in the subdifferential of a convex function. This implies that $S(\gamma) \setminus S(\gamma^*)$ can only contain either (x_2, y_3) *or* (x_3, y_2). However, given the construction of γ^*, the fact that $S(\gamma^*)$ is "jumping diagonally" from (x_2, y_2) to (x_3, y_3) means that $\mu_1 + \mu_2 = \nu_1 + \nu_2$. So there is no mass left for γ to try something different: if γ activates (x_2, y_3) (i.e., if γ sends a fraction of μ_2 to y_3), then γ must activate (x_3, y_2) (sending a corresponding fraction of μ_3 to y_2 to compensate the first modification), and this violates cyclical monotonicity. Therefore, γ^* is a unique optimal plan for $\mathbf{K}_c(\mu, \nu)$. (b): A geometric representation of a potential f such that $S(\gamma^*) \subset \partial f$. Notice that $\partial f(x_1) = [y_1, y_2]$ (with $y_1 = f'(x_1^-)$ and $y_2 = f'(x_1^+)$), $\partial f(x_2) = \{f'(x_2) = y_2\}$, $\partial f(x_3) = [y_3, y_5]$ (with $y_3 = f'(x_3^-)$, $y_5 = f'(x_3^+)$ and y_4 in the interior of $\partial f(x_3)$), and $\partial f(x_4) = \{f'(x_4) = y_5\}$. Notice that we have large freedom in accommodating the y_js as elements of the subdifferentials $\partial f(x_i)$; in particular, we can find other convex functions g with $S(\gamma^*) \subset \partial g$ and such that $f - g$ is not constant.

No matter what values of μ_i and ν_j are chosen, all the admissible transport plans will have the same cost. If, say, $\mu_i = \nu_j = 1/2$ for all i and j, then all the plans

$$\gamma_{11}^t = t, \qquad \gamma_{12}^t = \frac{1}{2} - t, \qquad \gamma_{21}^t = \frac{1}{2} - t, \qquad \gamma_{22}^t = t,$$

corresponding to $t \in [0, 1/2]$ are optimal. If $t \in (0, 1/2)$, then $S(\gamma^t)$ contains all the four possible pairs $S = \{(x_i, y_j)\}_{i,j}$. How many convex functions (modulo additive constants) can contain S in their subdifferential? Just one. Indeed, the slopes $y_1 = (-1, 0)$ and $y_2 = (1, 0)$ correspond to the affine functions $\ell(p) = a - p^1$ and $m(p) = b + p^1$ for $a, b \in \mathbb{R}$. The only way for $y_1, y_2 \in \partial f(x_1) \cap \partial f(x_2)$ is that the set $\{\ell = m\}$ contains both $x_1 = (0, 1)$ and $x_2 = (0, -1)$. Hence, we must have $a = b$ and, modulo additive constants, there exists a unique convex potential; see Figure 2.5.

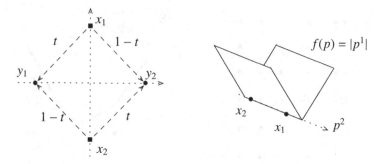

Figure 2.5 An example of discrete transport problem with quadratic cost where we have nonuniqueness of optimal transport plans, but where there is a unique (up to additive constants) convex potential such that the statement in Remark 2.12 holds.

2.5 The Discrete Monge Problem

We close this chapter with a brief discussion of the discrete Monge problem. The following theorem is our main result in this direction.

Theorem 2.15 *If $\{x_i\}_{i=1}^L$ and $\{y_j\}_{j=1}^L$ are families of L distinct points, and μ and ν denote the discrete measures*

$$\mu = \sum_{i=1}^L \frac{\delta_{x_i}}{L}, \qquad \nu = \sum_{j=1}^L \frac{\delta_{y_j}}{L}, \tag{2.40}$$

then for every optimal transport plan γ in $\mathbf{K}_c(\mu,\nu)$ there is a transport map T from μ to ν with the same transport cost as γ; in particular, T is optimal in $\mathbf{M}_c(\mu,\nu)$ and $\mathbf{M}_c(\mu,\nu) = \mathbf{K}_c(\mu,\nu)$.

Remark 2.16   It is interesting to notice that, by a perturbation argument, given an arbitrary pair of discrete probability measures (μ,ν) (i.e., μ and ν satisfy (2.1) and (2.2)), we can find a sequence $\{(\mu^\ell,\nu^\ell)\}_\ell$ of discrete probability measures such that $\mu^\ell \overset{*}{\rightharpoonup} \mu$ and $\nu^\ell \overset{*}{\rightharpoonup} \nu$ as $\ell \to \infty$, and each (μ^ℓ,ν^ℓ) satisfies the assumptions of Theorem 2.15 (with some $L = L_\ell$ in (2.40)). In a first approximation step, we can reduce to the case when all the weights μ_i and ν_j are rational numbers. Writing these weights with a common denominator L, we find $m_i, n_j \in \{1, \ldots, L\}$ such that

$$\mu_i = \frac{m_i}{L}, \qquad \nu_j = \frac{n_j}{L}, \qquad L = \sum_{i=1}^N m_i = \sum_{j=1}^M n_j.$$

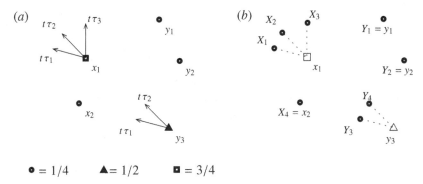

$\bullet = 1/4 \qquad \blacktriangle = 1/2 \qquad \blacksquare = 3/4$

Figure 2.6 The second step in the approximation procedure of Remark 2.16: (a) The starting measures $\mu = (3/4)\delta_{x_1} + (1/4)\delta_{x_2}$ and $\nu = (1/4)(\delta_{y_1} + \delta_{y_2}) + (1/2)\delta_{y_3}$; notice that there is no transport map between these two measures; (b) The measures $\mu^t = (1/4)\sum_{h=1}^4 \delta_{X_h}$ and $\nu^t = (1/4)\sum_{h=1}^4 \delta_{Y_h}$ resulting from the approximation by splitting with $t > 0$. As $t \to 0^+$, these measure weak-star converge to μ and ν respectively. Notice that there is a transport map from μ^t to ν^t corresponding to every permutation of $\{1, \ldots, 4\}$. One of them is optimal for the Monge problem with quadratic cost from μ^t to ν^t.

Then, in a second approximation step, we consider a set of L-distinct unit vectors $\{\tau_k\}_{k=1}^L$ in $\mathbb{R}^n$ and notice that, on the one hand,

$$\sum_{k=1}^{m_i} \frac{1}{L} \delta_{x_i + t\tau_k} \xrightarrow{*} \frac{m_i}{L} \delta_{x_i} \qquad \text{as } t \to 0^+,$$

while, on the other hand, for every sufficiently small but positive t, both

$$\{X_h\}_{h=1}^L = \{x_i + t\tau_k : 1 \le i \le N, 1 \le k \le m_i\},$$
$$\{Y_h\}_{h=1}^L = \{y_j + t\tau_k : 1 \le j \le M, 1 \le k \le n_j\},$$

consist of L-many distinct points. By combining these two approximation steps, we find a sequence $\{(\mu_j, \nu_j)\}_j$ with the required properties; see Figure 2.6.

Proof of Theorem 2.15 We notice that, by (2.40),

$$\Gamma(\mu, \nu) = \left\{ \{\gamma_{ij}\} : \gamma_{ij} = \frac{b_{ij}}{L}, \quad b = \{b_{ij}\} \in \mathcal{B}_L \right\}, \tag{2.41}$$

where $\mathcal{B}_L \subset \mathbb{R}^{L \times L}$ is the set of the $L \times L$-**bistochastic matrices** $b = \{b_{ij}\}$, i.e., for every $i, j, b_{ij} \in [0, 1]$ and $\sum_i b_{ij} = \sum_j b_{ij} = 1$. Therefore, by the Choquet theorem (2.33), if γ is an optimal plan in $\mathbf{K}_c(\mu, \nu)$, then $b = \{b_{ij} = L\gamma_{ij}\}$ is an extremal point of $\mathcal{B}_L$. The latter are characterized as follows (a result known as **the Birkhoff theorem**):

$$\text{permutation matrices are the extremal points of } \mathcal{B}_L, \qquad (2.42)$$

where $b = \{b_{ij}\}$ is a $L \times L$-**permutation matrix** if $b_{ij} = \delta_{j,\sigma(i)}$ for a permutation σ of $\{1,\ldots,L\}$. The proof of Birkhoff's theorem is very similar to the proof of Theorem 2.5 and goes as follows. It is enough to prove that if $b \in \mathcal{B}_L$ and $b_{i(1)j(1)} \in (0,1)$ for some pair $(i(1),j(1))$, then b is not an extremal point of $\mathcal{B}_L$. Indeed, by $\sum_j b_{i(1)j} = 1$ we can find, on the $i(1)$-row of b, an entry $b_{i(1)j(2)} \in (0,1)$ with $j(2) \neq j(1)$. We can then find, in the $j(2)$-column of b, an entry $b_{i(2)j(2)} \in (0,1)$ with $i(2) \neq i(1)$. If we iterate this procedure, there is a first step $k \geq 3$ such that either $j(k) = j(1)$ or $i(k) = i(1)$. In the first case we have identified an *even* number of entries b_{ij} of b; in the second case, discarding the entry $b_{i(1)j(1)}$, we have also identified an *even* number of entries b_{ij} of b; in both cases, these entries are arranged into a closed loop in the matrix representation of b, and belong to $(0,1)$. We can exploit this cyclical structure to define a family of *variations* b^t of b: considering, for notational simplicity, the case when $j(3) = j(1)$, these variations take the form

$$b^t_{i(1)j(1)} = b_{i(1)j(1)} + t, \qquad b^t_{i(1)j(2)} = b_{i(1)j(2)} - t,$$
$$b^t_{i(2)j(1)} = b_{i(2)j(1)} - t, \qquad b^t_{i(2)j(2)} = b_{i(2)j(2)} + t,$$

and $b^t_{ij} = b_{ij}$ otherwise. In this way there is $t_0 \in (0,1)$ such that $b^t = \{b^t_{ij}\} \in \mathcal{B}_L$ whenever $|t| \leq t_0$. In particular, by $b = (b^{t_0} + b^{-t_0})/2$, we see that b is not an extremal point of $\mathcal{B}_L$ and conclude the proof of (2.42).

Having proved that for every discrete optimal transport plan γ in $\mathbf{K}_c(\mu,\nu)$ there is a permutation σ of $\{1,\ldots,L\}$ such that $\gamma_{ij} = \delta_{\sigma(i),j}/L$, we can define a map $T : \{x_i\}_{i=1}^L \to \mathbb{R}^n$ by setting $T(x_i) = y_{\sigma(i)}$. By construction, $T_{\#}\mu = \nu$, with

$$\mathbf{M}_c(\mu,\nu) \leq \int_{\mathbb{R}^n} c(x,T(x))\,d\mu(x) = \text{Cost}(\gamma) = \mathbf{K}_c(\mu,\nu) \leq \mathbf{M}_c(\mu,\nu),$$

where we have used (2.7) and the general inequality $\mathbf{K}_c(\mu,\nu) \leq \mathbf{M}_c(\mu,\nu)$. $\square$

3

The Kantorovich Problem

In this chapter we finally present the general formulation of the Kantorovich problem. In Section 3.1 we extend the notion of transport plan, introduced in the discrete setting in Section 2.1, to the case of general probability measures. We then formulate the Kantorovich problem (Section 3.2), prove the existence of minimizers (Section 3.3), the c-cyclical monotonicity of supports of optimal transport plans (Section 3.4), the Kantorovich duality theorem (Section 3.5), and a few auxiliary, general results on c-cyclical monotonicity (Section 3.6). We close the chapter with Section 3.7, where we examine the Kantorovich duality theorem in the cases of the linear and quadratic transport costs.

3.1 Transport Plans

In Section 2.1, given two discrete probability measures

$$\mu = \sum_{i=1}^{N} \mu_i \, \delta_{x_i}, \qquad \nu = \sum_{j=1}^{M} \nu_j \, \delta_{y_j},$$

we have introduced discrete transport plans from μ to ν as matrices $\gamma = \{\gamma_{ij}\} \in \mathbb{R}^{N \times M}$, with the idea that the entry $\gamma_{ij} \in [0,1]$ represents the amount of mass at x_i to be transported to y_j. We have then imposed the transport conditions

$$\mu_i = \sum_{j=1}^{M} \gamma_{ij}, \qquad \nu_j = \sum_{i=1}^{N} \gamma_{ij}$$

(all the mass sitting at x_i is transported to ν, and all the mass needed at y_j arrives from μ) to express the condition that γ transports μ into ν. When working with arbitrary probability measures $\mu, \nu \in \mathcal{P}(\mathbb{R}^n)$, it is just natural to consider probability measures γ on $\mathbb{R}^n \times \mathbb{R}^n$, rather than matrices, with the idea that $\gamma(E \times F)$ is the amount of mass sitting at the Borel set $E \subset \mathbb{R}^n$ that,

35

under the instructions contained in γ, has to be transported to the Borel set $F \subset \mathbb{R}^n$. The conditions expressing that γ is a **transport plan** from μ to ν are then

$$\gamma(E \times \mathbb{R}^n) = \mu(E), \qquad \gamma(\mathbb{R}^n \times F) = \nu(F), \qquad \forall E, F \in \mathcal{B}(\mathbb{R}^n), \qquad (3.1)$$

or, using the notation of push-forward of measures (see Appendix A.4),

$$\mathbf{p}_\# \gamma = \mu, \qquad \mathbf{q}_\# \gamma = \nu, \tag{3.2}$$

where $\mathbf{p}, \mathbf{q} : \mathbb{R}^n \times \mathbb{R}^n \to \mathbb{R}^n$ are the projection maps defined, for $(x, y) \in \mathbb{R}^n \times \mathbb{R}^n$, by $\mathbf{p}(x, y) = x$ and $\mathbf{q}(x, y) = y$. The **family of the transport plans from μ to ν** is then denoted by

$$\Gamma(\mu, \nu) = \left\{ \gamma \in \mathcal{P}(\mathbb{R}^n \times \mathbb{R}^n) : \mathbf{p}_\# \gamma = \mu, \qquad \mathbf{q}_\# \gamma = \nu \right\}. \tag{3.3}$$

It is easily seen that if μ and ν satisfy (2.1) and (2.2), and $\gamma = \{\gamma_{ij}\} \in \Gamma(\mu, \nu) \subset \mathbb{R}^{N \times M}$ for $\Gamma(\mu, \nu)$ defined as in (2.4), then

$$\gamma = \sum_{i,j} \gamma_{ij} \, \delta_{(x_i, y_j)} \in \Gamma(\mu, \nu)$$

for $\Gamma(\mu, \nu)$ defined as in (3.3). In other words, the notation set in Chapter 2 is compatible with the one just introduced now. It is also convenient to notice that (3.1) can be equivalently formulated in terms of test functions, by saying that

$$\int_{\mathbb{R}^n \times \mathbb{R}^n} \varphi(x) \, d\gamma(x, y) = \int_{\mathbb{R}^n} \varphi \, d\mu, \qquad \int_{\mathbb{R}^n \times \mathbb{R}^n} \varphi(y) \, d\gamma(x, y) = \int_{\mathbb{R}^n} \varphi \, d\nu, \tag{3.4}$$

whenever $\varphi \in C_b^0(\mathbb{R}^n)$ or $\varphi : \mathbb{R}^n \to [0, \infty]$ is a Borel function.

The notion of transport plan just introduced is deeper than it may seem at first sight. There is a subtle indeterminacy in the transport instructions contained in $\gamma \in \Gamma(\mu, \nu)$. Indeed, given $x \in \mathbb{R}^n$, we cannot really answer the question "Where are we taking the mass sitting at x?" All we can say is that, given $r > 0$, we must take a $\gamma(B_r(x) \times B_r(y))$ amount of the mass stored by μ in $B_r(x)$ and transport it entirely inside $B_r(y)$. The stochastic character of transport plans will be further investigated in Section 17.3. For the moment, it is sufficient to clarify that transport maps can indeed be seen as transport plans.

Proposition 3.1 (Transport maps induce transport plans) *If $\mu, \nu \in \mathcal{P}(\mathbb{R}^n)$ and T transports μ (i.e., μ is concentrated on a Borel set $E \subset \mathbb{R}^n$ and $T : E \to \mathbb{R}^n$ is a Borel map), then T is a transport map from μ to ν ($T_\# \mu = \nu$) if and only if*

$$\gamma_T = (\mathrm{id} \times T)_\# \mu \tag{3.5}$$

is a transport plan from μ to ν. Here, $(\mathrm{id} \times T) : \mathbb{R}^n \times \mathbb{R}^n \to \mathbb{R}^n$ is defined by

$$(\mathrm{id} \times T)(x) = (x, T(x)), \qquad \forall x \in \mathbb{R}^n.$$

Proof Since $\mu \in \mathcal{P}(\mathbb{R}^n)$, it is obvious that $\gamma_T \in \mathcal{P}(\mathbb{R}^n \times \mathbb{R}^n)$. Moreover, using the general fact that *the push-forward by the composition is the composition of the push-forwards*, we see that $\mathbf{p}_\# \gamma_T = \mu$ and $\mathbf{q}_\# \gamma_T = T_\# \mu$: indeed, if $F \subset E$ is a Borel set, then

$$(\mathbf{p}_\# \gamma_T)(F) = ((\mathbf{id} \times T)_\# \mu)(F \times \mathbb{R}^n)$$
$$= \mu(\{x : (x, T(x)) \in F \times \mathbb{R}^n\}) = \mu(F),$$
$$(\mathbf{q}_\# \gamma_T)(F) = ((\mathbf{id} \times T)_\# \mu)(\mathbb{R}^n \times F) = \mu(\{x : (x, T(x)) \in \mathbb{R}^n \times F\})$$
$$= \mu(T^{-1}(F)) = (T_\# \mu)(F).$$

As a consequence, (3.2) holds if and only if $T_\# \mu = \nu$. □

Remark 3.2 We notice that *transport plans always exist*, since

$$\mu \times \nu \in \Gamma(\mu, \nu), \qquad \forall \mu, \nu \in \mathcal{P}(\mathbb{R}^n).$$

Indeed, if $\gamma = \mu \times \nu$ and $E \in \mathcal{B}(\mathbb{R}^n)$, then

$$\mathbf{p}_\# \gamma(E) = (\mu \times \nu)(\mathbf{p}^{-1}(E)) = (\mu \times \nu)(E \times \mathbb{R}^n) = \mu(E)\,\nu(\mathbb{R}^n) = \mu(E),$$

and, similarly, $\mathbf{q}_\# \gamma = \nu$. The fact that $\gamma = \mu \times \nu$ is always a transport plan is somehow reflected in the fact that, as a transport plan, $\mu \times \nu$ is indeed giving the most generic transport instructions (and, thus, the less likely transport-cost efficient ones!). Indeed, $\gamma(E \times F) = \mu(E)\,\nu(F)$ means that the amount of mass stored by μ at E has to be distributed uniformly along all possible "receiving sites" F for ν (i.e., those sets with $\nu(F) > 0$).

3.2 Formulation of the Kantorovich Problem

A **(transport) cost** is simply a Borel function $c : \mathbb{R}^n \times \mathbb{R}^n \to [0, \infty]$, so $c(x, y)$ represents the cost to transport a unit mass from location x to location y. Although our focus will be mostly on the cases of the **linear cost** (appearing in Monge's original problem (1.2)) $c(x, y) = |x - y|$ and the **quadratic cost** $c(x, y) = |x - y|^2$, in this section we will work with general cost functions. This choice has the advantage of not obscuring the true nature of some very general facts about transport problems, which have nothing to do with the linear or the quadratic cost; moreover, there are many examples of transport costs (beyond the linear and the quadratic case) that have theoretical and practical importance.

Given a transport plan $\gamma \in \Gamma(\mu, \nu)$, the **transport cost of** γ is given by

$$\int_{\mathbb{R}^n \times \mathbb{R}^n} c(x, y)\, d\gamma(x, y), \tag{3.6}$$

and the **Kantorovich problem** (with general cost c, from μ to ν) is the minimization problem

$$\mathbf{K}_c(\mu, \nu) = \inf\left\{ \int_{\mathbb{R}^n \times \mathbb{R}^n} c(x, y)\, d\gamma(x, y) : \gamma \in \Gamma(\mu, \nu)\right\}. \qquad (3.7)$$

Minimizers in $\mathbf{K}_c(\mu, \nu)$ are called **optimal transport plans**. Of course, using the notion of transport map, we can formulate a Monge problem for every cost function c by setting

$$\mathbf{M}_c(\mu, \nu) = \inf\left\{ \int_{\mathbb{R}^n \times \mathbb{R}^n} c(x, T(x))\, d\mu(x) : T_\# \mu = \nu\right\}, \qquad (3.8)$$

and we always have

$$\mathbf{M}_c(\mu, \nu) \geq \mathbf{K}_c(\mu, \nu). \qquad (3.9)$$

To prove (3.9) we notice that by Proposition 3.1, if T is a transport map from μ to ν and $\gamma_T = (\mathbf{id} \times T)_\# \mu$, then $\gamma_T \in \Gamma(\mu, \nu)$, and thus

$$\mathbf{K}_c(\mu, \nu) \leq \int_{\mathbb{R}^n \times \mathbb{R}^n} c(x, y)\, d\gamma_T(x, y) = \int_{\mathbb{R}^n} c(x, T(x))\, d\mu(x).$$

The crucial consequence of (3.9) is that it suggests an approach to the Monge problem: first, prove the existence of an optimal transport plan in $\mathbf{K}_c(\mu, \nu)$; second, when possible, show that optimal transport plans in $\mathbf{K}_c(\mu, \nu)$ are induced by transport maps as in (3.5). Of course, when this strategy works, it implies that (3.9) holds as an identity, therefore the interest of the following two remarks (see also Remark 4.1 in Chapter 4):

Remark 3.3 ($\mathbf{M}_c > \mathbf{K}_c$ may happen) If $c \in L^1(\mu \times \nu)$ (so that $\mathbf{K}_c(\mu, \nu) < \infty$), but there are no transport maps from μ to ν (so that $\mathbf{M}_c(\mu, \nu) = +\infty$), then (3.9) is obviously a strict inequality. In Remark 2.1 we have already seen that this can happen in the framework of discrete transport problems. More generally, we expect similar problems to emerge whenever μ has atoms, namely, whenever there is $x \in \mathbb{R}^n$ such that $\mu(\{x\}) > 0$, and thus it may be necessary to "split $T(x)$ into more than one value" to achieve the transport condition toward a given ν.

Remark 3.4 ($\mathbf{M}_c = \mathbf{K}_c$ if μ has no atoms) However, under very general conditions on the cost function, *if μ has no atoms* (i.e., if $\mu(\{x\}) = 0$ for every $x \in \mathbb{R}^n$), *then (3.9) holds as an equality*. The idea behind this result is that, when μ has no atoms, then every $\gamma \in \Gamma(\mu, \nu)$ can be approximated by maps-induced transport plans γ_T with the desired precision in transport cost. The formalization of this idea requires deeper technical tools than what is advisable to employ at this stage of our discussion, which will be postponed until Chapter 17; see, in particular, Theorem 17.11 therein.

Remark 3.5 (Symmetric costs)   Notice that we are not requiring the cost function to be symmetric, that is, to satisfy $c(x,y) = c(y,x)$ for every $x,y \in \mathbb{R}^n$. When this is the case, like for the linear or the quadratic cost, we have

$$\mathbf{K}_c(\mu,\nu) = \mathbf{K}_c(\nu,\mu). \tag{3.10}$$

Indeed, if we let $R : \mathbb{R}^n \times \mathbb{R}^n \to \mathbb{R}^n \times \mathbb{R}^n$ be the reflection map $R(x,y) = R(y,x)$, then $\gamma \in \Gamma(\mu,\nu)$ if and only if $R_\# \gamma \in \Gamma(\nu,\mu)$, and

$$\int_{\mathbb{R}^n \times \mathbb{R}^n} c(x,y)\, d[R_\# \gamma](x,y) = \int_{\mathbb{R}^n \times \mathbb{R}^n} c(y,x)\, d\gamma(x,y).$$

In particular, if c is symmetric, then γ and $R_\# \gamma$ have the same transport cost, and thus (3.10) follows.

Remark 3.6   Whenever μ and ν are discrete measures (see (2.1) and (2.2)) and we denote by γ both a discrete transport plan $\{\gamma_{ij}\} \in \mathbb{R}^{N \times M}$ from μ to ν and its realization as transport plan $\sum_{i,j} \gamma_{ij} \delta_{(x_i, y_j)} \in \mathcal{P}(\mathbb{R}^n \times \mathbb{R}^n)$, we have the cost function $\mathrm{Cost}(\gamma)$ defined in (2.5) and the one defined in (3.6) agree, since

$$\int_{\mathbb{R}^n \times \mathbb{R}^n} c\, d\gamma = \sum_{ij} c(x_i, y_j)\, \gamma_{ij}.$$

In particular, it is possible to see the theory developed in Chapter 2 as a particular case of the theory developed in this chapter.

We close this section with a useful remark, where we summarize the relations between spt μ, spt ν, and spt γ when $\gamma \in \Gamma(\mu,\nu)$.

Remark 3.7 (On the supports of transport plans)   If $\mu,\nu \in \mathcal{P}(\mathbb{R}^n)$ and $\gamma \in \Gamma(\mu,\nu)$, then we have

$$\mathrm{spt}\,\gamma \subset \mathrm{spt}\,\mu \times \mathrm{spt}\,\nu. \tag{3.11}$$

Moreover,

$$\mu \text{ is concentrated on } \mathbf{p}(\mathrm{spt}\,\gamma) \text{ and } \nu \text{ on } \mathbf{q}(\mathrm{spt}\,\gamma), \tag{3.12}$$

and, finally,

if spt ν is compact, then $\forall x \in \mathrm{spt}\,\mu$ there exists $y \in \mathrm{spt}\,\nu$ s.t. $(x,y) \in \mathrm{spt}\,\gamma.$
$$\tag{3.13}$$

To prove (3.11), we notice that (A.1) gives

$$\mathrm{spt}\,\gamma = \left\{ (x,y) \in \mathbb{R}^n \times \mathbb{R}^n : \gamma(B_r(x) \times B_r(y)) > 0 \quad \forall r > 0 \right\}, \tag{3.14}$$

so that (3.11) follows by $\gamma(B_r(x) \times \mathbb{R}^n) = \mu(B_r(x))$ and $\gamma(\mathbb{R}^n \times B_r(y)) = \nu(B_r(y))$. *To prove* (3.12), we notice that if $\mu(\mathbb{R}^n \setminus \mathbf{p}(\mathrm{spt}\,\gamma)) > 0$, then $\gamma(A) > 0$ for the open set $A = [\mathbb{R}^n \setminus \mathbf{p}(\mathrm{spt}\,\gamma)] \times \mathbb{R}^n$. Thus, there exists $(x,y) \in A \cap \mathrm{spt}\,\gamma$

so that $x \in \mathbf{p}(\mathrm{spt}\,\gamma)$, a contradiction to $x \in A$. Before proving (3.13) we notice that the property fails without assuming the compactness of ν: for example, let

$$\gamma = \sum_{j=1}^{\infty} \frac{1}{2^j} \delta_{(1/j,j)}, \qquad \mu = \sum_{j=1}^{\infty} \frac{1}{2^j} \delta_{1/j}, \qquad \nu = \sum_{j=1}^{\infty} \frac{1}{2^j} \delta_j,$$

then $0 \in \mathrm{spt}\,\mu$, but there is no $y \in \mathrm{spt}\,\nu$ such that $(0,y) \in \mathrm{spt}\,\gamma$. *To prove* (3.13): If $x \in \mathrm{spt}\,\mu$, then for every $r > 0$ we have $\gamma(B_r(x) \times \mathbb{R}^n) > 0$. The measure $\lambda(E) = \gamma(B_r(x) \times E)$ must have non-empty support, so that for every $r > 0$ we can find $y(r) \in \mathbb{R}^n$ such that $\gamma(B_r(x) \times B_s(y(r))) > 0$ for every $s > 0$. Notice that $y(r) \in \mathrm{spt}\,\nu$ since $0 < \gamma(B_r(x) \times B_s(y(r))) \leq \gamma(\mathbb{R}^n \times B_s(y(r))) = \nu(B_s(y(r)))$ for every $s > 0$. By compactness of $\mathrm{spt}\,\nu$, given $r_j \to 0^+$ we can find $y_j \in \mathrm{spt}\,\nu$ with $y_j \to y$ and $y \in \mathrm{spt}\,\nu$ such that $\gamma(B_{r_j}(x) \times B_s(y_j)) > 0$ for every $s > 0$. In particular, for every $\rho > 0$ we can find j large enough so that $\rho > r_j$, $B_{\rho/2}(y_j) \subset B_\rho(y)$, and thus

$$\gamma(B_\rho(x) \times B_\rho(y)) \geq \gamma(B_{r_j}(x) \times B_{\rho/2}(y_j)) > 0,$$

thus proving that $(x,y) \in \mathrm{spt}\,\gamma$ with $y \in \mathrm{spt}\,\nu$.

3.3 Existence of Minimizers in the Kantorovich Problem

As in the discrete case, the existence of minimizers in the Kantorovich problem is easily obtained as a reflection of the fact that we are minimizing the linear functional defined in (3.6) over the convex set $\Gamma(\mu,\nu)$. The main issue to be addressed is the compactness of $\Gamma(\mu,\nu)$, since now, at variance with the discrete case, we have an infinite-dimensional competition class. This is the content of the following theorem, where we prove the existence of minimizers in the Kantorovich problem under the sole assumption that the cost c is lower semicontinuous.

Theorem 3.8 *If $c : \mathbb{R}^n \times \mathbb{R}^n \to [0, \infty]$ is lower semicontinuous, then for every $\mu, \nu \in \mathcal{P}(\mathbb{R}^n)$ there exists a minimizer γ of $\mathbf{K}_c(\mu,\nu)$.*

Proof If $\mathbf{K}_c(\mu,\nu) = +\infty$, then every $\gamma \in \Gamma(\mu,\nu)$ is a minimizer in $\mathbf{K}_c(\mu,\nu)$. Therefore, we assume that $\mathbf{K}_c(\mu,\nu) < +\infty$ and consider a minimizing sequence $\{\gamma_j\}_j$ for $\mathbf{K}_c(\mu,\nu)$, i.e.,

$$\mathbf{K}_c(\mu,\nu) = \lim_{j \to \infty} \int_{\mathbb{R}^n \times \mathbb{R}^n} c \, d\gamma_j, \qquad \gamma_j \in \Gamma(\mu,\nu). \tag{3.15}$$

Since $\gamma_j(\mathbb{R}^n \times \mathbb{R}^n) = 1$ for every j, by the compactness theorem for Radon measures there exists a Radon measure γ on $\mathbb{R}^n \times \mathbb{R}^n$ such that, up to extracting subsequences, $\gamma_j \overset{*}{\rightharpoonup} \gamma$ as $j \to \infty$. Aiming at applying the narrow convergence criterion of Proposition A.1, we now show that for every $\varepsilon > 0$ we can find $K_\varepsilon \subset \mathbb{R}^n \times \mathbb{R}^n$ compact with

$$\sup_{j \in \mathbb{N}} \gamma_j ((\mathbb{R}^n \times \mathbb{R}^n) \setminus K_\varepsilon) < \varepsilon. \tag{3.16}$$

Indeed, since both $\mu(\mathbb{R}^n)$ and $\nu(\mathbb{R}^n)$ are finite, we can find compact sets K^μ and K^ν in $\mathbb{R}^n$ such that $\mu(\mathbb{R}^n \setminus K^\mu) < \varepsilon/2$ and $\nu(\mathbb{R}^n \setminus K^\nu) < \varepsilon/2$; since $K_\varepsilon = K^\mu \times K^\nu$ is compact in $\mathbb{R}^n \times \mathbb{R}^n$ and since

$$(\mathbb{R}^n \times \mathbb{R}^n) \setminus K_\varepsilon \subset \left[\left(\mathbb{R}^n \setminus K^\mu \right) \times \mathbb{R}^n \right] \cup \left[\mathbb{R}^n \times \left(\mathbb{R}^n \setminus K^\nu \right) \right],$$

by $\gamma_j \in \Gamma(\mu, \nu)$ we find

$$\gamma_j ((\mathbb{R}^n \times \mathbb{R}^n) \setminus K_\varepsilon) \leq \gamma_j \left[\left(\mathbb{R}^n \setminus K^\mu \right) \times \mathbb{R}^n \right] + \gamma_j \left[\mathbb{R}^n \times \left(\mathbb{R}^n \setminus K^\nu \right) \right]$$
$$= \mu(\mathbb{R}^n \setminus K^\mu) + \nu(\mathbb{R}^n \setminus K^\nu) < \varepsilon.$$

Having proved (3.16), we deduce from Proposition A.1 and from $\gamma_j(\mathbb{R}^n \times \mathbb{R}^n) = 1$ for every j that $\gamma(\mathbb{R}^n \times \mathbb{R}^n) = 1$ and that γ_j narrowly converges to γ on $\mathbb{R}^n \times \mathbb{R}^n$. In particular, since $\varphi \in C_b^0(\mathbb{R}^n)$ implies $\varphi \circ \mathbf{p} \in C_b^0(\mathbb{R}^n \times \mathbb{R}^n)$, we find that

$$\int_{\mathbb{R}^n} \varphi \, d\mu = \int_{\mathbb{R}^n \times \mathbb{R}^n} (\varphi \circ \mathbf{p}) \, d\gamma_j \to \int_{\mathbb{R}^n \times \mathbb{R}^n} (\varphi \circ \mathbf{p}) \, d\gamma \qquad \text{as } j \to \infty,$$

that is

$$\int_{\mathbb{R}^n \times \mathbb{R}^n} \varphi(x) \, d\gamma(x, y) = \int_{\mathbb{R}^n} \varphi \, d\mu \qquad \forall \varphi \in C_b^0(\mathbb{R}^n).$$

This shows $\mathbf{p}_\# \gamma = \mu$, and, similarly, we have $\mathbf{q}_\# \gamma = \nu$. To complete the proof we notice that, since c is lower semicontinuous on $\mathbb{R}^n \times \mathbb{R}^n$, the super level sets $\{c > t\}$ of c are open sets. Thus, by (A.8) we have

$$\liminf_{j \to \infty} \gamma_j (\{c > t\}) \geq \gamma(\{c > t\}) \qquad \forall t \in \mathbb{R},$$

so, by applying the layer cake formula (A.2) twice and by Fatou's lemma, we find

$$\mathbf{K}_c(\mu, \nu) = \lim_{j \to \infty} \int_{\mathbb{R}^n \times \mathbb{R}^n} c \, d\gamma_j = \lim_{j \to \infty} \int_0^\infty \gamma_j(\{c > t\}) \, dt$$
$$\geq \int_0^\infty \liminf_{j \to \infty} \gamma_j(\{c > t\}) \, dt \geq \int_0^\infty \gamma(\{c > t\}) \, dt$$
$$= \int_{\mathbb{R}^n \times \mathbb{R}^n} c \, d\gamma \geq \mathbf{K}_c(\mu, \nu),$$

where in the last inequality we have used the fact that $\gamma \in \Gamma(\mu, \nu)$. This proves that γ is a minimizer of $\mathbf{K}_c$. $\qquad \square$

3.4 c-Cyclical Monotonicity with General Measures

Next, in analogy with Theorem 2.5 for the discrete transport problem, we relate the property of being an optimal transport plan to c-cyclical monotonicity.

Theorem 3.9 *If $c : \mathbb{R}^n \times \mathbb{R}^n \to [0, \infty)$ is a continuous function, $\mu, \nu \in \mathcal{P}(\mathbb{R}^n)$ are such that $\mathbf{K}_c(\mu, \nu) < \infty$, and γ is an optimal transport plan in $\mathbf{K}_c(\mu, \nu)$, then*

$$\sum_{i=1}^{N} c(x_i, y_i) \leq \sum_{i=1}^{N} c(x_{i+1}, y_i)$$

whenever $\{(x_i, y_i)\}_{i=1}^{N} \subset \operatorname{spt} \gamma$ and $x_{N+1} = x_1$. In particular, $\operatorname{spt} \gamma$ is c-cyclically monotone.

Remark 3.10 (Non-crossing condition for the linear cost)   If γ is an optimal transport plan in $\mathbf{K}_c(\mu, \nu)$ with $c(x, y) = |x - y|$ and $(x, y), (x', y') \in \operatorname{spt} \gamma$ (so that, in particular, $x, x' \in \operatorname{spt} \mu$ and $y, y' \in \operatorname{spt} \nu$ by (3.11)), then the c-cyclical monotonicity of γ with respect to the linear cost implies

$$|x - y| + |x' - y'| \leq |x - y'| + |x' - y|. \tag{3.17}$$

Geometrically, this means that if the segments $[x, y] = \{(1-t)\, x + t\, y : t \in [0, 1]\}$ and $[x', y']$ have nontrivial intersection, then they either have one common endpoint (i.e., $x = x'$ and/or $y = y'$) or have the same orientation (i.e., $(y - x)/|y - x| = (y' - x')/|y' - x'|$). This *non-crossing condition*, which was already met in (1.18), will play a crucial role in solving the Monge problem in Chapter 18; see (18.21).

Proof of Theorem 3.9   As in the proof of the Theorem 2.5, given $\{(x_i, y_i)\}_{i=1}^{N} \subset \operatorname{spt} \gamma$ and setting $x_{N+1} = x_1$, we want to create variations of γ by *not transporting* an ε of the mass that γ transports from x_i to y_i and by sending instead to each to y_{i+1} an ε of the mass stored by μ at x_i. The difference is that now we are not working with discrete measures, so we cannot really identify objects like "the mass sent by γ from x_i to y_i" but have rather to work with $\gamma(B_r(x_i) \times B_r(y_i))$ for a value of r positive and small. This said, for every $r > 0$ and $(x_i, y_i) \in \operatorname{spt} \gamma$ we set

$$\varepsilon_i = \gamma(B_r(x_i) \times B_r(y_i)) > 0,$$

$$\gamma_i = \frac{1}{\varepsilon_i}\, \gamma \llcorner [B_r(x_i) \times B_r(y_i)] \in \mathcal{P}(\mathbb{R}^n \times \mathbb{R}^n),$$

$$\mu_i = \mathbf{p}_\# \gamma_i \in \mathcal{P}(\mathbb{R}^n),$$

$$\nu_i = \mathbf{q}_\# \gamma_i \in \mathcal{P}(\mathbb{R}^n),$$

so that γ_i encodes "what γ does near (x_i, y_i)," while $\mu_{i+1} \times \nu_i$ is a good approximation of the operation of "taking the mass transported by γ from x_{i+1} to y_{i+1}, and sending it to y_i instead." With this insight, we define our family of competitors parameterized over $0 < \varepsilon < \min_{i=1,...,N} \varepsilon_i$ by setting

$$\gamma^\varepsilon = \gamma - \frac{\varepsilon}{N} \sum_{i=1}^{N} \gamma_i + \frac{\varepsilon}{N} \sum_{i=1}^{N} \mu_{i+1} \times \nu_i.$$

We first check that $\gamma^\varepsilon \in \mathcal{P}(\mathbb{R}^n \times \mathbb{R}^n)$: indeed,

$$\gamma^\varepsilon(\mathbb{R}^n \times \mathbb{R}^n) = 1 - \frac{\varepsilon}{N} \sum_{i=1}^{N} \gamma_i(\mathbb{R}^n \times \mathbb{R}^n) + \frac{\varepsilon}{N} \sum_{i=1}^{N} \mu_{i+1}(\mathbb{R}^n) \nu_i(\mathbb{R}^n) = 1,$$

while for every $A \subset \mathbb{R}^n \times \mathbb{R}^n$ we have $\gamma^\varepsilon(A) \geq 0$, since $\varepsilon/\varepsilon_i \leq 1$, and thus

$$\gamma^\varepsilon(A) \geq \gamma(A) - \frac{\varepsilon}{N} \sum_{i=1}^{N} \frac{\gamma(A \cap (B_r(x_i) \times B_r(y_i)))}{\varepsilon_i} \geq \gamma(A) - \frac{1}{N} \sum_{i=1}^{N} \gamma(A) = 0.$$
$$(3.18)$$

Finally, denoting by ω a uniform modulus of continuity for c at the points $\{(x_i, y_i)\}_{i=1}^{N} \cup \{(x_{i+1}, y_i)\}_{i=1}^{N}$, we notice that since γ_i is concentrated on $B_r(x_i) \times B_r(y_i)$ and $\mu_{i+1} \times \nu_i$ is concentrated on $B_r(x_{i+1}) \times B_r(y_i)$, we have

$$0 \leq \int_{\mathbb{R}^n \times \mathbb{R}^n} c \, d\gamma^\varepsilon - \int_{\mathbb{R}^n \times \mathbb{R}^n} c \, d\gamma$$

$$= -\frac{\varepsilon}{N} \sum_{i=1}^{N} \int_{\mathbb{R}^n \times \mathbb{R}^n} c \, d\gamma_i + \frac{\varepsilon}{N} \sum_{i=1}^{N} \int_{\mathbb{R}^n \times \mathbb{R}^n} c \, d[\mu_{i+1} \times \nu_i]$$

$$\leq -\frac{\varepsilon}{N} \sum_{i=1}^{N} \left(c(x_i, y_i) - \omega(r) \right) + \frac{\varepsilon}{N} \sum_{i=1}^{N} \left(c(x_{i+1}, y_i) + \omega(r) \right).$$

Dividing by $\varepsilon > 0$, we find that

$$\sum_{i=1}^{N} c(x_i, y_i) \leq 2 N \omega(r) + \sum_{i=1}^{N} c(x_{i+1}, y_i),$$

and letting $r \to 0^+$ we conclude the proof. $\qquad\square$

3.5 Kantorovich Duality for General Transport Costs

Having proven c-cyclical monotonicity of spt γ to be necessary for optimality in $\mathbf{K}_c(\mu, \nu)$, we now discuss its sufficiency. The idea is to repeat the analysis performed in the case of the discrete transport problem with quadratic cost in Section 2.4. In that case the crucial step for proving sufficiency, discussed in the implication "(iii) implies (i)" in Theorem 2.10, was exploiting the relation of cyclical monotonicity to the notion of subdifferential of a convex function, and the relation between subdifferentials and equality cases in the Fenchel–Legendre inequality. We thus set ourselves to the task of introducing an appropriate notion of convexity related to the cost c and then develop related concepts of subdifferential and Fenchel–Legendre duality, with the final goal of repeating the argument of Theorem 2.10.

The appropriate definitions are correctly guessed by recalling how convexity is related to the quadratic cost transport problem, and in particular how the Euclidean scalar product, which plays a crucial role in the definitions of subdifferential and of Fenchel–Legendre transform, enters into the proof of Theorem 2.10. In this vein, we can notice that the identity

$$\int_{\mathbb{R}^n \times \mathbb{R}^n} |x-y|^2 \, d\gamma(x,y) = \int_{\mathbb{R}^n} |x|^2 \, d\mu(x) + \int_{\mathbb{R}^n} |y|^2 \, d\nu(y) - 2 \int_{\mathbb{R}^n \times \mathbb{R}^n} x \cdot y \, d\gamma(x,y)$$

implies that $\mathbf{K}_c(\mu, \nu)$ with $c(x,y) = |x - y|^2$ is actually equivalent to $\mathbf{K}_c(\mu, \nu)$ with $c(x,y) = -x \cdot y$ and then make the educated guess that we should introduce notions of c-convexity, of c-subdifferential, and of c-Fenchel–Legendre transform, by systematically replacing the scalar product $x \cdot y$ with $-c(x,y)$ in the corresponding definitions for convex functions from Section 2.3. We thus proceed to consider the following definitions:

c-**convexity:** Following (2.29), we say that $f : \mathbb{R}^n \to \mathbb{R} \cup \{+\infty\}$ is c-**convex** if there is a function $\alpha : \mathbb{R}^n \to \mathbb{R} \cup \{-\infty\}$ such that

$$f(x) = \sup \{\alpha(y) - c(x,y) : y \in \mathbb{R}^n\} \qquad \forall x \in \mathbb{R}^n. \tag{3.19}$$

We set $\text{Dom}(f) = \{f < \infty\}$ and notice that if c is continuous, then every c-convex function is lower semicontinuous on $\mathbb{R}^n$. When $c(x,y) = -x \cdot y$, we are just defining convex, lower-semicontinuous functions on $\mathbb{R}^n$, so that $f = f^{**}$. In the convex case we are taking of course $\alpha(y) = -f^*(y)$. Since f^* takes values in $\mathbb{R} \cup \{+\infty\}$, we have allowed α to take values in $\mathbb{R} \cup \{-\infty\}$.

c-**subdifferential of a c-convex function:** If f is c-convex, then, in analogy with (2.18) and (2.21) (and noticing that, if $c(x,y) = -x \cdot y$, then $y \cdot (z - x) = c(x,y) - c(z,y)$ for every $x, y, z \in \mathbb{R}^n$), we set

$$\partial_c f(x) = \left\{ y \in \mathbb{R}^n : f(z) \geq f(x) + c(x,y) - c(z,y) \quad \forall z \in \mathbb{R}^n \right\}, \tag{3.20}$$

$$\partial_c f = \bigcup_{x \in \mathbb{R}^n} \{x\} \times \partial_c f(x) \subset \mathbb{R}^n \times \mathbb{R}^n. \tag{3.21}$$

Clearly, if the cost c is continuous and f is c-convex (and thus lower semicontinuous), then $\partial_c f$ is closed in $\mathbb{R}^n \times \mathbb{R}^n$.

c-Fenchel-Legendre transform: If f is c-convex, then, in analogy with (2.28), we set

$$f^c(y) = \sup\{-c(x,y) - f(x) : x \in \mathbb{R}^n\} \qquad \forall y \in \mathbb{R}^n. \tag{3.22}$$

Finally, we recall that $S \subset \mathbb{R}^n \times \mathbb{R}^n$ is c-**cyclically monotone** if for every $\{(x_i, y_i)\}_{i=1}^N \subset S$ one has

$$\sum_{i=1}^N c(x_i, y_i) \le \sum_{i=1}^N c(x_{i+1}, y_i) \qquad \text{where } x_{N+1} = x_1. \tag{3.23}$$

The importance of definitions (3.19), (3.20), (3.21), and (3.22) is established by the following two results.

Theorem 3.11 (c-Rockafellar theorem) *Let $c : \mathbb{R}^n \times \mathbb{R}^n \to \mathbb{R}$ be given. A set $S \subset \mathbb{R}^n \times \mathbb{R}^n$ is c-cyclically monotone if and only if there exists a c-convex function $f : \mathbb{R}^n \to \mathbb{R} \cup \{+\infty\}$ such that*

$$S \subset \partial_c f. \tag{3.24}$$

Moreover, in the "only if" implication one can take f to be lower semicontinuous as soon as c is continuous.

Proposition 3.12 (c-Fenchel inequality) *Let $c : \mathbb{R}^n \times \mathbb{R}^n \to \mathbb{R}$ be given. If $f : \mathbb{R}^n \to \mathbb{R} \cup \{+\infty\}$ is a c-convex function, then*

$$c(x,y) \ge -f(x) - f^c(y) \qquad \forall(x,y) \in \mathbb{R}^n \times \mathbb{R}^n, \tag{3.25}$$
$$c(x,y) = -f(x) - f^c(y) \qquad \text{if and only if} \quad y \in \partial_c f(x). \tag{3.26}$$

Before proving Theorem 3.11 and Proposition 3.12, we show how they are used in the analysis of the Kantorovich problem, and, in particular, in proving that c-cyclical monotonicity of supports is effectively characterizing optimal transports plans, and in showing the validity of the duality formula (3.27) (which generalizes (2.36)):

Theorem 3.13 (Kantorovich theorem) *Let $c : \mathbb{R}^n \times \mathbb{R}^n \to [0, \infty)$ be a continuous function, $\mu, \nu \in \mathcal{P}(\mathbb{R}^n)$, and $c \in L^1(\mu \times \nu)$. Then, for every $\gamma \in \Gamma(\mu, \nu)$, the following three statements are equivalent:*

(i) *γ is an optimal transport plan in $\mathbf{K}_c(\mu, \nu)$;*
(ii) *spt γ is c-cyclically monotone in $\mathbb{R}^n \times \mathbb{R}^n$;*
(iii) *there exists a (lower semicontinuous) c-convex function $f : \mathbb{R}^n \to \mathbb{R} \cup \{+\infty\}$ such that spt $\gamma \subset \partial_c f$.*

Moreover, there exists a c-convex function $f : \mathbb{R}^n \to \mathbb{R} \cup \{+\infty\}$ *such that* $f \in L^1(\mu)$, $f^c \in L^1(\nu)$, *and*

$$\mathbf{K}_c(\mu, \nu) = \sup_{(\alpha, \beta) \in \mathcal{A}} \int_{\mathbb{R}^n} \alpha \, d\mu + \int_{\mathbb{R}^n} \beta \, d\nu \qquad (3.27)$$

$$= \int_{\mathbb{R}^n} (-f) \, d\mu + \int_{\mathbb{R}^n} (-f^c) \, d\nu,$$

where $\mathcal{A}$ *is the family of those pairs* $(\alpha, \beta) \in L^1(\mu) \times L^1(\nu)$ *such that*

$$\alpha(x) + \beta(y) \le c(x, y) \qquad \forall (x, y) \in \mathbb{R}^n \times \mathbb{R}^n. \qquad (3.28)$$

Remark 3.14 (Kantorovich dual problem in Economics) The maximization problem appearing in (3.27) is known as the **Kantorovich dual problem**, and (3.27) has the following economic interpretation. Let c_{ship} be the transport cost paid by a shipping company, and let c be the one that an individual would pay by transporting mass by their own means. Expressing both costs in the same currency, if the shipping company is well organized, then we arguably have $c_{\text{ship}}(x, y) \le c(x, y)$ (with a large gap if $|x - y|$ is large enough). The shipping company wants to establish prices $\alpha(x)$ and $\beta(y)$ for the services of, respectively, picking up a unit mass at a location x and of delivering a unit mass at location y. As soon as condition (3.28) holds, it will be convenient for any individual to delegate their transport needs to the shipping company. An optimal pair $(-f, -f^c)$ in (3.27) represents the maximal prices the shipping company can ask to be convenient for individuals. The actual prices will need to be lower than that to attract customers, but still such that $\alpha(x) + \beta(y) > c_{\text{ship}}(x, y)$ in order to keep the company profitable.

Proof of Theorem 3.13   Theorem 3.9 shows that (i) implies (ii), while Theorem 3.11 shows that (ii) implies (iii). We now show that if $\gamma \in \Gamma(\mu, \nu)$ and there exists a c-convex function $f : \mathbb{R}^n \to \mathbb{R} \cup \{+\infty\}$ such that spt $\gamma \subset \partial_c f$, then γ is an optimal transport plan in $\mathbf{K}_c(\mu, \nu)$. To this end let us first notice that

$$\int_{\mathbb{R}^n \times \mathbb{R}^n} c \, d\bar{\gamma} \ge \int_{\mathbb{R}^n} \alpha \, d\mu + \int_{\mathbb{R}^n} \beta \, d\nu$$

for every $\bar{\gamma} \in \Gamma(\mu, \nu)$ and $(\alpha, \beta) \in \mathcal{A}$, i.e.,

$$\mathbf{K}_c(\mu, \nu) \ge \sup_{(\alpha, \beta) \in \mathcal{A}} \int_{\mathbb{R}^n} \alpha \, d\mu + \int_{\mathbb{R}^n} \beta \, d\nu.$$

Therefore, to show that γ is an optimal transport plan in $\mathbf{K}_c(\mu, \nu)$ (and, at the same time, that (3.27) holds), it suffices to prove that $(-f, -f^c) \in \mathcal{A}$ and

$$\int_{\mathbb{R}^n \times \mathbb{R}^n} c \, d\gamma = \int_{\mathbb{R}^n} (-f) \, d\mu + \int_{\mathbb{R}^n} (-f^c) \, d\nu. \qquad (3.29)$$

To this end, let us recall that, by (3.25), $(\alpha, \beta) = (-f, -f^c)$ satisfies (3.28), while spt $\gamma \subset \partial^c f$ implies that if $(x, y) \in$ spt γ, then $y \in \partial_c f(x)$ and thus, thanks to (3.26), that

$$c(x, y) = -f(x) - f^c(y), \qquad \forall (x, y) \in \text{spt } \gamma. \tag{3.30}$$

Now, by $c \in L^1(\mu \times \nu)$ it follows that for ν-a.e. $y \in \mathbb{R}^n$, $\int_{\mathbb{R}^n} c(x, y) \, d\mu(x) < \infty$; similarly, by (3.12), ν is concentrated on $\mathbf{q}(\text{spt } \gamma)$, while $\mathbf{q}(\text{spt } \gamma) \subset \{f^c < \infty\}$ thanks to (3.30); therefore, for ν-a.e. $y \in \mathbb{R}^n$ we have both $\int_{\mathbb{R}^n} c(x, y) \, d\mu(x) < \infty$ and $f^c(y) \in \mathbb{R}$; we fix such a value of y and integrate $|f(x)| \le c(x, y) + |f^c(y)|$ in $d\mu(x)$ over $x \in \mathbf{p}(\text{spt } \gamma)$ to find that

$$\int_{\mathbb{R}^n} |f| \, d\mu \le \int_{\mathbb{R}^n} c(x, y) \, d\mu(x) + |f^c(y)| < \infty,$$

thus proving $f \in L^1(\mu)$; we then easily deduce that $f^c \in L^1(\nu)$ and conclude that $(-f, -f^c) \in \mathcal{A}$ and satisfies (3.29), thus completing the proof of the theorem. $\qquad\qquad\square$

The proofs of Theorem 3.11 and Proposition 3.12 are identical to those of their convex counterparts. Indeed, in proving Rockafellar's theorem, Fenchel's inequality and Fenchel's identity we have never made use of the bilinearity of the scalar product, nor of its continuity, homogeneity, or symmetry. For the sake of clarity we include anyway the details.

Proof of Theorem 3.11 *Proof that* (3.24) *implies c-cyclical monotonicity*: Given a finite subset $\{(x_i, y_i)\}_{i=1}^N$ of S, (3.24) implies that $y_i \in \partial_c f(x_i)$; in particular, $\partial_c f(x_i)$ is non-empty, $f(x_i) < \infty$, and

$$f(x) \ge f(x_i) + c(x_i, y_i) - c(x, y_i), \qquad \forall x \in \mathbb{R}^n.$$

Testing this inequality at $x = x_{i+1}$ (with $x_{N+1} = x_1$) and summing up over $i = 1, \ldots, N$ gives

$$\sum_{i=1}^N f(x_{i+1}) \ge \sum_{i=1}^N f(x_i) + c(x_i, y_i) - c(x_{i+1}, y_i),$$

and since $\sum_{i=1}^N f(x_{i+1}) = \sum_{i=1}^N f(x_i)$ we deduce that (3.23) holds.

Proof that c-cyclical monotonicity implies (3.24): Since S is non-empty we can pick $(x_0, y_0) \in S$ and (by replacing every term of the form $-z \cdot w$ with $c(z, w)$ in (2.24)) we define for $z \in \mathbb{R}^n$

$$f(z) = \sup \left\{ -c(z, y_N) + c(x_N, y_N) + \sum_{i=1}^{N-1} c(x_i, y_i) - c(x_{i+1}, y_i) \right. \tag{3.31}$$

$$\left. -c(x_1, y_0) + c(x_0, y_0) : \{(x_i, y_i)\}_{i=1}^N \subset S \right\}.$$

If we set $\alpha(y) = -\infty$ for $y \notin \mathbf{q}(\mathrm{spt}\,\gamma)$, and

$$\alpha(y) = \sup\left\{c(x_N, y_N) + \sum_{i=0}^{N-1} c(x_i, y_i) - c(x_{i+1}, y_i) : \{(x_i, y_i)\}_{i=1}^{N} \subset S, y_N = y\right\},$$

if $y \in \mathbf{q}(\mathrm{spt}\,\gamma)$ (so that the above supremum is taken over a non-empty set), then (3.31) takes the form

$$f(z) = \sup\{\alpha(y) - c(z, y) : y \in \mathbb{R}^n\} \qquad \forall z \in \mathbb{R}^n, \tag{3.32}$$

showing that f is, indeed, c-convex (and that f is lower semicontinuous as soon as c is continuous).

We claim that $f(x_0) = 0$ (so that f is not identically equal to $+\infty$). Indeed, by testing (3.31) with $\{(x_1, y_1)\} = \{(x_0, y_0)\}$ at $z = x_0$, we find $f(x_0) \geq 0$, while $f(x_0) \leq 0$ is equivalent to show that, for every $\{(x_i, y_i)\}_{i=1}^{N} \subset S$ we have

$$c(x_N, y_N) - c(x_0, y_N) + \sum_{i=1}^{N-1} c(x_i, y_i) - c(x_{i+1}, y_i) + c(x_0, y_0) - c(x_1, y_0) \leq 0.$$
$$\tag{3.33}$$

This follows by applying (3.23) to $\{(x_i, y_i)\}_{i=0}^{N} \subset S$.

Finally, we prove that $S \subset \partial_c f$. Indeed, let $(x_*, y_*) \in S$ and let $t \in \mathbb{R}$ be such that $t < f(x_*)$. By definition of $f(x_*)$, we can find $\{(x_i, y_i)\}_{i=1}^{N} \subset S$ such that

$$-c(x_*, y_N) + c(x_N, y_N) + \sum_{i=0}^{N-1} c(x_i, y_i) - c(x_{i+1}, y_i) \geq t. \tag{3.34}$$

Letting $x_{N+1} = x_*$ and $y_{N+1} = y_*$ and testing (3.31) with $\{(x_i, y_i)\}_{i=1}^{N+1} \subset S$, we find that, for every $z \in \mathbb{R}^n$,

$$f(z) \geq -c(z, y_{N+1}) + c(x_{N+1}, y_{N+1}) + \sum_{i=0}^{N} c(x_i, y_i) - c(x_{i+1}, y_i)$$

$$= -c(z, y_*) + c(x_*, y_*) - c(x_*, y_N) + c(x_N, y_N) + \sum_{i=0}^{N-1} c(x_i, y_i) - c(x_{i+1}, y_i)$$

$$\geq -c(z, y_*) + c(x_*, y_*) + t, \tag{3.35}$$

where in the last inequality we have used (3.34). Since $f(z)$ is finite at $z = x_0$, by letting $t \to f(x_*)^-$ in (3.35) first with $z = x_0$ we see that $x_* \in \mathrm{Dom}(f)$, and then, by taking the same limit for an arbitrary z, we see that $y_* \in \partial_c f(x_*)$. $\quad\square$

Proof of Proposition 3.12 Let $f : \mathbb{R}^n \to \mathbb{R} \cup \{+\infty\}$ be a c-convex function. The validity of (3.25) is immediate from the definition of f^c, so we just have to prove that

$$c(x, y) = -f(x) - f^c(y) \qquad \text{if and only if} \qquad y \in \partial_c f(x).$$

On the one hand, if $c(x, y) = -f(x) - f^c(y)$ for some $x, y \in \mathbb{R}^n$, then by using $f^c(y) \geq -c(z, y) - f(z)$ for every $z \in \mathbb{R}^n$ we deduce

$$c(x, y) \leq -f(x) + c(z, y) + f(z) \qquad \forall z \in \mathbb{R}^n, \tag{3.36}$$

i.e., $y \in \partial_c f(x)$. Conversely, if $y \in \partial_c f(x)$ for some $x, y \in \mathbb{R}^n$, then (3.36) holds, and the arbitrariness of z and the definition of $f^c(y)$ give $c(x, y) \leq -f(x) - f^c(y)$. Since the opposite inequality holds true, we deduce $c(x, y) = -f(x) - f^c(y)$. $\qquad\qquad\square$

3.6 Two Additional Results on c-Cyclical Monotonicity

We now prove two additional general results concerning c-cyclical monotonicity and transport problems. The first one, Theorem 3.15, is the generalization of Theorem 2.12 to general origin and final measures. The second one, Theorem 3.16, concerns the existence of c-cyclically monotone plans in $\Gamma(\mu, \nu)$ even when $\mathbf{K}_c(\mu, \nu)$ is not finite.

Theorem 3.15 (Set of optimal transport plans) *Let $c : \mathbb{R}^n \times \mathbb{R}^n \to [0, \infty)$ be a continuous function, let $\mu, \nu \in \mathcal{P}(\mathbb{R}^n)$ and assume that $c \in L^1(\mu \times \nu)$. Let*

$$\Gamma_{\mathrm{opt}}(\mu, \nu, c)$$

be the family of optimal transport plans in $\mathbf{K}_c(\mu, \nu)$. Then $\Gamma_{\mathrm{opt}}(\mu, \nu, c)$ is convex and compact in the narrow convergence of Radon measures. Moreover,

$$S = \bigcup \left\{ \mathrm{spt}\, \gamma : \gamma \in \Gamma_{\mathrm{opt}}(\mu, \nu, c) \right\} \text{ is c-cyclically monotone,}$$

and in particular there exists a (lower-semicontinuous and) c-convex function $f : \mathbb{R}^n \to \mathbb{R} \cup \{+\infty\}$ whose c-subdifferential $\partial_c f$ contains the supports of all the optimal plans in $\mathbf{K}_c(\mu, \nu)$.

Proof Given $\gamma \in \Gamma_{\mathrm{opt}}(\mu, \nu, c)$, by Theorem 3.13, there exists a c-convex function $f : \mathbb{R}^n \to \mathbb{R} \cup \{+\infty\}$ such that $\mathrm{spt}\, \gamma \subset \partial_c f$; moreover, since c is continuous and f is lower semicontinuous, we have that $\partial_c f$ is closed. Now, if $\gamma' \in \Gamma_{\mathrm{opt}}(\mu, \nu, c)$, then by (3.27) we have

$$\int_{\mathbb{R}^n \times \mathbb{R}^n} c(x, y) \, d\gamma'(x, y) = \int_{\mathbb{R}^n} (-f) \, d\mu + \int_{\mathbb{R}^n} (-f^c) \, d\nu,$$

so that Proposition 3.12 implies that γ' is concentrated on $\partial_c f$. Since $\partial_c f$ is closed, this implies that $\mathrm{spt}\, \gamma' \subset \partial_c f$. $\qquad\qquad\square$

Transport plans with prescribed marginals and whose support is c-cyclically monotone can be constructed even *outside* of the context of a well-posed transport problem, i.e., even when the assumption $\mathbf{K}_c(\mu, \nu) < \infty$ is dropped.

The proof is based on an approximation argument and on the remark that $K_c(\mu, \nu) < \infty$ always holds if μ and ν are discrete measures.

Theorem 3.16 *Let* $c : \mathbb{R}^n \times \mathbb{R}^n \to [0, \infty)$ *be a continuous function. If* $\mu, \nu \in \mathcal{P}(\mathbb{R}^n)$, *then there exist* $\gamma \in \Gamma(\mu, \nu)$ *and a c-convex function* $f : \mathbb{R}^n \to \mathbb{R} \cup \{+\infty\}$ *such that* spt γ *is c-cyclically monotone and contained in* $\partial_c f$.

Proof Step one: We prove that every probability measure on $\mathbb{R}^n$ can be approximated by discrete probability measures in the narrow convergence. This is usually proved, in much greater generality, by means of the Banach–Alaoglu theory – but, in the spirit of our presentation, we adopt here a more concrete argument. For $R > 0$ and for some integer $M \geq 2$, let $Q_R = (-R/2, R/2)^n$ and let $\{Q_R^i\}_{i=1}^N$ be a covering of Q_R by $N = (R/M)^n$ many (neither open or closed) cubes of side length R/M. Denote by x_R^i the center of Q_R^i, set $\lambda_R^i = \mu(Q_R^i)/\mu(Q_R)$, and let

$$\mu_{R,M} = \sum_{i=1}^N \lambda_R^i \, \delta_{x_R^i} \in \mathcal{P}(\mathbb{R}^n).$$

Then, for every $\varphi \in C_c^0(\mathbb{R}^n)$, denoting by ω_φ a modulus of continuity for φ, we have

$$\left| \int_{\mathbb{R}^n} \varphi \, d\mu - \int_{Q_R} \varphi \, d\mu \right| \leq \|\varphi\|_{C^0(\mathbb{R}^n)} \, \mu(\mathbb{R}^n \setminus Q_R),$$

$$\left| \int_{Q_R} \varphi \, d\mu - \int_{\mathbb{R}^n} \varphi \, d\mu_{R,N} \right| \leq \sum_{i=1}^N \int_{Q_R^i} |\varphi - \varphi(x_R^i)| \, d\mu + |\varphi(x_R^i)| \left| \mu(Q_R^i) - \frac{\mu(Q_R^i)}{\mu(Q_R)} \right|$$

$$\leq \omega_\varphi \left(\sqrt{n} \frac{R}{M} \right) \mu(Q_R) + \|\varphi\|_{C^0(\mathbb{R}^n)} \frac{1 - \mu(Q_R)}{\mu(Q_R)}.$$

Setting $R = \sqrt{M}$ and letting $M \to \infty$, we have proved that

$$\mu_{\sqrt{M},M} \overset{*}{\rightharpoonup} \mu \qquad \text{as } M \to \infty.$$

Since each $\mu_{\sqrt{M},M}$ is a probability measure, by Proposition A.1, the convergence of $\mu_{\sqrt{M},M}$ to μ is narrow.

Step two: Now let $\mu, \nu \in \mathcal{P}(\mathbb{R}^n)$, and let $\{\mu^j\}_{j=1}^\infty$ and $\{\nu^j\}_{j=1}^\infty$ be discrete probability measures such that

$$\mu^j \overset{n}{\to} \mu, \qquad \nu^j \overset{n}{\to} \nu, \qquad \text{as } j \to \infty.$$

By the theory developed in this section, for each j there is an optimal plan γ^j in $K_c(\mu^j, \nu^j) < \infty$, and spt γ^j is c-cyclically monotone. Up to extracting a subsequence, we have $\gamma^j \overset{*}{\rightharpoonup} \gamma$ for some Radon measure γ on $\mathbb{R}^n \times \mathbb{R}^n$. By narrow convergence of μ^j and ν^j to μ and ν respectively, we have that

$$\lim_{R\to\infty} \sup_j \mu^j(\mathbb{R}^n \setminus B_R) + \nu^j(\mathbb{R}^n \setminus B_R) = 0.$$

In particular, by arguing as in the proof of (3.16), we find that $\gamma^j \overset{n}{\rightharpoonup} \gamma$, and thus that $\gamma \in \mathcal{P}(\mathbb{R}^n)$. By (A.12), for every $(x,y) \in \operatorname{spt}\gamma$ there are $(x^j, y^j) \in \operatorname{spt}\gamma^j$ such that $(x^j, y^j) \to (x,y)$ as $j \to \infty$: since the c-cyclical monotonicity inequality for a continuous cost c is a closed condition, we deduce the c-cyclical monotonicity of $\operatorname{spt}\gamma$ from that of $\operatorname{spt}\gamma^j$. By Theorem 3.11, there exists a c-convex function $f : \mathbb{R}^n \to \mathbb{R} \cup \{+\infty\}$ such that $\operatorname{spt}\gamma \subset \partial_c f$. Moreover, the narrow convergence of μ^j, ν^j and γ^j to μ, ν and γ respectively implies that $\gamma \in \Gamma(\mu, \nu)$. □

3.7 Linear and Quadratic Kantorovich Dualities

We now examine some immediate consequences of the Kantorovich duality for the transport problems with linear and quadratic costs. The corresponding results, see Theorem 3.17 and Theorem 3.20, will be the starting points for the detailed analysis on the corresponding Monge problems in Part IV and Part II, respectively. It will be convenient to work with the set of **probability measures with finite p-moment**,

$$\mathcal{P}_p(\mathbb{R}^n) = \left\{ \mu \in \mathcal{P}(\mathbb{R}^n) : \int_{\mathbb{R}^n} |x|^p \, d\mu(x) < \infty \right\}, \qquad 1 \le p < \infty,$$

so that the Kantorovich problem corresponding to $c(x,y) = |x - y|^p$, denoted by $\mathbf{K}_p(\mu, \nu)$, is such that $\mathbf{K}_p(\mu, \nu) < \infty$ for every $\mu, \nu \in \mathcal{P}_p(\mathbb{R}^n)$ – indeed, if $\mu, \nu \in \mathcal{P}_p(\mathbb{R}^n)$, then $\mu \times \nu$ defines a transport plan from μ to ν with finite p-cost. This remark is useful because the finiteness of $\mathbf{K}_c(\mu, \nu)$ is a necessary assumption to apply Kantorovich's theory.

Theorem 3.17 *If $\mu, \nu \in \mathcal{P}_1(\mathbb{R}^n)$, then $\mathbf{K}_1(\mu, \nu)$ admits optimal transport plans, and there exists a Lipschitz function $f : \mathbb{R}^n \to \mathbb{R}$ with $\operatorname{Lip}(f) \le 1$, called a **Kantorovich potential from μ to ν**, such that $\gamma \in \Gamma(\mu, \nu)$ is an optimal transport plan in $\mathbf{K}_1(\mu, \nu)$ if and only if*

$$f(y) = f(x) + |x - y| \qquad \forall (x,y) \in \operatorname{spt}\gamma. \tag{3.37}$$

Moreover, the Kantorovich duality formula (3.27) holds in the form

$$\mathbf{K}_1(\mu, \nu) = \sup \left\{ \int_{\mathbb{R}^n} g \, d\nu - \int_{\mathbb{R}^n} g \, d\mu : g : \mathbb{R}^n \to \mathbb{R}, \operatorname{Lip}(g) \le 1 \right\}, \tag{3.38}$$

and any Kantorovich potential f achieves the supremum in (3.38).

Remark 3.18 A special situation where all Kantorovich potentials agree up to additive constants is described in Remark 19.5.

Remark 3.19 It is easily seen (arguing along the lines of the following proof) that if f is a Kantorovich potential from μ to ν, then $f^c = -f$ and $-f$ is a Kantorovich potential from ν to μ.

Proof of Theorem 3.17   In the following we set $c(x,y) = |x - y|$.

Step one: By Theorem 3.15, there exists a c-convex function $f : \mathbb{R}^n \to \mathbb{R} \cup \{+\infty\}$ such that $\gamma \in \Gamma(\mu, \nu)$ is an optimal transport plan for $\mathbf{K}_1(\mu, \nu)$ if and only if (3.37) holds. The c-convexity of f means that

$$f(x) = \sup\{\alpha(y) - c(x,y) : y \in \mathbb{R}^n\} \qquad \forall x \in \mathbb{R}^n, \tag{3.39}$$

where $\alpha : \mathbb{R}^n \to \mathbb{R} \cup \{-\infty\}$. We notice that $\mathrm{Dom}(f) \neq \emptyset$: indeed, $\partial_c f \neq \emptyset$ since there is an optimal plan γ for $\mathbf{K}_1(\mu, \nu)$ (Theorem 3.8), and therefore $\emptyset \neq \mathrm{spt}\,\gamma \subset \partial_c f$ by Theorem 3.15. In turn $\mathrm{Dom}(f) \neq \emptyset$ implies that f is 1-Lipschitz.

Indeed, let us notice as a general fact that if f is c-convex (with $c(x,y) = |x - y|$), then $\mathrm{Dom}(f) \neq \emptyset$ if and only if $\mathrm{Lip}(f) \leq 1$ (and, in particular, f is everywhere finite). Indeed, if f is c-convex and $f(x_0) < \infty$, then

$$\alpha(y) \leq f(x_0) + |y - x_0| \qquad \forall y \in \mathbb{R}^n, \tag{3.40}$$

and in particular

$$\alpha(y) - |x - y| \leq f(x_0) + |y - x_0| - |x - y| \leq f(x_0) + |x - x_0|,$$

so that, taking the supremum over $y \in \mathbb{R}^n$, $f(x) \leq f(x_0) + |x - x_0|$ and $\mathrm{Dom}(f) = \mathbb{R}^n$. Since $f(x) < \infty$ we can interchange roles and prove, by the same argument, that $f(x_0) \leq f(x) + |x - x_0|$, thus showing that $\mathrm{Lip}(f) \leq 1$. Conversely, if $\mathrm{Lip}(f) \leq 1$, then f is c-convex: indeed (3.39) holds trivially with $\alpha(y) = f(y)$ for $y \in \mathbb{R}^n$.

The fact that f is a 1-Lipschitz function implies that

$$y \in \partial_c f(x) \qquad \text{if and only if} \qquad f(y) = f(x) + |x - y|. \tag{3.41}$$

In other words, $y \in \partial_c f(x)$ if and only if y "saturates" the 1-Lipschitz condition of f with respect to x. In one direction, if $y \in \partial_c f(x)$, then (3.20) implies that

$$f(z) \geq f(x) + |x - y| - |z - y| \qquad \forall z \in \mathbb{R}^n.$$

Setting $z = y$ one finds $f(y) \geq f(x) + |x - y|$; since the opposite inequality follows by $\mathrm{Lip}(f) \leq 1$, we find $f(y) = f(x) + |y - x|$. Conversely, if $f(y) = f(x) + |x - y|$ and $z \in \mathbb{R}^n$, then by $\mathrm{Lip}(f) \leq 1$ we find

$$f(z) \geq f(y) - |z - y| = f(x) + |x - y| - |z - y|,$$

and thus, by arbitrariness of z, $y \in \partial_c f(x)$.

Step two: We now prove (3.38). First of all we trivially have

$$\mathbf{K}_1(\mu, v) \geq \sup \left\{ \int_{\mathbb{R}^n} g\, dv - \int_{\mathbb{R}^n} g\, d\mu : u : \mathbb{R}^n \to \mathbb{R}, \mathrm{Lip}(g) \leq 1 \right\}.$$

At the same time if γ is optimal in $\mathbf{K}_1(\mu, v)$ and f is a Kantorovich potential from μ to v, then f belongs to the competition class of the supremum problem since $\mathrm{Lip}(f) \leq 1$, while (3.37) gives

$$\mathbf{K}_1(\mu, v) = \int_{\mathbb{R}^n \times \mathbb{R}^n} |x - y|\, d\gamma(x, y)$$

$$= \int_{\mathbb{R}^n \times \mathbb{R}^n} f(y) - f(x)\, d\gamma(x, y) = \int_{\mathbb{R}^n} f\, dv - \int_{\mathbb{R}^n} f\, d\mu,$$

thus proving (3.38). $\qquad\square$

Let us now consider the case of the quadratic cost $c(x, y) = |x - y|^2/2$. It is easily seen that, (i) a function $f : \mathbb{R}^n \to \mathbb{R} \cup \{+\infty\}$ is c-convex if and only if it is lower semicontinuous on $\mathbb{R}^n$ and $x \mapsto f(x) + |x|^2/2$ is convex on $\mathbb{R}^n$; (ii) for every $x \in \mathbb{R}^n$, denoting by ∂ the subdifferential of a convex function,

$$\partial_c f(x) = \partial \left[f + \frac{|\cdot|^2}{2} \right](x);$$

(iii) if f is a c-convex function, then

$$f^c(y) = \left[f + \frac{|\cdot|^2}{2} \right]^*(y) - \frac{|y|^2}{2}. \tag{3.42}$$

It is thus immediate to deduce the following result.

Theorem 3.20 *If $\mu, v \in \mathcal{P}_2(\mathbb{R}^n)$, then $\mathbf{K}_2(\mu, v)$ admits optimal transport plans, and there exists a convex, lower semicontinuous function $f : \mathbb{R}^n \to \mathbb{R} \cup \{+\infty\}$ such that $\gamma \in \Gamma(\mu, v)$ is an optimal transport plan in $\mathbf{K}_2(\mu, v)$ if and only if*

$$\mathrm{spt}\, \gamma \subset \partial f,$$

and such that

$$\mathbf{K}_2(\mu, v) = \int_{\mathbb{R}^n} \frac{|x|^2}{2}\, d\mu + \int_{\mathbb{R}^n} \frac{|y|^2}{2}\, dv - \int_{\mathbb{R}^n} f\, d\mu - \int_{\mathbb{R}^n} f^*\, dv. \tag{3.43}$$

Moreover, the Kantorovich duality formula (3.27) holds in the form

$$\mathbf{K}_2(\mu, v) = \sup \left\{ \int_{\mathbb{R}^n} (-g)\, d\mu + \int_{\mathbb{R}^n} (-g^c)\, dv : g \in C_b^0(\mathbb{R}^n) \cap \mathrm{Lip}(\mathbb{R}^n) \right\}, \tag{3.44}$$

where $c(x, y) = |x - y|^2/2$.

Remark 3.21   It is qualitatively clear that the notion of subdifferential in the case of the linear cost, highlighted in (3.41), is much less stringent than the notion of subdifferential for the quadratic cost. Although just at an heuristic level, this remark points in the right direction in indicating that the linear cost problem is rougher than the quadratic cost problem – an indication that will be confirmed under every viewpoint: uniqueness of optimal plans, existence, uniqueness and regularity of transport maps, and so on.

Remark 3.22 (Improved Kantorovich duality)   Identity (3.44) points to the problem of the continued validity of the Kantorovich duality formula (3.27) when one restricts the competition class of the dual problem to only include functions more regular than Borel measurable: this may of course require attention since, by restricting the competition class of a supremum, we are possibly decreasing its value. We will address this problem only for the quadratic cost, since this is the case of the improved Kantorovich duality formula needed (or at least, convenient for technical reasons) in the proof of the Brenier–Benamou formula (see Chapter 15, Theorem 15.6). In particular, step two and step three of the proof of Theorem 3.20, which are devoted to the proof of (3.44), can be safely skipped on a first reading.

Proof of Theorem 3.20   *Step one*: Since $\mu, \nu \in \mathcal{P}_2(\mathbb{R}^n)$ implies $c(x, y) = |x - y|^2 \in L^1(\mu \times \nu)$, by Theorem 3.15, there exists a convex, lower-semicontinuous function $g : \mathbb{R}^n \to \mathbb{R} \cup \{+\infty\}$ such that every optimal transport plan γ in $\mathbf{K}_2(\mu, \nu)$ satisfies $\operatorname{spt} \gamma \subset \partial^c g$. Conversely by Theorem 3.13, if $\operatorname{spt} \gamma \subset \partial^c g$, then γ is optimal in $\mathbf{K}_2(\mu, \nu)$. Moreover,

$$\mathbf{K}_2(\mu, \nu) = \int_{\mathbb{R}^n} (-g) \, d\mu + \int_{\mathbb{R}^n} (-g^c) \, d\nu, \tag{3.45}$$

thanks to (3.27). By the above remarks, setting $f(x) = g(x) + |x|^2/2$ for $x \in \mathbb{R}^n$, $\operatorname{spt} \gamma \subset \partial^c g$ is equivalent to $\operatorname{spt} \gamma \subset \partial f$, where $f : \mathbb{R}^n \to \mathbb{R} \cup \{+\infty\}$ is convex and lower semicontinuous on $\mathbb{R}^n$, and (3.43) follows by $g(x) = f(x) - |x|^2/2$ and by (3.42).

Step two: To approach the proof of (3.44), we start making the following remark: if c is a nonnegative, bounded, and Lipschitz continuous cost function, $\mu, \nu \in \mathcal{P}(\mathbb{R}^n)$ are such that $\mathbf{K}_c(\mu, \nu) < \infty$, γ is an optimal plan in $\mathbf{K}_c(\mu, \nu)$, and g is the c-convex function associated by Theorem 3.11 to $S = \operatorname{spt} \gamma$ and to a given $(x_0, y_0) \in \mathbb{R}^n \times \mathbb{R}^n$ (see (3.31)), then both g and g^c are bounded and Lipschitz continuous. Indeed, by (3.32), g is Lipschitz continuous (as the supremum of a family of uniformly Lipschitz functions) and bounded from below (since c is bounded from above). From $g^c(y) = \sup\{-c(x, y) - g(x) : x \in \mathbb{R}^n\}$, we thus see that g^c is bounded from above (as g is bounded from below and

$c \geq 0$), bounded from below (as $g(x_0) = 0$, $g^c(y) \geq -c(x,y) \geq -\sup c$), and Lipschitz continuous (again, as a supremum of uniformly Lipschitz continuous functions). From $g(x) = g^{cc}(x) = \sup\{-c(x,y) - g^c(y) : y \in \mathbb{R}^n\}$ and since g^c is bounded from above, we also see that g is bounded from below.

Step three: Now let $c(x,y) = |x - y|^2/2$ and let $\{c_j\}_j$ be such that $c_j = c$ on $B_j^{2n} = \{(x,y) \in \mathbb{R}^n \times \mathbb{R}^n : |x|^2 + |y|^2 < j^2\}$, $c_j \uparrow c$ on $\mathbb{R}^n \times \mathbb{R}^n$, and c_j is bounded and Lipschitz continuous on $\mathbb{R}^n \times \mathbb{R}^n$. By step two and by Theorem 3.13, if $\mu, \nu \in \mathcal{P}_2(\mathbb{R}^n)$, then we have

$$\mathbf{K}_{c_j}(\mu,\nu) = \sup_{(\alpha,\beta) \in \mathcal{A}_j} \int_{\mathbb{R}^n} \alpha \, d\mu + \int_{\mathbb{R}^n} \beta \, d\nu, \tag{3.46}$$

where $\mathcal{A}_j$ is the class of those functions $\alpha, \beta \in C_b^0(\mathbb{R}^n) \cap \mathrm{Lip}(\mathbb{R}^n)$ such that $\alpha(x) + \beta(y) \leq c_j(x,y)$ for every $(x,y) \in \mathbb{R}^n \times \mathbb{R}^n$. If γ_j is an optimal transport plan for $\mathbf{K}_{c_j}(\mu,\nu)$, then by $\gamma_j \in \Gamma(\mu,\nu)$, and up to extracting subsequences, there is $\gamma \in \Gamma(\mu,\nu)$ such that $\gamma_j \overset{n}{\rightharpoonup} \gamma$. In particular, for each $(x,y) \in \mathrm{spt}\,\gamma$, there is $(x_j, y_j) \in \mathrm{spt}\,\gamma_j$ such that $(x_j, y_j) \to (x,y)$ as $j \to \infty$, and we can use this fact, together with the observation that testing the c-cyclical monotonicity of $\mathrm{spt}\,\gamma$ requires the consideration of only finitely many points in $\mathrm{spt}\,\gamma$ at a time, to see that the c_j-cyclical monotonicity of $\mathrm{spt}\,\gamma_j$ implies the c-cyclical monotonicity of $\mathrm{spt}\,\gamma$, and thus that, thanks to Theorem 3.13,

$$\mathbf{K}_2(\mu,\nu) = \int_{\mathbb{R}^n \times \mathbb{R}^n} c \, d\gamma. \tag{3.47}$$

Now, since $c = c_j$ on B_j^{2n}, setting $V_r = (\mathbb{R}^n \times \mathbb{R}^n) \setminus B_r^{2n}$ we see that, if $r < j$ is such that $\gamma(\partial B_r^{2n}) = 0$ ($\mathcal{L}^1$-a.e. $r < j$ has this property), then

$$\left| \int_{\mathbb{R}^n \times \mathbb{R}^n} c \, d\gamma - \int_{\mathbb{R}^n \times \mathbb{R}^n} c_j \, d\gamma_j \right| \leq \left| \int_{B_r^{2n}} c \, d\gamma - \int_{B_r^{2n}} c \, d\gamma_j \right| + \int_{V_r} c \, d\gamma + \int_{V_r} c_j \, d\gamma_j,$$

where the first term converges to zero as $j \to \infty$ by $\gamma(\partial B_r^{2n}) = 0$, while

$$\int_{V_r} c_j \, d\gamma_j \leq \int_{V_r} |x|^2 + |y|^2 \, d\gamma_j = \int_{\mathbb{R}^n} |x|^2 \, d\mu + \int_{\mathbb{R}^n} |y|^2 \, d\nu - \int_{B_r^{2n}} |x|^2 + |y|^2 \, d\gamma_j$$

so that, by $\gamma_j \overset{n}{\rightharpoonup} \gamma$ and lower semicontinuity on open sets,

$$\limsup_{j \to \infty} \int_{V_r} c_j \, d\gamma_j \leq \int_{\mathbb{R}^n} |x|^2 \, d\mu + \int_{\mathbb{R}^n} |y|^2 \, d\nu - \int_{B_r^{2n}} |x|^2 + |y|^2 \, d\gamma = \int_{V_r} |x|^2 + |y|^2 \, d\gamma.$$

In summary,

$$\limsup_{j \to \infty} \left| \int_{\mathbb{R}^n \times \mathbb{R}^n} c \, d\gamma - \int_{\mathbb{R}^n \times \mathbb{R}^n} c_j \, d\gamma_j \right| \leq 2 \int_{V_r} |x|^2 + |y|^2 \, d\gamma,$$

where the latter quantity converges to zero as $r \to \infty$, since $\gamma \in \mathcal{P}_2(\mathbb{R}^n \times \mathbb{R}^n)$. Combining this with (3.47) and (3.46) we see that

$$\mathbf{K}_2(\mu, \nu) = \lim_{j \to \infty} \sup_{(\alpha, \beta) \in \mathcal{A}_j} \int_{\mathbb{R}^n} \alpha \, d\mu + \int_{\mathbb{R}^n} \beta \, d\nu$$

$$\leq \sup \left\{ \int_{\mathbb{R}^n} (-g) \, d\mu + \int_{\mathbb{R}^n} (-g^c) \, d\nu : g \in C_b^0(\mathbb{R}^n) \cap \mathrm{Lip}(\mathbb{R}^n) \right\} \leq \mathbf{K}_2(\mu, \nu),$$

where in the first inequality we have used $c_j \leq c$ to notice that every $(\alpha, \beta) \in \mathcal{A}_j$ satisfies $\alpha(x) + \beta(y) \leq c(x, y)$ whenever $(x, y) \in \mathbb{R}^n \times \mathbb{R}^n$, while in the second inequality we have used (3.27). $\qquad \square$

PART II

Solution of the Monge Problem with Quadratic Cost: The Brenier–McCann Theorem

4

The Brenier Theorem

With In this chapter we start our analysis of the Monge problem $\mathbf{M}_c(\mu, \nu)$ for the quadratic cost $c(x, y) = |x - y|^2$, namely, of

$$\mathbf{M}_2(\mu, \nu) = \inf \left\{ \int_{\mathbb{R}^n \times \mathbb{R}^n} |x - T(x)|^2 \, d\mu(x) : T_\# \mu = \nu \right\}.$$

As noticed in Remark 2.1 the competition class for this problem could be empty, and, as explained in Chapter 1, even assuming to be in a situation where the existence of transport maps is not in doubt (e.g., because we assume from the onset that $\nu = T_\# \mu$ for some map T that transports μ), it is not clear how to prove the existence of minimizers. At the same time, in Theorem 3.20, given $\mu, \nu \in \mathcal{P}_2(\mathbb{R}^n)$, we have proved the existence of optimal transport plans in the quadratic Kantorovich problem

$$\mathbf{K}_2(\mu, \nu) = \inf \left\{ \int_{\mathbb{R}^n \times \mathbb{R}^n} |x - y|^2 \, d\gamma(x, y) : \gamma \in \Gamma(\mu, \nu) \right\}.$$

Since $\mathbf{K}_2(\mu, \nu) \le \mathbf{M}_2(\mu, \nu)$, a natural strategy for proving the existence of minimizers of $\mathbf{M}_2(\mu, \nu)$ is to show that for every optimal plan γ in $\mathbf{K}_2(\mu, \nu)$ there is a transport map T from μ to ν such that $\gamma = \gamma_T = (\mathbf{id} \times T)_\# \mu$ (cf. Proposition 3.1). Theorem 3.20 strongly suggests that such a map T should be *the gradient of a convex function*. Indeed, by Theorem 3.20, if $\mu, \nu \in \mathcal{P}_2(\mathbb{R}^n)$, then there is a convex, lower-semicontinuous function $f : \mathbb{R}^n \to \mathbb{R} \cup \{+\infty\}$ such that every optimal transport plan γ in $\mathbf{K}_2(\mu, \nu)$ satisfies

$$\operatorname{spt} \gamma \subset \partial f = \{(x, y) : y \in \partial f(x)\}. \tag{4.1}$$

For a generic x, $\partial f(x)$ is a set, not a point; however, by Rademacher's theorem (see Appendix A.10), the local Lipschitz property of convex functions, and Proposition 2.7, we know that $\partial f(x) = \{\nabla f(x)\}$ for $\mathcal{L}^n$-a.e. $x \in \operatorname{Dom}(f)$. In particular, if $\mu \ll \mathcal{L}^n$, there is hope that ∇f could be an optimal transport map from μ to ν. Moreover, the fact that (4.1) holds with the same f for *every* optimal transport plan γ suggests that the uniqueness of minimizers should hold (both in $\mathbf{K}_2(\mu,\nu)$ and in $\mathbf{M}_2(\mu,\nu)$). The main goal of this chapter will be thus proving the *existence and uniqueness* of an optimal transport map in $\mathbf{M}_2(\mu,\nu)$ under the assumptions that $\mu, \nu \in \mathcal{P}_2(\mathbb{R}^n)$ with $\mu \ll \mathcal{L}^n$, a result known as the **Brenier theorem**. In light of the discussion in Chapter 1, this is indeed a remarkable result.

We close with an important remark. Notice that a necessary condition for the above strategy to hold is that $\mathbf{M}_2(\mu,\nu) = \mathbf{K}_2(\mu,\nu)$. As already anticipated in Remark 3.4, this identity holds as soon as μ has no atoms (see Theorem 17.11). From this viewpoint, the assumption $\mu \ll \mathcal{L}^n$ in the Brenier theorem could seem quite restrictive, but this is actually not the case: indeed, in the following remark, we construct $\mu, \nu \in \mathcal{P}_2(\mathbb{R}^2)$ such that $\mu \ll \mathcal{H}^1 \llcorner \ell$ for a line ℓ in $\mathbb{R}^2$ (in particular, μ has no atoms), but $\mathbf{M}_2(\mu,\nu)$ *does not admit minimizers*. Therefore, the assumption $\mu \ll \mathcal{L}^n$, although not sharp (as we shall see in Theorem 6.1), is definitely appropriate.

Remark 4.1 ($\mathbf{K}_c = \mathbf{M}_c$ but $\mathbf{M}_c$ has no minimizers) Let us consider the segments $\sigma_t = [0,1] \times \{t\} \subset \mathbb{R}^2$ corresponding to $t \in \mathbb{R}$ and define

$$\mu = \mathcal{H}^1 \llcorner \sigma_0, \qquad \nu = (1/2)\, \mathcal{H}^1 \llcorner (\sigma_d \cup \sigma_{-d}), \qquad d > 0.$$

We claim that for every transport cost of the form $c(x,y) = h(|x - y|)$, with $h : [0,\infty) \to [0,\infty)$ continuous, strictly increasing and such that $h(0) = 0$, we have $\mathbf{K}_c(\mu,\nu) = \mathbf{M}_c(\mu,\nu) = h(d)$ and

$$\int_{\mathbb{R}^n} c(x,T(x))\, d\mu(x) > h(d) \quad \text{whenever } T_{\#}\mu = \nu,$$

thus showing that no minimizer exists in $\mathbf{M}_c(\mu,\nu)$ even if the infima of the Monge and Kantorovich problems agree. To prove this, let us start by noticing that, since h is strictly increasing,

$$x \in \sigma_0,\, y \in \sigma_{\pm d} \quad \Rightarrow \quad \begin{array}{l} h(|y - x|) \geq h(d), \\ \text{with } = \text{ if and only if } y = x \pm d\, e_2. \end{array} \tag{4.2}$$

In particular, if $\gamma \in \Gamma(\mu,\nu)$ and $(x,y) \in \mathrm{spt}\,\gamma$, then, by (3.11), $(x,y) \in \mathrm{spt}\,\mu \times \mathrm{spt}\,\nu$, so that $h(|x - y|) \geq d$ and $\mathbf{K}_c(\mu,\nu) \geq h(d)$. At the same time, setting $S_1(x) = x + d\, e_2$ and $S_2(x) = x - d\, e_2$, the transport plan

$$\gamma_0 = (1/2)\, \{(\mathbf{id} \times S_1)_{\#}\mu + (\mathbf{id} \times S_2)_{\#}\mu\},$$

(which corresponds to the action of *two* distinct transport maps) is optimal in $\mathbf{K}_c(\mu,\nu)$, as $\int_{\mathbb{R}^2 \times \mathbb{R}^2} c\, d\gamma_0 = h(d)$. Now, by formalizing the construction described in Figure 4.1, we find $\mathbf{M}_c(\mu,\nu) = h(d) = \mathbf{K}_c(\mu,\nu)$. In particular,

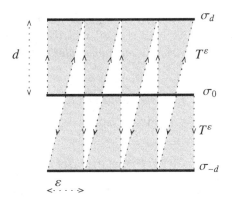

Figure 4.1 Divide σ_0 into an even number of intervals of length $\varepsilon/2$ and denote by E_1^ε the union of the "odd" intervals and by E_2^ε the union of the "even" intervals. Define a map T^ε so that T^ε stretches by a factor 2 each odd interval and then translate it upwards by $d\,e_2$ into σ_d; and that stretches by a factor 2 each even interval, and then translate it downwards by $-d\,e_2$ into σ_{-d}. In this way T^ε is a transport map with the property that $d^2 \le |x - T^\varepsilon(x)|^2 \le d^2 + (\varepsilon/2)^2$ at each $x \in \sigma_0$. In particular, as soon as h is continuous, $\int_{\mathbb{R}^2} h(|x - T^\varepsilon(x)|)\,d\mu(x) \to h(d)^+$ as $\varepsilon \to 0^+$.

should there be a minimizer T of $\mathbf{M}_c(\mu, \nu)$, then γ_T would be optimal in $\mathbf{K}_c(\mu, \nu)$, and thus we would have $h(|x-y|) = h(d)$ for γ_T-a.e. $(x, y) \in \mathbb{R}^2 \times \mathbb{R}^2$. By $\gamma_T = (\mathbf{id} \times T)_\# \mu$, we would find $h(|x - T(x)|) = h(d)$ for μ-a.e. $x \in \mathbb{R}^2$, and then, by (4.2), there would be a Borel partition $\{E_1, E_2\}$ of σ_0 (modulo $\mathcal{H}^1$-null sets) such that $T(x) = x + d\,e_2$ on E_1 and $T(x) = x - d\,e_2$ on E_2. Thus,

$$T_\# \mu = \mathcal{H}^1 \llcorner (E_1 + d\,e_2) + \mathcal{H}^1 \llcorner (E_2 - d\,e_2),$$

so that $\nu = T_\# \mu$ would imply the contradiction

$$(1/2)\,\mathcal{H}^1 \llcorner \sigma_d = \mathcal{H}^1 \llcorner (E_1 + d\,e_2), \qquad (1/2)\,\mathcal{H}^1 \llcorner \sigma_{-d} = \mathcal{H}^1 \llcorner (E_2 - d\,e_2).$$

4.1 The Brenier Theorem: Statement and Proof

We now state and prove the Brenier theorem. Since we will work many times with probability measures that are absolutely continuous with respect to $\mathcal{L}^n$, it is convenient to set $\mathcal{P}_{\mathrm{ac}}(\mathbb{R}^n) = \{\mu \in \mathcal{P}(\mathbb{R}^n) : \mu \ll \mathcal{L}^n\}$ and

$$\mathcal{P}_{p,\mathrm{ac}}(\mathbb{R}^n) = \mathcal{P}_p(\mathbb{R}^n) \cap \mathcal{P}_{\mathrm{ac}}(\mathbb{R}^n), \qquad 1 \le p < \infty.$$

By (A.21) and (A.22) one sees that $\mathcal{P}_{\mathrm{ac}}(\mathbb{R}^n)$ and $\mathcal{P}_{p,\mathrm{ac}}(\mathbb{R}^n)$ are dense in the narrow convergence, respectively, in $\mathcal{P}(\mathbb{R}^n)$ and $\mathcal{P}_p(\mathbb{R}^n)$.

Theorem 4.2 (Brenier theorem) *If $\mu \in \mathcal{P}_{2,\mathrm{ac}}(\mathbb{R}^n)$ and $v \in \mathcal{P}_2(\mathbb{R}^n)$, then there is a convex function $f : \mathbb{R}^n \to \mathbb{R} \cup \{+\infty\}$ such that, if F denotes the set of differentiability points of f and γ is an optimal plan in $\mathbf{K}_2(\mu, v)$, then*

$$\Omega = \mathrm{Int}\,\mathrm{Dom}(f) \neq \emptyset, \tag{4.3}$$

$$\mu \text{ is concentrated on } F, \tag{4.4}$$

$$\gamma = (\mathbf{id} \times \nabla f)_{\#}\mu, \tag{4.5}$$

$$(\nabla f)_{\#}\mu = v, \tag{4.6}$$

$$\mathrm{Cl}(\nabla f(F \cap \mathrm{spt}\,\mu)) = \mathrm{spt}\,v. \tag{4.7}$$

In particular, $(\mathbf{id} \times \nabla f)_{\#}\mu$ is the unique optimal transport plan in $\mathbf{K}_2(\mu, v)$, ∇f is an optimal transport map in $\mathbf{M}_2(\mu, v)$ with

$$\mathbf{K}_2(\mu, v) = \mathbf{M}_2(\mu, v) = \int_{\mathbb{R}^n} |\nabla f(x) - x|^2 \, d\mu(x), \tag{4.8}$$

and S is optimal in $\mathbf{M}_2(\mu, v)$ if and only if $S = \nabla f$ μ-a.e. on $\mathbb{R}^n$.

Remark 4.3 The map ∇f is called **the Brenier map from μ to v**. It is the **unique optimal transport map** in $\mathbf{M}_2(\mu, v)$ if one agrees to identify maps that transport μ when they agree μ-a.e. It is of course natural to do so, given that any two maps that differ on a μ-negligible set end up transporting μ at the same cost and pushing it forward to the same target measure.

Proof of the Brenier theorem By Theorem 3.20 there is a convex and lower-semicontinuous function $f : \mathbb{R}^n \to \mathbb{R} \cup \{+\infty\}$ such that if γ is an optimal transport plan in $\mathbf{K}_2(\mu, v)$, then $\mathrm{spt}\,\gamma \subset \partial f$. In step one we prove (4.3) and (4.4). In step two we prove (4.5), (4.6), (4.8), and the uniqueness statement for optimal transport maps. Finally, in step three, we prove (4.7).

Step one: We prove (4.3) and (4.4). By (2.31), $\mathrm{spt}\,\gamma \subset \partial f$ gives

$$x \cdot y = f(x) + f^*(y) \qquad \forall (x, y) \in \mathrm{spt}\,\gamma. \tag{4.9}$$

Since $f^* : \mathbb{R}^n \to \mathbb{R} \cup \{+\infty\}$ and $x \cdot y \in \mathbb{R}$, (4.9) implies that $f(x) < \infty$ for γ-a.e. $(x, y) \in \mathbb{R}^n \times \mathbb{R}^n$, which in turn implies $f(x) < \infty$ for μ-a.e. $x \in \mathbb{R}^n$, thanks to

$$\mu(\{f = +\infty\}) = \gamma(\{f = +\infty\} \times \mathbb{R}^n).$$

Having proved that $f < \infty$ μ-a.e. on $\mathbb{R}^n$, we deduce (4.3): indeed, if $\Omega = \emptyset$, then $\mathrm{Dom}(f)$ has affine dimension $k \leq n - 1$, and therefore $\mathcal{L}^n(\{f < \infty\}) = 0$; in particular, $f < \infty$ μ-a.e. on $\mathbb{R}^n$ implies that μ is concentrated on a set of null Lebesgue measure, and thus, by $\mu \ll \mathcal{L}^n$, that $\mu(\mathbb{R}^n) = 0$, a contradiction to $\mu(\mathbb{R}^n) = 1$. This proves (4.3), and since Ω is $\mathcal{L}^n$-equivalent to

Dom(f), we deduce that μ is concentrated on Ω. By Rademacher's theorem, f is differentiable $\mathcal{L}^n$-a.e. on Ω. If F is the differentiability set of f in Ω, then $\mathcal{L}^n(\Omega \setminus F) = 0$, and by $\mu \ll \mathcal{L}^n$ we conclude that f is differentiable μ-a.e. on $\mathbb{R}^n$, that is (4.4). In particular, (4.4) ensures that ∇f transports μ, and thus that $(\nabla f)_{\#}\mu$ is well-defined.

Step two: We prove (4.5), (4.6), (4.8), and the uniqueness statement for optimal transport maps. We first notice that

$$\text{if } (x,y) \in \operatorname{spt}\gamma \text{ and } x \in F, \text{ then } y = \nabla f(x), \tag{4.10}$$

(indeed, $(x,y) \in \operatorname{spt}\gamma$ implies $y \in \partial f(x)$ where $\partial f(x) = \{\nabla f(x)\}$ thanks to $x \in F$ – recall (2.19)), which can be equivalently reformulated as

$$\operatorname{spt}\gamma \cap (F \times \mathbb{R}^n) \subset \{(x, \nabla f(x)) : x \in F\}. \tag{4.11}$$

Now, if $\varphi \in C_b^0(\mathbb{R}^n \times \mathbb{R}^n)$, then by using (4.11) and the fact that $\gamma((\mathbb{R}^n \setminus F) \times \mathbb{R}^n) = \mu(\mathbb{R}^n \setminus F) = 0$ (recall (4.4)), we find that

$$\int_{\mathbb{R}^n \times \mathbb{R}^n} \varphi(x,y)\, d\gamma(x,y) = \int_{(F \times \mathbb{R}^n) \cap \operatorname{spt}\gamma} \varphi(x,y)\, d\gamma(x,y)$$

$$= \int_{\mathbb{R}^n \times \mathbb{R}^n} 1_F(x)\, \varphi(x, \nabla f(x))\, d\gamma(x,y).$$

$$(\text{by } (3.4)) = \int_F \varphi(x, \nabla f(x))\, d\mu(x)$$

$$= \int_{\mathbb{R}^n \times \mathbb{R}^n} \varphi(x,y)\, d[(\mathbf{id} \times \nabla f)_{\#}\mu](x,y).$$

By the arbitrariness of φ we deduce (4.5). By (4.5), $(\mathbf{id} \times \nabla f)_{\#}\mu \in \Gamma(\mu, \nu)$, and thus Proposition 3.1 implies $(\nabla f)_{\#}\mu = \nu$, that is, (4.6). By (4.6) and (4.5),

$$\mathbf{M}_2(\mu, \nu) \le \int_{\mathbb{R}^n} |\nabla f - x|^2\, d\mu = \int_{\mathbb{R}^n \times \mathbb{R}^n} |x - y|^2\, d\gamma = \mathbf{K}_2(\mu, \nu),$$

where in last identity we have used first the optimality of γ. By the general inequality $\mathbf{K}_2 \le \mathbf{M}_2$ we deduce (4.8). Finally, let S be an optimal transport map in $\mathbf{M}_2(\mu, \nu)$, so that $S : E \to \mathbb{R}^n$ for a Borel set $E \subset \mathbb{R}^n$ on which μ is concentrated. Setting $\gamma_S = (\mathbf{id} \times S)_{\#}\mu$, we notice that

$$\gamma_S \text{ is concentrated on } G(\nabla f) := \{(x, \nabla f(x)) : x \in F\}, \tag{4.12}$$

$$\gamma_S \text{ is concentrated on } G(S) := \{(x, S(x)) : x \in E\}. \tag{4.13}$$

Indeed, by (4.8) and since S is optimal in $\mathbf{M}_2(\mu, \nu)$, we have that γ_S is an optimal plan in $\mathbf{K}_2(\mu, \nu)$, so that we can apply (4.11) and $\mu(\mathbb{R}^n \setminus F) = 0$ to γ and deduce (4.12). At the same time, $G(S) = (\mathbf{id} \times S)^{-1}(\mathbb{R}^n \times \mathbb{R}^n)$, so that (4.13) is immediate. Now, should $X = \{x \in E \cap F : S(x) \ne \nabla f(x)\}$ be such that $0 < \mu(X) = \gamma_S(X \times \mathbb{R}^n)$, then, by (4.13), the set $Y = G(S) \cap (X \times \mathbb{R}^n) = \{(x, S(x)) : x \in X\}$ is such that $\gamma_S(Y) > 0$. However, $Y \cap G(\nabla f) = \emptyset$, so that $\gamma_S(Y) > 0$ is in contradiction with (4.12). By $\mu(X) = 0$ and since μ is concentrated on $E \cap F$, we deduce that $S = \nabla f$ μ-a.e. on $\mathbb{R}^n$.

Step three: We finally prove (4.7). If $y \in \operatorname{spt} \nu$ and $r > 0$, then

$$\mu((\nabla f)^{-1}(B_r(y))) = \nu(B_r(y)) > 0,$$

and thus there exists $x \in (\operatorname{spt} \mu) \cap F$ such that $|\nabla f(x) - y| < r$: hence, the inclusion $\supset$ in (4.7) follows. To prove the opposite inclusion, let us recall from Proposition 2.7 that for every $x \in F$ and $\varepsilon > 0$ there exists $\delta > 0$ such that

$$\partial f(B_\delta(x)) \subset B_\varepsilon(\nabla f(x)). \tag{4.14}$$

By (4.14), we see that if $x \in F$, then for every $\varepsilon > 0$ we can find $\delta > 0$ such that, thanks to $(\nabla f)_{\#}\mu = \nu$,

$$\nu(B_\varepsilon(\nabla f(x))) \geq \nu(\partial f(B_\delta(x))) \geq \nu(\nabla f(F \cap B_\delta(x)))$$
$$= \mu\big(\{z \in F : \nabla f(z) \in \nabla f(F \cap B_\delta(x))\}\big) \geq \mu(F \cap B_\delta(x)) = \mu(B_\delta(x)).$$

In particular, if $x \in (\operatorname{spt} \mu) \cap F$, then $\mu(B_\delta(x)) > 0$ and the arbitrariness of ε give $\nabla f(x) \in \operatorname{spt} \nu$. We have thus proved $\nabla f(F \cap \operatorname{spt} \mu) \subset \operatorname{spt} \nu$, and since $\operatorname{spt} \nu$ is closed, (4.7) follows.

4.2 Inverse of a Brenier Map and Fenchel–Legendre Transform

Of course, the case when **both** μ and ν are absolutely continuous with respect to $\mathcal{L}^n$ is of particular importance. In this case, in addition to the Brenier map ∇f from μ to ν, we have the Brenier map ∇g from ν to μ. In the following theorem we prove that ∇g is the inverse of ∇f (in the proper a.e. sense), and that the Fenchel–Legendre transform f^* of f is always a suitable choice for the convex potential g.

Theorem 4.4 *If $\mu, \nu \in \mathcal{P}_{2,\mathrm{ac}}(\mathbb{R}^n)$ and ∇f is the Brenier map from μ to ν, then ∇f^* is the Brenier map from ν to μ. Moreover, the following properties hold:*

(i) *for ν-a.e. $y \in \mathbb{R}^n$, f is differentiable at $x = \nabla f^*(y)$, and*

$$\nabla f(\nabla f^*(y)) = y;$$

(ii) *for μ-a.e. $x \in \mathbb{R}^n$, f^* is differentiable at $y = \nabla f(x)$, and*

$$\nabla f^*(\nabla f(x)) = x.$$

Proof *Step one*: Let $R(x,y) = (y,x)$. As already noticed in Remark 3.5, if $\gamma \in \Gamma(\mu, \nu)$, then $R_\# \gamma \in \Gamma(\nu, \mu)$ and the two plans have the same quadratic cost. Therefore, $\mathbf{K}_2(\mu, \nu) = \mathbf{K}_2(\nu, \mu)$. If now $\gamma = (\mathbf{id} \times \nabla f)_\# \mu$ is optimal in $\mathbf{K}_2(\mu, \nu)$, then $\gamma^* = R_\# \gamma$ is optimal in $\mathbf{K}_2(\nu, \mu)$. By (4.10), for γ-a.e. $(x, y) \in \mathbb{R}^n \times \mathbb{R}^n$ we have $y \in \partial f(x)$. Since $y \in \partial f(x)$ if and only if $x \in \partial f^*(y)$, we conclude that for γ^*-a.e. $(x, y) \in \mathbb{R}^n \times \mathbb{R}^n$ we have $x \in \partial f^*(y)$. Since $\nu \ll \mathcal{L}^n$, we can now repeat the argument in step two of the proof of Theorem 4.2 to show that $\gamma^* = (\mathbf{id} \times \nabla f^*)_\# \nu$. This shows that ∇f^* is the Brenier map from ν to μ.

Step two: We prove statements (i) and (ii). Let F and F^* be the sets of differentiability points of f and f^*, so that $\mu(\mathbb{R}^n \setminus F) = \nu(\mathbb{R}^n \setminus F^*) = 0$. By $(\nabla f)_\# \mu = \nu$ we see that

$$1 = \nu(\mathbb{R}^n) = \nu(F^*) = \mu((\nabla f)^{-1}(F^*)),$$

so that μ-a.e. $x \in \mathbb{R}^n$ satisfies $x \in F$ and $\nabla f(x) \in F^*$. We already proved that $(\mathbf{id} \times \nabla f^*)_\# \nu$ is optimal in $\mathbf{K}_2(\nu, \mu)$, so that

$$[(\nabla f^*) \times \mathbf{id}]_\# \nu = R_\# [(\mathbf{id} \times \nabla f^*)_\# \nu]$$

is optimal in $\mathbf{K}_2(\mu, \nu)$. By the uniqueness of optimal transport plans in the Brenier theorem we deduce that

$$(\mathbf{id} \times \nabla f)_\# \mu = [(\nabla f^*) \times \mathrm{Id}]_\# \nu.$$

In particular, if $\varphi : \mathbb{R}^n \times \mathbb{R}^n \to [0, \infty]$ is a Borel function, then

$$\int_F \varphi(x, \nabla f(x)) \, d\mu(x) = \int_{F^*} \varphi(\nabla f^*(y), y) \, d\nu(y).$$

By testing this identity with $\varphi(x, y) = 1_{F^*}(y) |x - \nabla f^*(y)|$ we conclude that

$$\int_F 1_{F^*}(\nabla f(x)) |x - \nabla f^*(\nabla f(x))| \, d\mu(x) = \int_{\mathbb{R}^n} |\nabla f^*(y) - \nabla f^*(y)| \, d\nu(y) = 0.$$

Since we already know that $1_{F^*}(\nabla f) = 1$ μ-a.e. on F, we conclude that $x = \nabla f^*(\nabla f(x))$ for μ-a.e. $x \in \mathbb{R}^n$.

4.3 Brenier Maps under Rigid Motions and Dilations

As an illustrative exercise concerning the uniqueness statement in the Brenier theorem, we briefly discuss the behavior of Brenier maps under rigid motions and dilations of the data measures. Translations, rotations, and volume-preserving scalings of $\mathbb{R}^n$ are closed operations in $\mathcal{P}_{2,\mathrm{ac}}(\mathbb{R}^n)$: more precisely, if $x_0 \in \mathbb{R}^n$, $Q \in \mathbf{O}(n)$ (the orthogonal group of $\mathbb{R}^n$, i.e., the group of linear isometries of $\mathbb{R}^n$), $\lambda > 0$ and $\mu = \rho \, d\mathcal{L}^n \in \mathcal{P}_{2,\mathrm{ac}}(\mathbb{R}^n)$, then the measures

$$\tau_{x_0}\mu(E) = \mu(E - x_0), \tag{4.15}$$

$$Q_\#\mu(E) = \mu(Q^{-1}(E)), \tag{4.16}$$

$$\mu^\lambda(E) = \mu(E/\lambda), \qquad E \in \mathcal{B}(\mathbb{R}^n), \tag{4.17}$$

belong to $\mathcal{P}_{2,\mathrm{ac}}(\mathbb{R}^n)$ and are such that

$$\mathrm{spt}\, \tau_{x_0}\mu = x_0 + \mathrm{spt}\,\mu, \qquad \mathrm{spt}\,(Q_\#\,\mu) = Q(\mathrm{spt}\,\mu), \qquad \mathrm{spt}\,\mu^\lambda = \lambda\,\mathrm{spt}\,\mu.$$

Let $f : \mathbb{R}^n \to \mathbb{R} \cup \{+\infty\}$ be convex such that ∇f transports μ, so that ∇f is the Brenier map from μ to $\nu = (\nabla f)_\#\mu$ by Theorem 4.2. We can easily obtain the Brenier maps from $\tau_{x_0}\mu$, $Q_\#\mu$, and μ^λ to, respectively, $\tau_{x_0}\nu$, $Q_\#\nu$, and ν^λ, by looking at the convex functions

$$x \mapsto f(x - x_0) + x \cdot x_0, \qquad x \mapsto f(Q^*(x)), \qquad x \mapsto \lambda^2 f(x/\lambda).$$

For example, if $\varphi \in C_c^0(\mathbb{R}^n)$, then $(\nabla f)_\#\mu = \nu$ gives

$$\int_{\mathbb{R}^n} \varphi \, d[\tau_{x_0}\nu] = \int_{\mathbb{R}^n} \varphi(y + x_0) \, d\nu(y) = \int_{\mathbb{R}^n} \varphi(\nabla f(x) + x_0) \, d\mu(x),$$

while at the same time if g is convex and such that $(\nabla g)_\#[\tau_{x_0}\mu] = \tau_{x_0}\nu$, then

$$\int_{\mathbb{R}^n} \varphi \, d[\tau_{x_0}\nu] = \int_{\mathbb{R}^n} \varphi(\nabla g) \, d[\tau_{x_0}\mu] = \int_{\mathbb{R}^n} \varphi(\nabla g(x + x_0)) \, d\mu(x),$$

so that we must have $\nabla g(x) = \nabla f(x - x_0) + x_0$ for μ-a.e. $x \in \mathbb{R}^n$; in particular, we may take $g(x) = f(x - x_0) + x \cdot x_0$ and (4.15) holds. The proofs of (4.16) and (4.17) are entirely analogous.

5

First Order Differentiability of Convex Functions

In this chapter we establish two basic facts concerning the first order differentiability of convex functions-a sharp dimensional estimate on the set of non-differentiability points (Theorem 5.1), and a non-smooth, convex version of the implicit function theorem (Theorem 5.3). These results will then be used in Chapter 6 to relax the condition of $\mathcal{L}^n$-absolute continuity on the origin measure in the Brenier theorem.

5.1 First Order Differentiability and Rectifiability

The $\mathcal{L}^n$-a.e. differentiability of convex functions was deduced in Section 2.3 as a direct consequence of Rademacher's theorem and the local Lipschitz bounds implied by convexity. This approach does not take full advantage of convexity, and indeed a bit of experimenting suggests that the set of non–differentiability points of a convex function should have, at most, codimension one; and, in Theorem 5.1 below, we prove indeed its countable $(n-1)$-rectifiability. Here we are using the terminology (see Appendix A.15 for more context) according to which a Borel set $M \subset \mathbb{R}^n$ is **countably k-rectifiable in $\mathbb{R}^n$** (where $0 \leq k \leq n-1$ is an integer), if there exist countably many Lipschitz maps $g_j : \mathbb{R}^k \to \mathbb{R}^n$ such that

$$M \subset \bigcup_{j \in \mathbb{N}} g_j(\mathbb{R}^k).$$

Theorem 5.1 (First order differentiability of convex functions)   *If* $f : \mathbb{R}^n \to \mathbb{R} \cup \{+\infty\}$ *is a convex function on* $\mathbb{R}^n$ *with set of differentiability points F, then*

$$\mathrm{Dom}(f) \setminus F$$

is a countably $(n-1)$-rectifiable subset of $\mathbb{R}^n$.

Proof   For a preliminary insight in the proof, see Figure 5.1.

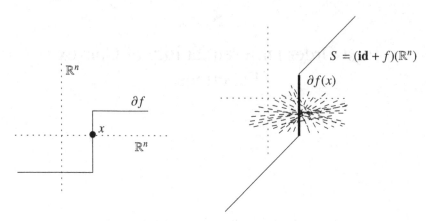

Figure 5.1 Proof of Theorem 5.1. On the left, the subdifferential ∂f of a convex function f. On the right, the graph S of $(\mathbf{id} + \partial f)$: By monotonicity of ∂f we can invert $(\mathbf{id} + \partial f)$ on S as a 1-Lipschitz map. Now, if $x \in \text{Dom}(f) \setminus F$, then $\partial f(x)$ has affine dimension at least 1, in particular $\partial f(x)$ intersects every element of an open ball in the metric space of affine hyperplanes (this ball is represented by a "cloud of lines" in the picture). By taking a countable dense subset in the space of affine hyperplanes and using the inverse map of $(\mathbf{id} + \partial f)$, we thus cover $\text{Dom}(f) \setminus F$ by countably many Lipschitz images of $\mathbb{R}^{n-1}$.

Step one: Let us consider the multi-valued map $(\mathbf{id} + \partial f) : \mathbb{R}^n \to \mathcal{K}$, i.e.,

$$(\mathbf{id} + \partial f)(x) = \left\{ x + y : y \in \partial f(x) \right\} \qquad x \in \mathbb{R}^n,$$

where $\mathcal{K}$ is the family of closed convex subsets of $\mathbb{R}^n$, and let

$$S = (\mathbf{id} + \partial f)(\mathbb{R}^n) = \left\{ z \in \mathbb{R}^n : z = x + y, x \in \mathbb{R}^n, y \in \partial f(x) \right\}.$$

We first claim that for every $z \in S$ there exists a unique $x \in \mathbb{R}^n$ so that $z = x + y$ for some $y \in \partial f(x)$. Indeed, if $z_1 = x_1 + y_1$ and $z_2 = x_2 + y_2$ are elements of S, then by cyclical monotonicity of ∂f we have

$$(y_1 - y_2) \cdot (x_1 - x_2) \geq 0, \tag{5.1}$$

and therefore

$$|x_1 - x_2|^2 \leq [(x_1 - x_2) + (y_1 - y_2)] \cdot (x_1 - x_2) = (z_1 - z_2) \cdot (x_1 - x_2)$$
$$\leq |z_1 - z_2|\,|x_1 - x_2|$$

so that $|x_1 - x_2| \leq |z_1 - z_2|$: In particular, if $z_1 = z_2$, then $x_1 = x_2$ and, necessarily, $y_1 = y_2$. Based on this claim we can define a map $T = (\mathbf{id} + \partial f)^{-1} : S \to \mathbb{R}^n$ so that $T(z) = x$ is the unique element of $\mathbb{R}^n$ with the property that $z - x \in \partial f(x)$, and the resulting map is 1-Lipschitz, i.e.,

$$|T(z_1) - T(z_2)| \le |z_1 - z_2| \qquad \forall z_1, z_2 \in S.$$

Finally, we extend T as a 1-Lipschitz map from $\mathbb{R}^n$ to $\mathbb{R}^n$.

Step two: We can assume that $\Omega = \text{Int Dom}(f) \ne \emptyset$: Otherwise $F = \emptyset$, $\text{Dom}(f)$ is contained in a hyperplane, and the theorem is proved. Since $\text{Dom}(f) \setminus \Omega$ is contained in the boundary of the convex set Ω – which is easily seen to be countably $(n-1)$-rectifiable – we can directly work with $\Omega \setminus F$. To conclude the argument, let us denote by $\mathbf{G}_n^{n-1}$ the metric space of $(n-1)$-dimensional linear subspaces of $\mathbb{R}^n$, and let $\mathscr{A}_n^{n-1} = \mathbb{R}^n \times \mathbf{G}_n^{n-1}$ represent the metric space of $(n-1)$-dimensional affine subspaces of $\mathbb{R}^n$. We let $Q = \{A_j\}_{j \in \mathbb{N}}$ denote a countable dense subset of $\mathscr{A}_n^{n-1}$. Now, if $x \in \Omega \setminus F$, then $\partial f(x)$ is a non-trivial convex set of affine dimension at least 1. In particular, there exists an open ball in $\mathscr{A}_n^{n-1}$ such that every affine plane A in this ball intersects $\partial f(x)$: By density, we can find $j \in \mathbb{N}$ such that $\partial f(x) \cap A_j \ne \emptyset$. Let $g_j : A_j \to \mathbb{R}^n$ be defined as the restriction of T to A_j: If $z \in A_j \cap \partial f(x)$, then $g_j(z) = x$. In summary, for every $x \in \Omega \setminus F$, there exists $j \in \mathbb{N}$ such that $x \in g_j(A_j)$. Since $A_j \equiv \mathbb{R}^{n-1}$ (and g_j is Lipschitz since T is), $\Omega \setminus F$ is countably $(n-1)$-rectifiable. $\qquad \square$

Remark 5.2 With the same proof, the set of those x such that $\partial f(x)$ has affine dimension greater or equal than k is proved to be countably $(n-k)$-rectifiable. In particular, $\partial f(x)$ has non-empty interior only at countably many $x \in \mathbb{R}^n$.

5.2 Implicit Function Theorem for Convex Functions

The implicit function theorem for C^1-functions makes crucial use of the continuity assumption of the gradients and is indeed false for Lipschitz functions.[1] In the following theorem we prove a version of the implicit function theorem for convex functions where the continuity of gradients is replaced by the continuity property of convex subdifferentials proved in Proposition 2.7.

Theorem 5.3 (Convex implicit function theorem) *If $f, g : \mathbb{R}^n \to \mathbb{R} \cup \{+\infty\}$ are convex functions, both differentiable at x and with $\nabla f(x) \ne \nabla g(x)$, then there exists $r > 0$ such that $B_r(x) \cap \{f = g\}$ is contained in the graph of a Lipschitz function of $(n-1)$-variables.*

Proof The idea is to follow the classical proof of the implicit function theorem in the case of the two smooth functions $f_\varepsilon = f \star \rho_\varepsilon$ and $g_\varepsilon = g \star \rho_\varepsilon$ defined by

---

[1] An example is given in [McC95]: let $n = 2$, $f(x) = x_1$ if $|x_1| \ge x_2^2$, $f(x) = 0$ if $|x_1| \le x_2^2/2$ and let $f(x)$ be defined by a Lipschitz extension elsewhere. In this example f is differentiable at $x = 0$ with $\nabla f(0) = e_1$, and $f(0) = g(0) = 0$, where $g \equiv 0$. Nevertheless, the level set $\{f = g\} = \{f = 0\}$ contains an open set in every neighborhood of the origin.

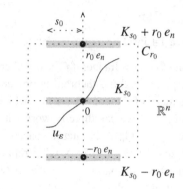

Figure 5.2 The notation used in the proof of Theorem 5.3.

ε-regularization of f and g. By exploiting the continuity of subdifferentials, i.e., the fact that if f is differentiable at x, then for every $\varepsilon > 0$ there exists $\delta > 0$ such that

$$(\partial f)(B_\delta(x)) \subset B_\varepsilon(\nabla f(x)), \tag{5.2}$$

we will be able to take the limit $\varepsilon \to 0^+$ in the implicit functions so constructed, thus proving the theorem. Without loss of generality, we set $x = 0$, let $h = f - g$, assume that

$$\nabla h(0) = 2\lambda e_n, \qquad \text{for } \lambda > 0, \tag{5.3}$$

and introduce the projection $\pi : \mathbb{R}^n \to \mathbb{R}^{n-1}$ defined by $\pi(y) = (y_1, \ldots, y_{n-1})$, and the corresponding cylinders and $(n-1)$-dimensional disks

$$C_r = \left\{ y \in \mathbb{R}^n : |\pi(y)| < r, |y_n| < r \right\}, \qquad K_r = \left\{ y \in \mathbb{R}^n : |\pi(y)| < r, y_n = 0 \right\};$$

see Figure 5.2. We claim that there exists $\varepsilon_0 > 0$ and $r_0 > 0$ such that

$$\nabla h_\varepsilon(y) \cdot e_n \geq \lambda \qquad \forall y \in C_{r_0}, \forall \varepsilon < \varepsilon_0, \tag{5.4}$$

where $h_\varepsilon = h \star \rho_\varepsilon$. Indeed, if (5.4) fails, then we can find $\varepsilon_j \to 0^+$ and $y_j \to 0$ such that

$$\nabla h_{\varepsilon_j}(y_j) \cdot e_n < \lambda.$$

If F is the set of differentiability points of f, then

$$\nabla f_{\varepsilon_j}(y_j) = \int_{B_{\varepsilon_j}(y_j)} \rho_{\varepsilon_j}(z - y_j) \nabla f(z) \, dz \in \mathrm{Conv}\left((\nabla f)(F \cap B_{\varepsilon_j}(y_j)) \right)$$
$$\subset \mathrm{Conv}\left(\partial f(B_{\varepsilon_j + |y_j|}(0)) \right),$$

so that, thanks to (5.2), we find $\nabla f_{\varepsilon_j}(y_j) \to \nabla f(0)$. Similarly, $\nabla g_{\varepsilon_j}(y_j) \to \nabla g(0)$, and thus $\nabla h(0) \cdot e_n \le \lambda$, a contradiction with (5.3). This proves (5.4). By (5.4), we find

$$h_\varepsilon(r_0\, e_n) \ge h_\varepsilon(0) + \lambda\, r_0, \qquad h_\varepsilon(-r_0\, e_n) \le h_\varepsilon(0) - \lambda\, r_0.$$

By combining these bounds with the basic uniform estimate

$$\sup_{\varepsilon < r_0} \|\nabla h_\varepsilon\|_{C^0(C_{r_0})} \le \|\nabla h\|_{L^\infty(C_{2r_0})}, \tag{5.5}$$

we find that there exists $s_0 \in (0, r_0)$ such that, for every $\varepsilon < e_0$,

$$h_\varepsilon \ge h_\varepsilon(0) + \frac{\lambda\, r_0}{2} \ge \frac{\lambda\, r_0}{4} \qquad \text{on } K_{s_0} + r_0\, e_n,$$

$$h_c \le h_\varepsilon(0) - \frac{\lambda\, r_0}{2} \le -\frac{\lambda\, r_0}{4} \qquad \text{on } K_{s_0} - r_0\, e_n,$$

(see, again, Figure 5.2), where we have also used the fact that $h_\varepsilon(0) \to h(0) = 0$ as $\varepsilon \to 0^+$. By continuity of h_ε, for every $z \in K_{s_0}$ and $\varepsilon < \varepsilon_0$, there exists a (unique, thanks to (5.4)) $u_\varepsilon(z) \in (-r_0, r_0)$ such that

$$h_\varepsilon(z, u_\varepsilon(z)) = 0, \qquad |u_\varepsilon(0)| < r_0. \tag{5.6}$$

Finally, we claim that

$$\mathrm{Lip}(u_\varepsilon; K_{s_0}) \le \frac{\|\nabla h\|_{L^\infty(C_{2r_0})}}{\lambda}. \tag{5.7}$$

Indeed, given $z_0, z_1 \in K_{s_0}$, the map

$$\ell_\varepsilon(t) = h_\varepsilon\big(z_t, (1-t)\, u_\varepsilon(z_0) + t\, u_\varepsilon(z_1)\big),$$
$$\text{where } z_t = (1-t)\, z_0 + t\, z_1,$$

is such that $\ell_\varepsilon(0) = \ell_\varepsilon(1) = 0$: therefore there exists $t_\varepsilon \in (0, 1)$ such that, setting

$$x_\varepsilon = \big(z_{t_\varepsilon}, (1 - t_\varepsilon)\, u_\varepsilon(z_0) + t_\varepsilon\, u_\varepsilon(z_1)\big),$$

then

$$0 = \ell'_\varepsilon(t_\varepsilon) = \pi[\nabla h_\varepsilon(x_\varepsilon)] \cdot (z_1 - z_0) + (u_\varepsilon(z_1) - u_\varepsilon(z_0))\, [\nabla h_\varepsilon(x_\varepsilon)] \cdot e_n.$$

Considering that $x_\varepsilon \in K_{s_0} \times (-r_0, r_0) \subset C_{r_0}$ and applying (5.4) and (5.5) again, we get

$$\lambda\, |u_\varepsilon(z_1) - u_\varepsilon(z_0))| \le |u_\varepsilon(z_1) - u_\varepsilon(z_0))|\, |\nabla h_\varepsilon(x_\varepsilon) \cdot e_n|$$
$$\le |\pi[\nabla h_\varepsilon(x_\varepsilon)]|\, |z_1 - z_0| \le \|\nabla h\|_{L^\infty(C_{2r_0})}\, |z_1 - z_0|,$$

i.e., we prove (5.7). By (5.7) and $|u_\varepsilon(0)| < r_0$ we deduce that $u_\varepsilon \to u$ uniformly in K_{s_0} to a Lipschitz function u. Given that $h_\varepsilon \to h$ uniformly in C_{r_0}, we conclude from $h_\varepsilon(z, u_\varepsilon(z)) = 0$ that $h(z, u(z)) = 0$ for every $z \in K_{s_0}$, that is

$$\text{Graph}(u; K_{s_0}) \subset \{f = g\}.$$

To finally show

$$\text{Graph}(u; K_{s_0}) = \{f = g\} \cap [K_{s_0} \times (-r_0, r_0)],$$

we obtain a contradiction from the existence of $w \in \{f = g\} \cap (K_{s_0} \times (-r_0, r_0))$ such that $w \neq (z, u(z))$ for $z = \pi[w]$. Indeed, if such point w exists and, say, $w_n > u(z)$, then $w_n > u_\varepsilon(z)$ for every ε small enough, and thus, thanks to (5.6) and (5.4),

$$h_\varepsilon(w) = h_\varepsilon(z, u_\varepsilon(z)) + (w_n - u_\varepsilon(z)) \int_0^1 \nabla h_\varepsilon(z, u_\varepsilon(z) + t(w_n - u_\varepsilon(z))) \cdot e_n \, dt$$

$$\geq \lambda(w_n - u_\varepsilon(z)),$$

which in turn, in the limit $\varepsilon \to 0^+$, gives $0 = h(w) \geq \lambda(w_n - u(z)) > 0$, a contradiction.   □

6

The Brenier–McCann Theorem

We now exploit the results presented in Chapter 5 to improve the Brenier theorem in several directions.

Theorem 6.1 (The Brenier–McCann theorem)   *If $\mu \in \mathcal{P}(\mathbb{R}^n)$ is such that*

$$\mu(S) = 0 \tag{6.1}$$

for every countably $(n-1)$-rectifiable set S in $\mathbb{R}^n$, then:

Existence: *For every $\nu \in \mathcal{P}(\mathbb{R}^n)$ there exists a lower semicontinuous and convex function $f : \mathbb{R}^n \to \mathbb{R} \cup \{+\infty\}$ with set of differentiability points F such that*

$$\Omega = \operatorname{Int} \operatorname{Dom}(f) \neq \emptyset,$$

$$\mu \text{ is concentrated on } F,$$

$$(\nabla f)_{\#}\mu = \nu,$$

$$\operatorname{Cl}(\nabla f(F \cap \operatorname{spt}\mu)) = \operatorname{spt}\nu.$$

Moreover, if $\operatorname{spt}\nu$ is compact, then we can construct f so that

$$\operatorname{Dom}(f) = \mathbb{R}^n, \qquad \nabla f(x) \in \operatorname{spt}\nu \text{ for every } x \in F. \tag{6.2}$$

Uniqueness: *If $f, g : \mathbb{R}^n \to \mathbb{R} \cup \{+\infty\}$ are convex functions such that $(\nabla f)_{\#}\mu = (\nabla g)_{\#}\mu$, then $\nabla f = \nabla g$ μ-a.e. on $\mathbb{R}^n$.*

Remark 6.2   The **existence statement** in Theorem 6.1 improves on Theorem 4.2 under two aspects: First, it drops the assumption that $\mu, \nu \in \mathcal{P}_2(\mathbb{R}^n)$; second, μ is no longer required to be absolutely continuous with respect to $\mathcal{L}^n$, but just to be null on countably $(n-1)$-rectifiable sets. The first improvement simply hinges on the use of Theorem 3.16 in place of Theorem 3.20. The second improvement is based on the first order differentiability result for convex functions established in Theorem 5.1. Notice in particular that, in light of Remark 4.1, **condition (6.1) is sharp**; i.e., if μ is positive on some $(n-1)$-dimensional set, there may be no optimal transport map. The **uniqueness statement** in Theorem 6.1 is formally the same as the uniqueness of Brenier maps proved in Theorem 4.2, but the proof is substantially different. The reason is, once again, that we may not have $|x - y|^2 \in L^1(\mu \times \nu)$ and thus may be unable to exploit the Kantorovich duality theorem (which is behind Theorem 3.15, and thus the uniqueness statements in the Brenier theorem). We will thus need an entirely new argument, which will be based on the implicit function theorem for convex functions (Theorem 5.3).

6.1 Proof of the Existence Statement

Step one: By Theorem 3.16 there are $\gamma \in \Gamma(\mu, \nu)$ and a convex function $f : \mathbb{R}^n \to \mathbb{R} \cup \{+\infty\}$ such that spt $\gamma \subset \partial f$. We now argue as in step one of the proof of the Brenier theorem, using (6.1) in place of $\mu \ll \mathcal{L}^n$. First, by spt $\gamma \subset \partial f$ we have $f(x) + f^*(y) = x \cdot y$ for every $(x, y) \in$ spt γ, and thus that $f < \infty$ μ a.e. on $\mathbb{R}^n$. In particular, μ is concentrated on $\text{Dom}(f)$. If $\Omega = \emptyset$, then $\text{Dom}(f)$ is countably $(n - 1)$-rectifiable, and, by (6.1), $0 = \mu(\text{Dom}(f)) \geq \mu(\text{spt } \mu) = 1$, a contradiction. Hence, $\Omega \neq \emptyset$ and, since $\text{Dom}(f) \backslash \Omega$ is locally the graph of a convex function over $\mathbb{R}^{n-1}$, $\text{Dom}(f) \setminus \Omega$ is in turn μ-negligible, thanks to (6.1). In summary, μ is concentrated in Ω, and since $\Omega \backslash F$ is countably $(n-1)$-rectifiable, thanks to Theorem 5.1, we conclude that $\mu(\mathbb{R}^n \setminus F) = 0$. We can now *verbatim* repeat the arguments in steps two and three of the proof of Theorem 4.2 to conclude that $\gamma = (\text{id} \times \nabla f)_\# \mu$, $(\nabla f)_\# \mu = \nu$, and $\text{Cl}(\nabla f(F \cap \text{spt } \mu)) = \text{spt } \nu$.

Step two: We prove that in the special case when spt ν is compact, one can find a Brenier map satisfying (6.2). To this end, it will suffice to set

$$g(x) = \sup_{y \in \text{spt } \nu} x \cdot y - f^*(y).$$

We notice that $g : \mathbb{R}^n \to \mathbb{R}$ is convex (in particular, $\text{Dom}(g) = \mathbb{R}^n$), and that if G denotes the set of differentiability points of g, then $\nabla g(x) \in$ spt ν for every $x \in G$. Indeed, since spt ν is compact and f^* is lower semicontinuous, for every $x \in \mathbb{R}^n$ there exists a maximizer $y(x) \in$ spt ν in the definition of $g(x)$: In particular, $y(x) \in \partial g(x)$, and if in addition $x \in G$ this gives $\nabla g(x) = y(x) \in$ spt ν. We are thus left to prove that μ is concentrated on G and that $(\nabla g)_\# \mu = \nu$. We first notice that, since f is convex and lower semicontinuous on $\mathbb{R}^n$,

$$f(x) = f^{**}(x) = \sup_{y \in \mathbb{R}^n} \{x \cdot y - f^*(y)\} \geq g(x) \qquad \forall x \in \mathbb{R}^n.$$

At the same time, if $x \in F \cap$ spt μ and $y = \nabla f(x)$, then $y \in$ spt ν, so that y is admissible in the definition of $g(x)$, and $x \cdot y - f^*(y) = f(x)$ by (2.31): Therefore

$$f \geq g \text{ on } \mathbb{R}^n \text{ and } f = g \text{ on } F \cap \text{spt } \mu. \tag{6.3}$$

Since, by definition, f is differentiable on F, (6.3) gives $F \subset G$ and $\nabla f = \nabla g$ on F. In particular, since μ is concentrated on F we find that ∇g transports μ, and then $(\nabla g)_\# \mu = \nu$ follows since for every Borel set $E \subset \mathbb{R}^n$ we have

$$(\nabla g)_\# \mu(E) = \mu(G \cap (\nabla g)^{-1}(E)) = \mu(F \cap (\nabla g)^{-1}(E)) = \mu(F \cap (\nabla f)^{-1}(E)) = \nu(E).$$

6.2 Proof of the Uniqueness Statement

We want to prove that if $f, g : \mathbb{R}^n \to \mathbb{R} \cup \{+\infty\}$ are convex functions such that $(\nabla f)_\# \mu = (\nabla g)_\# \mu$, then $\nabla f = \nabla g$ μ-a.e. on $\mathbb{R}^n$. Indeed, let F and G denote the sets of differentiability of f and g: we are going to prove that $\nabla f = \nabla g$ on spt $\mu \cap F \cap G$. Arguing by contradiction, let $x_0 \in (\text{spt } \mu) \cap F \cap G$ be such that $\nabla f(x_0) \neq \nabla g(x_0)$. There is no loss of generality in assuming that $f(x_0) = g(x_0)$. By Theorem 5.3 there exist $v \in \mathbb{S}^{n-1}$, $\delta_0 > 0$, and a Lipschitz function $u : v^\perp \to \mathbb{R}$ such that $u(0)v = x_0$ and

$$B_{\delta_0}(x_0) \cap \{f > g\} = B_{\delta_0}(x_0) \cap \{z + tv : z \in v^\perp, t > u(z)\},$$

where $v^\perp = \{z \in \mathbb{R}^n : z \cdot v = 0\}$. By (6.1) we have $\mu(B_{\delta_0}(x_0) \cap \{f = g\}) = 0$, which, combined with the fact that $\mu(B_\delta(x_0)) > 0$ for every $\delta > 0$, implies that for every $\delta \in (0, \delta_0)$ we have

either $\quad \mu(B_\delta(x_0) \cap \{f > g\}) > 0 \quad$ or $\quad \mu(B_\delta(x_0) \cap \{f < g\}) > 0.$

$$(6.4)$$

To find a contradiction to the fact that ∇f and ∇g define the same push-forward of μ, we need to relate the behaviors of ∇f and ∇g on $\{f > g\}$ and $\{f < g\}$. To this end, we first prove that if $x \in G$, then

$$\nabla g(x) \in \partial f(\{f > g\}) \quad \Rightarrow \quad f(x) > g(x). \tag{6.5}$$

Indeed, since $\nabla g(x) \in \partial f(z)$ for some z such that $f(z) > g(z)$, for every $w \in \mathbb{R}^n$ we have

$$f(w) \geq f(z) + \nabla g(x) \cdot (w - z) > g(z) + \nabla g(x) \cdot (w - z)$$
$$\geq g(x) + \nabla g(x) \cdot (z - x) + \nabla g(x) \cdot (w - z)$$

i.e.,

$$f(w) > g(x) + \nabla g(x) \cdot (w - x) \quad \forall w \in \mathbb{R}^n. \tag{6.6}$$

Taking $w = x$ proves (6.5). We can prove a stronger localized version of (6.5) near x_0, namely we can show the existence of $\eta > 0$ such that if $x \in G$, then

$$\nabla g(x) \in \partial f(\{f > g\}) \quad \Rightarrow \quad \begin{cases} f(x) > g(x), \\ |x - x_0| \geq \eta. \end{cases} \tag{6.7}$$

Indeed, if (6.7) fails, then we can find $x_j \to x_0$ with $\nabla g(x_j) \in \partial f(z_j)$ for $z_j \in \{f > g\}$, so that by (6.6)

$$f(w) > g(x_j) + \nabla g(x_j) \cdot (w - x_j) \qquad \forall j, \forall w \in \mathbb{R}^n;$$

since $\nabla g(x_j) \to \nabla g(x_0)$ by the continuity of convex subdifferentials, we find

$$f(w) \geq A^g_{x_0}(w) \qquad \forall w \in \mathbb{R}^n. \tag{6.8}$$

Since $f(x_0) = g(x_0) = A^g_{x_0}(x_0)$, we conclude that $\nabla g(x_0) \in \partial f(x_0) = \{\nabla f(x_0)\}$, a contradiction. This proves (6.7). Therefore, by repeating the argument with the roles of f and g inverted, we have proved the existence of $\eta > 0$, depending on f, g and x_0, such that

$$(\nabla g)^{-1}\big(\partial f(\{f > g\})\big) \subset \{f > g\} \setminus B_\eta(x_0), \tag{6.9}$$

$$(\nabla f)^{-1}\big(\partial g(\{g > f\})\big) \subset \{g > f\} \setminus B_\eta(x_0).$$

Assuming without loss of generality that $\eta < \delta_0$ and that the first condition in (6.4) holds for $\delta = \eta$, i.e., that

$$\mu(B_\eta(x_0) \cap \{f > g\}) > 0,$$

we now conclude that proof. Indeed, setting

$$Z = \partial f(\{f > g\})$$

and taking into account that we trivially have

$$F \cap \{f > g\} = \big\{y \in F : \nabla f(y) \in Z\big\} = (\nabla f)^{-1}(Z), \tag{6.10}$$

we conclude that, by $(\nabla g)_\# \mu = \nu$, (6.9), $\mu(\mathbb{R}^n \setminus F) = 0$, and (6.10), one has

$$\begin{aligned}
\nu(Z) &\leq \mu(\{f > g\} \setminus B_\eta(x_0)) \\
&= \mu(F \cap \{f > g\} \setminus B_\eta(x_0)) \\
&\leq \mu((\nabla f)^{-1}(Z) \cap \{f > g\} \setminus B_\eta(x_0)) \\
&= \mu((\nabla f)^{-1}(Z) \cap \{f > g\}) - \mu((\nabla f)^{-1}(Z) \cap \{f > g\} \cap B_\eta(x_0)) \\
&= \mu((\nabla f)^{-1}(Z)) - \mu(F \cap \{f > g\} \cap B_\eta(x_0)) \\
&= \nu(Z) - \mu(\{f > g\} \cap B_\eta(x_0)) \\
&< \nu(Z)
\end{aligned}$$

where in the last identity we have used $(\nabla f)_\# \mu = \nu$ and $\mu(\mathbb{R}^n \setminus F) = 0$, and where in the last inequality we have used (6.4) with $\delta = \eta$. This contradiction proves that $\nabla f = \nabla g$ on $\operatorname{spt} \mu \cap F \cap G$, and thus μ-a.e.

7

Second Order Differentiability of
Convex Functions

In Chapter 1 we have introduced two different formulations of the Monge problem: one based on the pointwise transport condition (1.1) (and limited to the case when $\mu, \nu \ll \mathcal{L}^n$), and one based on the notion of transport map (and covering the case of arbitrary mass distributions). It thus makes sense to ask if the Brenier maps constructed in the solution of the Monge problem with quadratic cost (between measures in $\mathcal{P}_{ac}(\mathbb{R}^n)$) satisfy the pointwise transport condition (1.1). This delicate issue, which is the subject of Chapter 8, calls into play the second order differentiability properties of convex functions discussed in this chapter. We first address second order differentiability from the distributional viewpoint (Section 7.1). Then, in Section 7.2, we discuss the existence, at a.e. point in $\mathbb{R}^n$, of second order Taylor expansions of convex functions (this is the celebrated Alexandrov's theorem; see Theorem 7.2) and also obtain first order expansions for their gradients (Theorem 7.3).

7.1 Distributional Derivatives of Convex Functions

Let us recall that if Ω is an open set in $\mathbb{R}^n$, then the space of functions of locally bounded variation in Ω with values in $\mathbb{R}^m$, $BV_{\mathrm{loc}}(\Omega; \mathbb{R}^m)$, consists of those vector-fields $T \in L^1_{\mathrm{loc}}(\Omega; \mathbb{R}^m)$, whose distributional derivative DT defines an $\mathbb{R}^{n \times m}$-valued Radon measure in Ω. We denote by $\nabla T \in L^1_{\mathrm{loc}}(\Omega; \mathbb{R}^{n \times m})$ the density of DT with respect to the Lebesgue measure, so that the Radon–Nikodym decomposition of DT takes the form

$$DT = \nabla T \, d\mathcal{L}^n + [DT]^s,$$

for an $\mathbb{R}^{n \times m}$-valued Radon measure $[DT]^s$ in Ω, which is singular with respect to the Lebesgue measure in Ω.

Theorem 7.1 (Distributional derivatives of a convex functions)   *Let $f : \mathbb{R}^n \to \mathbb{R} \cup \{+\infty\}$ be convex with $\Omega = \operatorname{Int} \operatorname{Dom}(f) \neq \emptyset$. Then f is locally Lipschitz continuous in Ω with*

$$\|\nabla f\|_{L^\infty(B_r(x))} \leq \frac{C(n)}{r^{n+1}} \|f\|_{L^1(B_{3r}(x))}, \qquad \forall B_{3r}(x) \subset\subset \Omega. \tag{7.1}$$

Moreover, the distributional Hessian $D^2 f$ of f satisfies

$$D^2 f \text{ is a } \mathbb{R}^{n\times n}_{\text{sym}}\text{-valued Radon measure in } \Omega, \tag{7.2}$$

such that, if we set

$$(D^2_{vw} f)[\varphi] = \int_\Omega f(x) \, v \cdot \nabla^2 \varphi(y)[w] \, dy \qquad \varphi \in C^\infty_c(\Omega), v, w \in \mathbb{R}^n, \tag{7.3}$$

then $D^2_{vv} f$ is a (nonnegative) Radon measure in Ω for every $v \in \mathbb{R}^n$. In particular, $\nabla f \in BV_{\text{loc}}(\Omega; \mathbb{R}^n)$.

Proof   Step one: In Section 2.3 we have already proved that $f \in W^{1,\infty}_{\text{loc}}(\Omega)$, with

$$\|\nabla f\|_{L^\infty(B_r)} \leq \frac{C(n)}{r} \operatorname{osc}(f; B_{2r}), \tag{7.4}$$

whenever B_r and B_{2r} denote concentric balls of radii r and $2r$ with $B_{2r} \subset\subset \Omega$. In this step we deduce (7.1) from (7.4). Indeed, if $x \in \Omega$, then $\partial f(x) \neq \emptyset$, and if $y \in \partial f(x)$, then $f(z) \geq f(x) + y \cdot (z - x)$ for every $z \in \mathbb{R}^n$: therefore if $x \in B_{2r}$ and $B_{3r} \subset\subset \Omega$, we can integrate this lower bound on $z \in B_r(x)$ to find

$$f(x) \leq \frac{1}{\omega_n r^n} \int_{B_r(x)} f + y \cdot \frac{1}{\omega_n r^n} \int_{B_r(x)} (x - z) \, dz$$

$$= \frac{1}{\omega_n r^n} \int_{B_r(x)} f \leq \frac{C(n)}{r^n} \int_{B_{3r}} f.$$

By taking the supremum over $x \in B_{2r}$ we have proved that

$$\sup_{B_{2r}} f \leq \frac{C(n)}{r^n} \int_{B_{3r}} f. \tag{7.5}$$

Similarly, for a.e. $y \in B_r(x)$ we have $f(x) \geq f(y) + \nabla f(y) \cdot (x - y)$: multiplying this inequality by $\varphi(y)$ for some nonnegative $\varphi \in C^\infty_c(B_r(x))$ with $|\nabla \varphi| \leq C/r$ and $0 < \int_{B_r(x)} \varphi \leq C r^n$, we find that

$$f(x) \int_{B_r(x)} \varphi \geq \int_{B_r(x)} \{f(y) + \nabla f(y) \cdot (x - y)\} \, \varphi(y) \, dy$$

$$= \int_{B_r(x)} f(y)\{\varphi(y) + \operatorname{div}(\varphi(y)(y - x))\} \, dy \geq -C(n) \int_{B_r(x)} |f|;$$

dividing by $\int_{B_r(x)} \varphi$ and combining the result with (7.5), we obtain (7.1).

Step two: Let us now consider the distributional Hessian $D^2 f$, defined on $C_c^\infty(\Omega)$ by

$$D^2 f(\varphi) = \int_\Omega f \nabla^2 \varphi, \qquad \forall \varphi \in C_c^\infty(\Omega).$$

We first prove that $D_{vv}^2 f$ defines a nonnegative distribution in Ω, and thus it is a Radon measure in Ω: indeed, denoting by f_ε the convolution of f with a standard regularizing kernel ρ_ε, we easily see that f_ε is convex in $\Omega_\varepsilon = \{x : \mathrm{dist}(x, \Omega^c) > \varepsilon\}$; therefore, if $\varphi \in C_c^\infty(\Omega)$ with $\varphi \geq 0$, then for ε small enough we have $\mathrm{spt}\, \varphi \subset\subset \Omega_\varepsilon$ and

$$0 \leq \int_{\mathrm{spt}\, \varphi} \varphi \nabla_{vv}^2 f_\varepsilon = \int_{\mathrm{spt}\, \varphi} f_\varepsilon \nabla_{vv}^2 \varphi,$$

so that, letting $\varepsilon \to 0^+$, $D_{vv}^2 f \geq 0$ on $C_c^\infty(\Omega)$, as claimed. Now, setting $D_{ij}^2 f = D_{e_i e_j}^2 f$, the identity

$$(e_i + e_j) \cdot A[e_i + e_j] = e_i \cdot A[e_i] + e_j \cdot A[e_j] + 2\, e_i \cdot A[e_j] \qquad \forall A \in \mathbb{R}_{\mathrm{sym}}^{n \times n}$$

applied to $A = \nabla^2 \varphi$ gives

$$2\, D_{ij}^2 f = D_{vv}^2 f - D_{ii}^2 f - D_{jj}^2 f, \qquad v = e_i + e_j,$$

thus proving that $D_{ij}^2 f$ is a signed Radon measure in Ω. This proves (7.2), and since $D^2 f = D(\nabla f)$, we also have $\nabla f \in BV_{\mathrm{loc}}(\Omega; \mathbb{R}^n)$. $\qquad\square$

7.2 Alexandrov's Theorem for Convex Functions

We now address the existence of second order Taylor expansions for convex functions, proving the following classical result of Alexandrov.

Theorem 7.2 (Alexandrov's theorem) *Let $f : \mathbb{R}^n \to \mathbb{R} \cup \{+\infty\}$ be convex on $\mathbb{R}^n$, and let F_2 denote the set of points of twice differentiability of f, namely, for every $x \in F_2$, f is differentiable at x with gradient $\nabla f(x)$ and there exists $\nabla^2 f(x) \in \mathbb{R}_{\mathrm{sym}}^{n \times n}$ such that*

$$f(x + v) = f(x) + \nabla f(x) \cdot v + \frac{1}{2} v \cdot \nabla^2 f(x)[v] + \mathrm{o}(|v|^2), \qquad (7.6)$$

as $v \to 0$ in $\mathbb{R}^n$. Then,

$$\mathcal{L}^n(\mathrm{Dom}(f) \setminus F_2) = 0,$$

and, for $\mathcal{L}^n$-a.e. $x \in F_2$ one can take $\nabla^2 f(x)$ to be the value at x of density of $D^2 f$ with respect to the Lebesgue measure.

Proof   Of course, there is nothing to prove unless $\Omega = \text{Int Dom}(f) \neq \emptyset$. Let $D^2 f = \nabla^2 f \, d\mathcal{L}^n \llcorner \Omega + [D^2 f]^s$ be the Radon–Nikodym decomposition of $D^2 f$ with respect to $\mathcal{L}^n \llcorner \Omega$, and let x be a Lebesgue point for both ∇f and $\nabla^2 f$, and a point such that $\nabla^2 f(x) = D_{\mathcal{L}^n}[D^2 f](x)$, so that, as $r \to 0^+$,

$$\int_{B_r(x)} |\nabla f - \nabla f(x)| = o_x(r^n), \tag{7.7}$$

$$\int_{B_r(x)} |\nabla^2 f - \nabla^2 f(x)| = o_x(r^n), \tag{7.8}$$

$$\lim_{r \to 0^+} \frac{D^2 f(B_r(x))}{\omega_n \, r^n} = \nabla^2 f(x); \tag{7.9}$$

see Appendix A.9. Notice that (7.8) and (7.9) imply

$$|[D^2 f]^s|(B_r(x)) = o_x(r^n), \tag{7.10}$$

as $r \to 0^+$. Without loss of generality, we set

$$x = 0, \qquad B_r(x) = B_r, \qquad \Phi(y) = f(y) - f(0) - \nabla f(0) \cdot y - \frac{1}{2} \, y \cdot \nabla^2 f(0)[y],$$

for $y \in \mathbb{R}^n$, and notice that the thesis follows by showing

$$\|\Phi\|_{C^0(B_r)} = o(r^2) \qquad \text{as } r \to 0^+. \tag{7.11}$$

We first prove an L^1-version of (7.11) and then conclude by an interpolation argument.

Step one: We prove that

$$\int_{B_r} |\Phi| = o(r^{n+2}), \qquad \text{as } r \to 0^+. \tag{7.12}$$

Given standard regularizing kernels ρ_ε, we set $f_\varepsilon = f \star \rho_\varepsilon$ and let

$$\Phi_\varepsilon(y) = f_\varepsilon(y) - f_\varepsilon(0) - \nabla f_\varepsilon(0) \cdot y - \frac{1}{2} \, y \cdot \nabla^2 f(0)[y], \qquad y \in \mathbb{R}^n.$$

(Notice that we are intentionally using $\nabla^2 f(0)$ and not $\nabla^2 f_\varepsilon(0)$ in defining Φ_ε.) Since f is continuous in Ω, $f_\varepsilon \to f$ locally uniformly in Ω; moreover, since 0 is a Lebesgue point of ∇f, we have $\nabla f_\varepsilon(0) \to \nabla f(0)$. Therefore,

$$\Phi_\varepsilon \to \Phi \qquad \text{locally uniformly on } \mathbb{R}^n.$$

In particular, if $\varphi \in C_c^0(\Omega)$, then

$$\int_{B_r} \varphi \, \Phi = \lim_{\varepsilon \to 0^+} \int_{B_r} \varphi \, \Phi_\varepsilon. \tag{7.13}$$

Notice that by applying Taylor's formula for f_ε we find

$$\Phi_\varepsilon(y) = \int_0^1 (1-s)\, y \cdot \nabla^2 f_\varepsilon(s\, y)[y]\, ds - \frac{1}{2}\, y \cdot \nabla^2 f(0)[y]$$

$$= \int_0^1 (1-s)\, y \cdot \left(\nabla^2 f_\varepsilon(s\, y) - \nabla^2 f(0)\right)[y]\, ds,$$

so that, setting

$$h_\varepsilon^\varphi(s) = \int_{B_{rs}} \varphi(z/s)\, z \cdot \left(\nabla^2 f_\varepsilon(z) - \nabla^2 f(0)\right)[z]\, dz \qquad s \in (0,1),$$

by the change of variables $s\, y = z$ we find

$$\int_{B_r} \varphi(y)\, \Phi_\varepsilon(y)\, dy = \int_0^1 \frac{(1-s)}{s^{n+2}}\, h_\varepsilon^\varphi(s)\, ds. \tag{7.14}$$

Now, in the weak-star convergence of $\mathbb{R}_{\text{sym}}^{n\times n}$-valued Radon measure on Ω, we have

$$\left(\nabla^2 f_\varepsilon(x) - \nabla^2 f(0)\, d\mathcal{L}^n \llcorner \Omega\right) = \left(D^2 f - \nabla^2 f(0)\, d\mathcal{L}^n \llcorner \Omega\right) \star \rho_\varepsilon$$

$$\overset{*}{\rightharpoonup} D^2 f - \nabla^2 f(0)\, d\mathcal{L}^n \llcorner \Omega, \qquad \text{as } \varepsilon \to 0^+.$$

Therefore, for every $\varphi \in C_c^0(\Omega)$ and $s \in (0,1)$, we have $h_\varepsilon^\varphi(s) \to h^\varphi(s)$ as $\varepsilon \to 0^+$, where h^φ is given by

$$h^\varphi(s) = \int_{B_{rs}} \varphi(z/s)\, z \cdot \left(\nabla^2 f(z) - \nabla^2 f(0)\right)[z]\, dz$$

$$+ \sum_{i,j=1}^n \int_{B_{rs}} \varphi(z/s)\, z_i\, z_j\, d[D^2 f]_{ij}^s;$$

here we have taken into account that $D^2 f = \nabla^2 f\, d[\mathcal{L}^n \llcorner \Omega] + [D^2 f]^s$ and have denoted by $[D^2 f]_{ij}^s$ the (i,j)-component of the $\mathbb{R}^{n\times n}$-valued Radon measure $[D^2 f]^s$. Setting for brevity $\mu = D^2 f - \nabla^2 f(0)\, d\mathcal{L}^n \llcorner \Omega$ and recalling the basic estimate for Radon measures

$$\int_{B_t} |\mu \star \rho_\varepsilon| \le C(n)\, \frac{\min\{t^n, \varepsilon^n\}}{\varepsilon^n}\, |\mu|(B_{t+\varepsilon}), \qquad \text{(see (A.23))},$$

we find that

$$|h_\varepsilon^\varphi(s)| \le \|\varphi\|_{C^0}\, (rs)^2 \int_{B_{rs}} |(D^2 f - \nabla^2 f(0)\, d\mathcal{L}^n) \star \rho_\varepsilon|(z)\, dz$$

$$\le C(n)\, \|\varphi\|_{C^0}\, (rs)^2\, \frac{\min\{\varepsilon^n, (rs)^n\}}{\varepsilon^n}\, |D^2 f - \nabla^2 f(0)\, d\mathcal{L}^n|(B_{rs+\varepsilon})$$

$$\le C(n)\, \|\varphi\|_{C^0}\, (rs)^2\, \frac{\min\{\varepsilon^n, (rs)^n\}}{\varepsilon^n}\, (rs + \varepsilon)^n \le C(n)\, \|\varphi\|_{C^0}\, (rs)^{n+2},$$

where in the penultimate inequality we have used (7.8) and (7.9). This proves that, uniformly in ε,

$$|h^{\varphi}_{\varepsilon}(s)| \le C(n)\,\|\varphi\|_{C^0}\,(rs)^{n+2}, \qquad \forall s \in (0,1), \qquad (7.15)$$

so that by (7.13), (7.14), and dominated convergence we get

$$\int_{B_r} \varphi\,\Phi = \lim_{\varepsilon \to 0^+} \int_{B_r} \varphi\,\Phi_{\varepsilon} = \lim_{\varepsilon \to 0^+} \int_0^1 \frac{(1-s)}{s^{n+2}}\,h^{\varphi}_{\varepsilon}(s)\,ds = \int_0^1 \frac{(1-s)}{s^{n+2}}\,h^{\varphi}(s)\,ds.$$

We now notice that, again by (7.8) and (7.9), we have

$$\int_0^1 \frac{(1-s)}{s^{n+2}}\,h^{\varphi}(s)\,ds \le r^2\,\|\varphi\|_{C^0}\left\{ \int_0^1 \frac{1-s}{s^n}\,ds \int_{B_{rs}} |\nabla^2 f(z) - \nabla^2 f(0)|\,dz \right.$$

$$\left. + \int_0^1 \frac{1-s}{s^n}\,|\,[D^2 f]^s\,|(B_{rs})\,ds \right\}$$

$$= \|\varphi\|_{C^0}\,o(r^{n+2}), \qquad \text{as } r \to 0^+,$$

where $o(r^{n+2})$ is independent from φ. We conclude that

$$\int_{B_r} |\Phi| = \sup\left\{ \int_{B_r} \varphi\,\Phi : \varphi \in C^0_c(\Omega), |\varphi| \le 1 \right\} = o(r^{n+2}), \qquad \text{as } r \to 0^+,$$

as claimed.

Step two: In this step we show how deduce (7.11) from (7.12), thus concluding the proof of the theorem. Let us start noticing that we can write $\Phi = \Phi_1 - \Phi_2$, where Φ_1 and Φ_2 are the convex functions defined by

$$\Phi_1(y) = f(y) - f(0) - \nabla f(0) \cdot y, \qquad \Phi_2(y) = \frac{1}{2}\,y \cdot \nabla^2 f(0)[y].$$

By applying (7.1) to both Φ_1 and Φ_2, we find

$$\|\nabla\Phi\|_{L^\infty(B_r)} \le \frac{C(n)}{r^{n+1}}\left\{ \|\Phi_1\|_{L^1(B_{2r})} + \|\Phi_2\|_{L^1(B_{2r})} \right\}$$

$$\le \frac{C(n)}{r^{n+1}}\left\{ \|\Phi\|_{L^1(B_{2r})} + 2\,\|\Phi_2\|_{L^1(B_{2r})} \right\},$$

where $\|\Phi_2\|_{C^0(B_{2r})} \le \Lambda r^2$ if $\Lambda = |\nabla^2 f(0)|$, so that

$$\|\nabla\Phi\|_{L^\infty(B_r)} \le C(n,\Lambda)\left\{ r + \frac{1}{r^{n+1}} \int_{B_{2r}} |\Phi| \right\}, \qquad \forall B_{3r} \subset\subset \Omega. \qquad (7.16)$$

Interestingly, we cannot deduce (7.11) directly from the basic interpolation of (7.16) and (7.12) obtained by looking at the first order Taylor's formula for Φ, namely,

$$|\Phi(y)| \le |y| \int_0^1 |\nabla\Phi(sy)|\,ds \le C\,|y|^2 + \frac{C}{|y|^n} \int_{B_{2|y|}} |\Phi|;$$

indeed, in this way, we just find $\|\Phi\|_{C^0(B_r)} = O(r^2)$. We rather argue as follows. For every ε small enough consider the bad set $K_\varepsilon = \{y \in B_r : |\Phi| \geq \varepsilon r^2\}$ and notice that by (7.12) we can find r_ε such that

$$\int_{B_r} |\Phi| \leq \varepsilon^{n+1} r^2 |B_r|, \qquad \forall r < r_\varepsilon,$$

and, correspondingly,

$$|B_r \cap K_\varepsilon| \leq \frac{1}{\varepsilon r^2} \int_{B_r} |\Phi| \leq \varepsilon^n |B_r|, \qquad \forall r < r_\varepsilon.$$

In particular, if $r < r_\varepsilon$ and $y \in B_{r/2}$, then $B_{2\varepsilon r}(y) \setminus K_\varepsilon \neq \emptyset$, for otherwise

$$\varepsilon^n |B_r| \geq |B_r \cap K_\varepsilon| \geq |B_{2\varepsilon r}(y)| = 2^n \varepsilon^n |B_r|,$$

a contradiction. Hence, for every $r < r_\varepsilon$ and $y \in B_{r/2}$ there exists z with $|z - y| < 2\varepsilon r$ and $|\Phi(z)| \leq \varepsilon r^2$, so that (7.16) gives

$$|\Phi(y)| \leq |\Phi(z)| + |z - y| \, \|\nabla\Phi\|_{L^\infty(B_r)}$$
$$\leq \varepsilon r^2 + 2\varepsilon r \left\{ Cr + \frac{C}{r^{n+1}} \int_{B_{2r}} |\Phi| \right\} \leq C\varepsilon r^2,$$

where in the last inequality we have again used (7.12). This proves (7.11). $\qquad\square$

We finally discuss the validity of a first order Taylor's expansion for the gradient of a convex function. In general, the validity of (7.6) does not imply that $\nabla f(x+v) - \nabla f(x) - \nabla^2 f(x)[v] = o(|v|)$ as $v \to 0$ (as it is seen, for example, by taking $f(x) = |x|^3 \sin(|x|^{-2})$ if $x \neq 0$, and $f(0) = 0$ with $n = 1$). For convex functions, however, the existence of a second order Taylor expansion at a point implies the existence of a first order expansion at the level of gradients, although the latter concept has to be properly formulated since the classical gradient is not everywhere defined.

Theorem 7.3 (Differentiability of the subdifferential) *If $f : \mathbb{R}^n \to \mathbb{R} \cup \{+\infty\}$ is convex and twice differentiable at x_0, then*

$$\lim_{x \to x_0} \sup_{y \in \partial f(x)} \left| \frac{y - \nabla f(x_0) - \nabla^2 f(x_0)[x - x_0]}{|x - x_0|} \right| = 0. \qquad (7.17)$$

Remark 7.4 Notice that, in turn, the validity of (7.17) with some $A \in \mathbb{R}^{n \times n}_{\text{sym}}$ in place of $\nabla^2 f(x_0)$ implies that

$$f(x) = f(x_0) + \nabla f(x_0) \cdot (x - x_0) + \frac{1}{2} (x - x_0) \cdot A[x - x_0] + o(|x - x_0|^2)$$

as $x \to x_0$ in $\mathbb{R}^n$ – and thus, that $A = \nabla^2 f(x_0)$. To prove this, let

$$Q(x) = f(x_0) + \nabla f(x_0) \cdot (x - x_0) + \frac{1}{2} (x - x_0) \cdot A[x - x_0], \qquad x \in \mathbb{R}^n,$$

pick any $v \in \mathbb{S}^n$, and integrate over $[x_0, x_0 + tv]$ to find that

$$|f(x_0 + tv) - Q(x_0 + tv)| \leq \int_0^t |\nabla f(x_0 + sv) - (\nabla f(x_0) + A[sv])|\, ds$$

$$\leq \int_0^t \sup_{y \in \partial f(x_0 + sv)} \left| \frac{y - y_0 - A[sv]}{|sv|} \right| |sv|\, ds$$

$$\leq \frac{t^2}{2} \sup_{x \in B_t(x_0)} \sup_{y \in \partial f(x)} \left| \frac{y - \nabla f(x_0) - A[x - x_0]}{|x - x_0|} \right|,$$

i.e., $|f(x_0 + tv) - Q(x_0 + tv)| = o(t^2)$ as $t \to 0^+$, uniformly on $v \in \mathbb{S}^n$.

Proof of Theorem 7.3   *Step one*: We show that pointwise convergence of convex functions implies locally uniform convergence of their subdifferentials. More formally, if f_j and f are convex functions on $\mathbb{R}^n$, $K \subset\subset \mathrm{IntDom}(f)$, and $f_j \to f$ pointwise on $\mathbb{R}^n$ as $j \to \infty$, then

$$\lim_{j \to \infty} \sup \left\{ \mathrm{dist}(y, \partial f(x)) : y \in \partial f_j(x), x \in K \right\} = 0. \qquad (7.18)$$

Indeed, if x is interior to the domain of f, then we can find an n-dimensional simplex Σ containing x and compactly contained in $\mathrm{Dom}(f)$. Every convex function will be bounded on Σ by its values at the $(n + 1)$-vertexes of Σ, and in turn its Lipschitz constant on a ball $B_r(x_0)$ such that $B_{2r}(x_0) \subset \Sigma$ will be bounded in terms of the oscillation of the convex function on Σ. Covering K by finitely many such simplexes we see that, for A an open neighborhood of K with $A \subset\subset \mathrm{IntDom}(f)$, we have $\mathrm{Lip}(f_j; A) \leq M < \infty$ for every j. In particular, $f_j \to f$ uniformly on K. Now let $x_j \in K$ and $y_j \in \partial f_j(x_j)$ be such that

$$\mathrm{dist}(y_j, \partial f(x_j)) + \frac{1}{j} \geq \sup_j,$$

where $\sup_j$ is an abbreviation for the j-dependent supremum appearing in (7.18). Since K is compact and $\mathrm{Lip}(f_j; A) \leq M$, we deduce that both x_j and y_j are bounded sequences and thus, up to extracting subsequences, have limits x_0 and y_0. Since $f_j \to f$ uniformly on K, we find that $f_j(x_j) \to f(x_0)$. By taking the limit $j \to \infty$ in the inequalities $f_j(z) \geq f_j(x_j) + y_j \cdot (z - x_j)$ for every $z \in \mathbb{R}^n$ we see that $y_0 \in \partial f(x_0)$. Since $\mathrm{dist}(y_j, \partial f(x_j)) \to \mathrm{dist}(y_0, \partial f(x_0)) = 0$ we conclude that $\sup_j \to 0$, as claimed.

Step two: We know that $\partial f(x_0) = \{\nabla f(x_0)\}$ and that there exists $A = \nabla^2 f(x_0) \in \mathbb{R}^{n \times n}_{\mathrm{sym}}$ such that, locally uniformly on $v \in \mathbb{R}^n$,

$$\frac{f(x_0 + hv) - f(x_0) - h\nabla f(x_0) \cdot v}{h^2} \to \frac{1}{2} v \cdot A[v] \qquad \text{as } h \to 0^+.$$

This fact can be reformulated by saying that the convex functions

$$g_h(v) = \frac{f(x_0 + hv) - f(x_0) - h\nabla f(x_0) \cdot v}{h^2}, \qquad g(v) = \frac{1}{2}v \cdot A[v],$$

are such that $g_h \to g$ uniformly on compact subsets of $\mathbb{R}^n$ as $h \to 0^+$. Clearly, $\nabla g(v) = A[v]$ for every $v \in \mathbb{R}^n$, while

$$\partial g_h(v) = \left\{ \frac{y - \nabla f(x_0)}{h} : y \in \partial f(x_0 + hv) \right\}.$$

By applying step one with $K = \{|v| \le 1\}$, and taking into account that $\partial g(v) = \{A[v]\}$, and thus $\text{dist}(y, \partial g(v)) = |y - A[v]|$, we conclude that

$$0 = \lim_{h \to 0^+} \sup \left\{ |w - A[v]| : w \in \partial g_h(v), |v| \le 1 \right\}$$

$$= \lim_{h \to 0^+} \sup \left\{ \left| \frac{y - \nabla f(x_0)}{h} - A[v] \right| : y \in \partial f(x_0 + hv), |v| \le 1 \right\}$$

$$= \lim_{x \to x_0} \sup_{y \in \partial f(x)} \left| \frac{y - \nabla f(x_0) - A[x - x_0]}{|x - x_0|} \right|,$$

as claimed.   □

8

The Monge–Ampère Equation for Brenier Maps

In this chapter we address the fundamental problem of the validity, for Brenier maps, of the ($\mathcal{L}^n$-a.e.) pointwise transport condition (1.1) – namely, we prove that, if $\mu = \rho \, d\mathcal{L}^n$, $\nu = \sigma \, d\mathcal{L}^n$, and ∇f is the Brenier map from μ to ν, then

$$\rho(x) = \sigma(\nabla f(x)) \det \nabla^2 f(x) \qquad \mathcal{L}^n\text{-a.e. on Dom}(f). \qquad (8.1)$$

This identity plays a crucial role in the applications of the quadratic Monge problem discussed in Part III. It also asserts that the convex potential f solves (in the interior of its domain) the **Monge–Ampère equation** $\det \nabla^2 f = h$ (with $h(x) = \rho(x)/\sigma(\nabla f(x))$). This is a type of equation with a rich regularity theory, which, under suitable assumptions on ρ and σ, can be exploited to infer regularity results for Brenier maps.

We cannot deduce (8.1) from $(\nabla f)_{\#}\mu = \nu$ as in Proposition 1.1, since there we were making use of the area formula for injective Lipschitz maps T, whereas $T = \nabla f$ is, in general, neither Lipschitz nor injective. The second order differentiability theory of convex functions developed in Chapter 7 provides the starting point to address these two issues and prove the main result of this chapter, Theorem 8.1. Notice that Theorem 8.1 can be seen as a further improvement on Theorem 6.1 in the special case when $\mu, \nu \in \mathcal{P}_{ac}(\mathbb{R}^n)$.

We set the following convenient notation. Given a convex function $f : \mathbb{R}^n \to \mathbb{R} \cup \{+\infty\}$, we denote by F_1 and F_2 the points of first and second differentiability of f, and set

$$F_{2,\text{inv}} = \{x \in F_2 : \nabla^2 f(x) \text{ is invertible}\}.$$

The corresponding sets for f^* are denoted by F_1^*, F_2^*, and $F_{2,\text{inv}}^*$. Moreover, given $w \in L^1_{\text{loc}}(\Omega)$, Ω an open set in $\mathbb{R}^n$, we denote by $\text{Leb}(w)$ the set of the Lebesgue points of w in Ω, see (A.15). With this notation in place we state the main result of this chapter.

Theorem 8.1 (Monge–Ampère equation for Brenier maps)   *If $\mu, \nu \in \mathcal{P}_{ac}(\mathbb{R}^n)$ with*

$$\mu = \rho\, d\mathcal{L}^n, \qquad \nu = \sigma\, d\mathcal{L}^n,$$

then there exists a convex function $f : \mathbb{R}^n \to \mathbb{R} \cup \{+\infty\}$ such that $\Omega = \mathrm{IntDom}(f) \neq \emptyset$, $(\nabla f)_\# \mu = \nu$, $\mathrm{Cl}(\nabla f(F_1 \cap \mathrm{spt}\,\mu)) = \mathrm{spt}\,\nu$, and

$$
\begin{aligned}
&\mu \text{ is concentrated on } X, \text{ where} \\
&X = F_{2,\mathrm{inv}} \cap (\nabla f)^{-1}(\mathrm{Leb}(\sigma)) \cap \mathrm{Leb}(\rho) \cap \{\rho > 0\}.
\end{aligned}
\tag{8.2}
$$

Moreover, for every $x \in X$ we have

$$\nabla f(x) \in \{\sigma > 0\}, \tag{8.3}$$

$$\sigma(\nabla f(x))\, \det \nabla^2 f(x) = \rho(x). \tag{8.4}$$

In particular, (8.3) and (8.4) hold μ-a.e. on $\mathbb{R}^n$. Finally, $\det \nabla^2 f \in L^1_{\mathrm{loc}}(\Omega)$ and the area formula

$$\int_{(\nabla f)(M)} \varphi = \int_M \varphi(\nabla f)\, \det \nabla^2 f, \tag{8.5}$$

holds for every Borel function $\varphi : \mathbb{R}^n \to [0, \infty]$, where

$$M = F_{2,\mathrm{inv}} \cap \mathrm{Leb}(\det \nabla^2 f).$$

Remark 8.2   The convex potential f in Theorem 8.1 is constructed as in Theorem 6.1 – in particular, if spt ν is compact, then we can further assert that $\mathrm{Dom}(f) = \mathbb{R}^n$ and that $\nabla f(x) \in \mathrm{spt}\,\nu$ for every $x \in F_1$.

We now discuss the strategy of the proof of Theorem 8.1. The first step consists of the following basic proposition, which holds in greater generality than Theorem 8.1, and is of independent interest.

Proposition 8.3   *If $\mu \in \mathcal{P}(\mathbb{R}^n)$, $f : \mathbb{R}^n \to \mathbb{R} \cup \{+\infty\}$ is convex, and ∇f transports μ, then for every Borel set E in $\mathbb{R}^n$ we have*

$$(\nabla f)_\# \mu(E) = \mu(\partial f^*(E)). \tag{8.6}$$

Proof   Indeed, recalling that F_1 denotes the differentiability set of f, the fact that ∇f transports μ means that μ is concentrated on F_1. At the same time,

$$
\begin{aligned}
F_1 \cap \partial f^*(E) &= \big\{ x \in F_1 : \exists\, y \in E \text{ s.t. } x \in \partial f^*(y) \big\} \\
&= \big\{ x \in F_1 : \exists\, y \in E \text{ s.t. } y \in \partial f(x) = \{\nabla f(x)\} \big\} \\
&= \{ x \in F_1 : \nabla f(x) \in E \} = (\nabla f)^{-1}(E),
\end{aligned}
$$

so that $\partial f^*(E)$ is μ-equivalent to $(\nabla f)^{-1}(E)$, as claimed. $\square$

With (8.6) at hand, we can illustrate the main idea behind the proof of (8.4). With μ and ν as in Theorem 8.1, if x is both a Lebesgue point of ρ and such that $\nabla f(x)$ is a Lebesgue point of σ, then, thanks to (8.6),

$$\frac{\sigma(\nabla f(x))}{\rho(x)} \approx \frac{\nu(B_r(\nabla f(x)))}{\omega_n \, r^n} \frac{\omega_n \, r^n}{\mu(B_r(x))} = \frac{\mu(\partial f^*(B_r(\nabla f(x))))}{\mu(B_r(x))}. \tag{8.7}$$

We thus need to understand how the subdifferential of the convex function f^* is deforming volumes: Of course, should f^* be twice differentiable at $y = \nabla f(x)$, we may hope that, in analogy with the smooth case,

$$\frac{\mu(\partial f^*(B_r(y)))}{\mu(B_r(x))} \approx \det \nabla^2 f^*(y). \tag{8.8}$$

By (8.7) and (8.8), we could prove (8.4) by further showing that

$$\nabla^2 f^*(y) \text{ is the inverse matrix of } \nabla^2 f(x). \tag{8.9}$$

With this scheme of proof in mind we now discuss these various issues separately: (8.9) is addressed in Section 8.1, (8.8) in Section 8.2, and finally the formal proof of Theorem 8.1 is presented in Section 8.3.

8.1 Convex Inverse Function Theorem

Theorem 8.4 *Let $f : \mathbb{R}^n \to \mathbb{R} \cup \{+\infty\}$ be convex and twice differentiable at $x_0 \in \mathbb{R}^n$. If $\nabla^2 f(x_0)$ is invertible (as an element of $\mathbb{R}^{n \times n}_{\mathrm{sym}}$), then f^* is twice differentiable at $y_0 = \nabla f(x_0)$ with $\nabla f^*(y_0) = x_0$ and*

$$\nabla^2 f^*(y_0) = [\nabla^2 f(x_0)]^{-1}. \tag{8.10}$$

Viceversa, if $\nabla^2 f(x_0)$ is not invertible, then f^* is not twice differentiable at y_0. Finally, if F_2 denotes the sets of points where f is twice differentiable and $F_{2,\mathrm{inv}} = \{x \in F_2 : \nabla^2 f(x) \text{ is invertible}\}$, then

$$F_2 \setminus F_{2,\mathrm{inv}} \subset \partial f^*(\mathrm{Dom}(f^*) \setminus F_2^*) \tag{8.11}$$

where F_2^* is the set of those $y_0 \in \mathbb{R}^n$ such that f^* is twice differentiable at y_0.

Proof Step one: We prove that if $A = \nabla^2 f(x_0)$ is invertible, then f^* is differentiable at $y_0 = \nabla f(x_0)$ with $\nabla f^*(y_0) = x_0$. Indeed, $y_0 \in \partial f(x_0)$ implies $x_0 \in \partial f^*(y_0)$. If f^* is not differentiable at y_0, then there is some $x_1 \in \partial f^*(y_0)$, $x_1 \neq x_0$. Since $\partial f^*(y_0)$ is a convex set, we deduce that $x_t = (1-t)x_0 + t \, x_1 \in \partial f^*(y_0)$ for every $t \in [0,1]$. Since f is twice differentiable at x_0, by Theorem 7.3 we have

$$0 = \lim_{x \to x_0} \sup_{y \in \partial f(x)} \frac{|y - y_0 - A[x - x_0]|}{|x - x_0|} \geq \lim_{t \to 0^+} \sup_{y \in \partial f(x_t)} \left| \frac{y - y_0}{t \, |x_1 - x_0|} - A \left[\frac{x_1 - x_0}{|x_1 - x_0|} \right] \right|.$$

Since $x_t \in \partial f^*(y_0)$ implies $y_0 \in \partial f(x_t)$ we can pick $y = y_0$ and conclude that $A[x_1 - x_0] = 0$, a contradiction to the invertibility of A.

Step two: We prove that if A is invertible, then f^* is twice differentiable at y_0 with $\nabla^2 f^*(y_0) = A^{-1}$. Thanks to Remark 7.4, it is enough to show that

$$\lim_{y \to y_0} \sup_{x \in \partial f^*(y)} \frac{|x - x_0 - A^{-1}[y - y_0]|}{|y - y_0|} = 0. \qquad (8.12)$$

Indeed by keeping into account that $x \in \partial f^*(y)$ if and only if $y \in \partial f(x)$, that $\partial f(x_0) = \{y_0\}$, that $\partial f^*(y_0) = \{x_0\}$ and the continuity of subdifferentials, Proposition 2.7, we find that

$$\lim_{y \to y_0} \sup_{x \in \partial f^*(y)} a(x, y) = \lim_{x \to x_0} \sup_{y \in \partial f(x)} a(x, y), \qquad a(x, y) = \frac{|x - x_0 - A^{-1}[y - y_0]|}{|y - y_0|},$$
$$(8.13)$$

whenever one of the two limits exists. Since A is invertible we have

$$\frac{x - x_0 - A^{-1}[y - y_0]}{|y - y_0|} = -A^{-1}\left(\frac{y - y_0 - A(x - x_0)|}{|x - x_0|}\right) \frac{|x - x_0|}{|y - y_0|}. \qquad (8.14)$$

We conclude the proof since Theorem 7.3 gives

$$\lim_{x \to x_0} \sup_{y \in \partial f(x)} \left| \frac{y - y_0 - A(x - x_0)|}{|x - x_0|} \right| = 0,$$

which in turn implies that

$$\lim_{x \to x_0} \sup_{y \in \partial f(x)} \frac{|x - x_0|}{|y - y_0|} = \lim_{x \to x_0} \sup_{y \in \partial f(x)} \frac{|x - x_0|}{|A(x - x_0)|} < \infty.$$

since A is invertible. Combining (8.13) and (8.14) with these facts we conclude the proof.

Step three: We prove that if A is not invertible, then f^* is not twice differentiable at $y_0 = \nabla f(x_0)$. We can directly consider the case when f^* is differentiable at y_0, in which case it must be $\nabla f^*(y_0) = x_0$. Under this assumption we prove that, for every $B \in \mathbb{R}^{n \times n}_{\text{sym}}$,

$$\limsup_{y \to y_0} \sup_{x \in \partial f^*(y)} \frac{|x - x_0 - B[y - y_0]|}{|y - y_0|} = \infty,$$

thus showing that f^* is not twice differentiable at y_0. Indeed, if $A[w] = 0$ for some $w \in \mathbb{S}^n$ then we can deduce from (7.17) the existence of $x_j \to x_0$ and $y_j \in \partial f(x_j)$ such that $|y_j - y_0| = o(|x_j - x_0|)$ as $j \to \infty$. Therefore, as $j \to \infty$,

$$\sup_{x \in \partial f^*(y_j)} \frac{|x - x_0 - B[y - y_0]|}{|y - y_0|} \geq \frac{|x_j - x_0|}{|y_j - y_0|} - \|B\| \to +\infty.$$

Step four: We finally prove (8.11). If $x \in F_2 \backslash F_{2,\mathrm{inv}} \subset F_1$, then $y = \nabla f(x)$ is well defined and $x \in \partial f^*(y)$, so that $\partial f^*(y)$ is non-empty, and hence $y \in \mathrm{Dom}(f^*)$. This shows $x \in \partial f^*(\mathrm{Dom}(f^*))$. If f^* is not differentiable at y, this also shows that $x \in \partial f^*(\mathrm{Dom}(f^*) \backslash F_1^*) \subset \partial f^*(\mathrm{Dom}(f^*) \backslash F_2^*)$. If, instead, f^* is differentiable at y, then the fact that $\nabla^2 f(x)$ is not invertible implies that f^* is not twice differentiable at $y = \nabla f(x)$ by step three, and thus $y \notin F_2^*$, as required. $\qquad\qquad\qquad\qquad\qquad\qquad\qquad\qquad\qquad\qquad\qquad\qquad\qquad\qquad\square$

8.2 Jacobians of Convex Gradients

Having in mind (8.8), we now discuss the volume deforming properties of convex subdifferentials. The first step is obtained by combining the differentiability theorem for convex subdifferentials, Theorem 7.3, with the convex inverse function theorem, Theorem 8.4 to prove a sort of Lipschitz property for convex subdifferentials at points of twice differentiability of a convex function. In the statement, we set $I_r(X) = \{x \in \mathbb{R}^n : \mathrm{dist}(x, X) < r\}$.

Theorem 8.5 (Lipschitz continuity of convex subdifferentials) *If $f : \mathbb{R}^n \to \mathbb{R} \cup \{+\infty\}$ is convex and twice differentiable at x_0 with $y_0 = \nabla f(x_0)$ and $A = \nabla^2 f(x_0)$, then*

$$\partial f(B_r(x_0)) \subset y_0 + r\, I_{\omega(r)}(A[B_1]), \qquad \forall B_r(x_0) \subset\subset \Omega, \tag{8.15}$$

where ω is an increasing function with $\omega(0^+) = 0$ and $\Omega = \mathrm{Int}\,\mathrm{Dom}(f)$. If, in addition, A is invertible and r is such that $B_{(1+\omega(\|A\|r))\,r}(x_0) \subset\subset \Omega$, then

$$\partial f(B_r(x_0)) \subset y_0 + r\, A[B_{1+\|A^{-1}\|\,\omega(r)}], \tag{8.16}$$

$$y_0 + A[B_r] \subset \partial f(B_{(1+\omega(\|A\|r))\,r}(x_0)). \tag{8.17}$$

Proof By (7.17) we have

$$\sup_{y \in \partial f(x)} |y - y_0 - A[x - x_0]| \le |x - x_0|\, \omega(|x - x_0|),$$

for a nonnegative increasing function $\omega : [0, \infty) \to (0, \infty)$ such that $\omega(0^+) = 0$. In particular, if $B_r(x_0) \subset\subset \Omega$, then

$$\partial f(B_r(x_0)) \subset y_0 + I_{\omega(r)\,r}(A[B_r]) = y_0 + r\, I_{\omega(r)}(A[B_1]).$$

that is (8.15). If now A is invertible, then we have

$$I_\varepsilon(A[B_1]) \subset A[B_{1+\|A^{-1}\|\varepsilon}], \qquad \forall \varepsilon > 0,$$

and thus (8.16) follows from (8.15). Moreover, by Theorem 8.4 we also have that y_0 is a point of twice differentiability of f^* with $\nabla^2 f^*(y_0) = A^{-1}$, so that

$$\sup_{x \in \partial f^*(y)} |x - x_0 - A^{-1}[y - y_0]| \le |y - y_0|\, \omega(|y - y_0|),$$

up to possibly increase the values of ω. Now, if $y \in y_0 + A[B_s]$, then $|y - y_0| \le \|A\|\, s$ and

$$\partial f^*(y_0 + A[B_s]) \subset x_0 + I_{\omega(\|A\| s)\, s} (A^{-1}[A[B_s]]) \subset B_{(1+\omega(\|A\| s))\, s}(x_0),$$

so that (8.17) follows thanks to the fact that

$$\partial f^*(F) \subset E \qquad \text{iff} \qquad F \subset \partial f(E), \tag{8.18}$$

whenever $E \subset \text{Dom}(f)$ and $F \subset \text{Dom}(f^*)$. $\qquad\square$

Next we combine Theorem 8.5 with the area formula for linear maps to obtain volume distortion estimates for subdifferentials. Let us recall indeed that, whenever $L : \mathbb{R}^n \to \mathbb{R}^n$ is a linear map,

$$\frac{\mathcal{L}^n(L[B_r])}{\omega_n\, r^n} = |\det L| \qquad \forall r > 0. \tag{8.19}$$

Theorem 8.6 (Jacobian of ∇f and volume distortion) *If $f : \mathbb{R}^n \to \mathbb{R} \cup \{+\infty\}$ is convex and $\Omega = \text{IntDom}(f)$, then:*

(i) *if f is twice differentiable at x_0, then*

$$\lim_{r \to 0^+} \frac{\mathcal{L}^n(\partial f(B_r(x_0)))}{\omega_n\, r^n} = \det \nabla^2 f(x_0). \tag{8.20}$$

(ii) *if f is twice differentiable at x_0, $\nabla^2 f(x_0)$ is invertible, and $y_0 = \nabla f(x_0) \in \text{Leb}(w)$ form some $w \in L^1_{\text{loc}}(B_\eta(y_0))$, $\eta > 0$, then*

$$\lim_{r \to 0^+} \frac{1}{\mathcal{L}^n(\partial f(B_r(x_0)))} \int_{\partial f(B_r(x_0))} |w - w(y_0)| = 0. \tag{8.21}$$

(iii) *if $\mu = u\, d\mathcal{L}^n$, ∇f transports μ, f is twice differentiable at x_0, $\nabla^2 f(x_0)$ is invertible, and x_0 is a Lebesgue point of u, then*

$$\lim_{r \to 0^+} \frac{[(\nabla f)_\# \mu](B_r(y_0))}{\omega_n\, r^n} = \frac{u(x_0)}{\det \nabla^2 f(x_0)}, \tag{8.22}$$

where $y_0 = \nabla f(x_0)$.

Proof *Step one*: We prove statement (i). Let $y_0 = \nabla f(x_0)$ and $A = \nabla^2 f(x_0)$. If A is not invertible, then $A[B_1]$ is contained in a $(n-1)$-dimensional plane, and therefore

$$\mathcal{L}^n(I_{\omega(r)}(A[B_1])) \le C(n, \|A\|)\, \omega(r).$$

By combining this estimate with (8.15) we find that

$$\mathcal{L}^n(\partial f(B_r(x_0))) \leq r^n\, C(n, \|A\|)\, \omega(r)$$

and (8.20) holds with $\det A = 0$. Assuming from now on that A is invertible, by (8.16), (8.19) and $|\det A| = \det A$, we have

$$\mathcal{L}^n(\partial f(B_r(x_0))) \leq r^n\, \mathcal{L}^n(A[B_{1+\|A^{-1}\|\,\omega(r)}]) = \omega_n\, r^n\, (1 + \|A^{-1}\|\,\omega(r))^n\, \det A,$$

that is,

$$\limsup_{r \to 0^+} \frac{\mathcal{L}^n(\partial f(B_r(x_0)))}{\omega_n\, r^n} \leq \det A. \tag{8.23}$$

Finally, by (8.17) and since $s \mapsto (1 + \omega(\|A\|\,s))s$ is strictly increasing on $s > 0$, we conclude that

$$\liminf_{r \to 0^+} \frac{\mathcal{L}^n(\partial f(B_r(x_0)))}{\omega_n\, r^n} = \liminf_{s \to 0^+} \frac{\mathcal{L}^n(\partial f(B_{(1+\omega(\|A\|\,s))\,s}(x_0)))}{\omega_n\, (1 + \omega(\|A\|\,s))\,s)^n}$$

$$\geq \liminf_{s \to 0^+} \frac{\mathcal{L}^n(A[B_s])}{\omega_n\, (1 + \omega(\|A\|\,s))\,s)^n} = \det A, \tag{8.24}$$

which combined with (8.23) proves (8.20) also when A is invertible.

Step two: We prove statement (ii). We start showing that $\{\partial f(B_r(x_0))\}_{r>0}$ shrinks nicely to $y_0 = \nabla f(x_0)$ with respect to $\mathcal{L}^n$ as $r \to 0^+$, in the sense that we can find $\rho(r) \to 0^+$ as $r \to 0^+$ such that

$$\partial f(B_r(x_0)) \subset B_{\rho(r)}(y_0) \quad \forall r < r_0, \qquad \liminf_{r \to 0^+} \frac{\mathcal{L}^n(\partial f(B_r(x_0)))}{\mathcal{L}^n(B_{\rho(r)}(y_0))} \geq c, \tag{8.25}$$

for some $r_0, c > 0$. Indeed (8.15) implies

$$\partial f(B_r(x_0)) \subset B_{(\|A\|+\omega(r))r}(y_0) \qquad \forall B_r(x_0) \subset\subset \Omega,$$

so that, setting $\rho(r) = (\|A\| + \omega(r))r$ and using (8.24) we find that (8.25) holds with $c = \det A/\|A\|^n$. If now $w \in L^1_{\mathrm{loc}}(B_\eta(y_0))$, $\eta > 0$, and y_0 is a Lebesgue point of w, then

$$\frac{\int_{\partial f(B_r(x_0))} |w - w(y_0)|}{\mathcal{L}^n(\partial f(B_r(x_0)))} \leq \frac{\omega_n\, \rho(r)^n}{\mathcal{L}^n(\partial f(B_r(x_0)))} \frac{1}{\omega_n\, \rho(r)^n} \int_{B_{\rho(r)}(y_0)} |w - w(y_0)|$$

so that (8.21) follows by (8.25).

Step three: We prove statement (iii). By Proposition 8.3, setting $v = (\nabla f)_{\#}\mu$ we have

$$v(B_r(y_0)) = \mu(\partial f^*(B_r(y_0))) \qquad y_0 = \nabla f(x_0).$$

Taking into account that $\mu = u \, d\mathcal{L}^n$ we have that

$$\frac{\nu(B_r(y_0))}{\omega_n \, r^n} = \frac{\int_{\partial f^*(B_r(y_0))} u}{\mathcal{L}^n(\partial f^*(B_r(y_0)))} \, \frac{\mathcal{L}^n(\partial f^*(B_r(y_0)))}{\omega_n \, r^n}.$$

By statement (i), if $y_0 \in F_2^*$, then the second factor converges to $\det \nabla^2 f^*(y_0)$, while by statement (ii), if x_0 is a Lebesgue point of u and $y_0 \in F_{2,\mathrm{inv}}^*$, then the first factor converges to $u(x_0)$: Therefore,

$$\lim_{r \to 0^+} \frac{\nu(B_r(y_0))}{\omega_n \, r^n} = u(x_0) \, \det \nabla^2 f^*(y_0),$$

whenever $x_0 \in \mathrm{Leb}(u)$ and $y_0 = \nabla f(x_0) \in F_{2,\mathrm{inv}}^*$. If $x_0 \in F_{2,\mathrm{inv}} \cap \mathrm{Leb}(u)$, then by Theorem 8.4, $y_0 = \nabla f(x_0) \in F_{2,\mathrm{inv}}^*$ with $\nabla^2 f^*(y_0) = [\nabla^2 f(x_0)]^{-1}$, so that we have completed the proof of (8.22). $\square$

8.3 Derivation of the Monge–Ampère Equation

Proof of Theorem 8.1   By assumption, $\mu = \rho \, d\mathcal{L}^n$ and $\nu = \sigma \, d\mathcal{L}^n$. By Theorem 6.1 there exists $f : \mathbb{R}^n \to \mathbb{R} \cup \{+\infty\}$ such that $\Omega = \mathrm{IntDom}(f) \neq \emptyset$, μ is concentrated on F_1, $(\nabla f)_\# \mu = \nu$, and $\mathrm{Cl}(\nabla f(F_1 \cap \mathrm{spt}\,\mu)) = \mathrm{spt}\,\nu$.

Step one: We show that μ is concentrated on $F_{2,\mathrm{inv}}$. Since μ is concentrated on F_1, $\mu \ll \mathcal{L}^n$ and $\mathcal{L}^n(F_1 \setminus F_2) = 0$ by Alexandrov's theorem, we have that μ is concentrated on F_2. Similarly, since f^* is twice differentiable $\mathcal{L}^n$-a.e. on its domain and $\nu \ll \mathcal{L}^n$ we have that $\nu(\mathrm{Dom}(f^*) \setminus F_2^*) = 0$. In particular, since Proposition 8.3 gives

$$\nu(E) = \mu(\partial f^*(E)) \qquad \forall E \in \mathcal{B}(\mathbb{R}^n), \tag{8.26}$$

we conclude that $0 = \mu(\partial f^*(\mathrm{Dom}(f^*) \setminus F_2^*))$. By (8.11) (that is to say, by $F_2 \setminus F_{2,\mathrm{inv}} \subset \partial f^*(\mathrm{Dom}(f^*) \setminus F_2^*)$), we finally deduce that f is twice differentiable with invertible Hessian μ-a.e. on F_2. Having already proved $\mu(\mathbb{R}^n \setminus F_2) = 0$, we conclude that $\mu(\mathbb{R}^n \setminus F_{2,\mathrm{inv}}) = 0$.

Step two: We show that μ is concentrated on $(\nabla f)^{-1}(\mathrm{Leb}(\sigma))$. Indeed, by $\nu \ll \mathcal{L}^n$, Rademacher's theorem, the Lebesgue points theorem, and (8.26) give

$$1 = \nu(\mathrm{Leb}(\sigma) \cap F_1^*) = \mu(\partial f^*(\mathrm{Leb}(\sigma) \cap F_1^*))$$
$$= \mu(F_1 \cap \partial f^*(\mathrm{Leb}(\sigma) \cap F_1^*)). \tag{8.27}$$

Since $x \in F_1$, $y \in F_1^*$ and $x = \nabla f^*(y)$ imply $y = \nabla f(x)$, we conclude that

$$F_1 \cap \partial f^*(\mathrm{Leb}(\sigma) \cap F_1^*) \subset \{x \in F_1 : \nabla f(x) \in \mathrm{Leb}(\sigma)\} = (\nabla f)^{-1}(\mathrm{Leb}(\sigma)),$$

and thus that μ is concentrated on $(\nabla f)^{-1}(\text{Leb}(\sigma))$. Combined with step one, this proves (8.2).

Step three: We prove (8.3) and (8.4). Indeed, let

$$X = F_{2,\text{inv}} \cap \text{Leb}(\rho) \cap (\nabla f)^{-1}(\text{Leb}(\sigma)) \cap \{\rho > 0\}.$$

If $x \in X$, then by Theorem 8.6-(iii) and since $x \in F_{2,\text{inv}} \cap \text{Leb}(\rho)$, we have

$$\frac{\rho(x)}{\det \nabla^2 f(x)} = \lim_{r \to 0^+} \frac{\nu(B_r(\nabla f(x)))}{\omega_n r^n} = \sigma(\nabla f(x)),$$

thanks to $\nabla f(x) \in \text{Leb}(\sigma)$. Since $\det \nabla^2 f(x) > 0$ and $\rho(x) > 0$, we find that $\sigma(\nabla f(x)) > 0$, so that (8.3) is proved.

Step four: We finally prove that $\det \nabla^2 f \in L^1_{\text{loc}}(\Omega)$ and that the area formula (8.5) holds true. Let us first show that

$$\int_K \det \nabla^2 f \le \mathcal{L}^n(\partial f(K)) \qquad \forall K \subset\subset \Omega. \tag{8.28}$$

Indeed, set $\Omega^* = \text{IntDom}(f^*)$ and consider the measure λ defined by

$$\lambda(E) = (\nabla f^*)_\#(\mathcal{L}^n \llcorner \Omega^*)(E) = \mathcal{L}^n(\partial f(E)) \qquad \forall E \in \mathcal{B}(\mathbb{R}^n),$$

where in the last identity we have used Proposition 8.3, the fact that ∇f^* transports $\mathcal{L}^n \llcorner \Omega^*$, and that $f^{**} = f$. Since $\nabla f \in L^\infty_{\text{loc}}(\Omega; \mathbb{R}^n)$, we easily see that if $K \subset\subset \Omega$, then $\partial f(K)$ is a bounded set in $\mathbb{R}^n$. Therefore, λ is locally finite, and thus a Radon measure, in Ω. By the Lebesgue–Besicovitch differentiation theorem there exist $h \in L^1_{\text{loc}}(\Omega)$ and a Radon measure $\lambda^s \perp \mathcal{L}^n \llcorner \Omega$ such that

$$\lambda = h \mathcal{L}^n \llcorner \Omega + \lambda^s, \qquad h(x) = \lim_{r \to 0^+} \frac{\lambda(B_r(x))}{\omega_n r^n}$$

for $\mathcal{L}^n$-a.e. $x \in \Omega$. At the same time, thanks to statement (a) and Alexandrov's theorem, for $\mathcal{L}^n$-a.e. $x \in \Omega$ the limit defining $h(x)$ is equal to $\det \nabla^2 f(x)$, so that $\det \nabla^2 f \in L^1_{\text{loc}}(\Omega)$, as claimed. To prove (8.5) we need to show that

$$(\nabla f)_\# (\det \nabla^2 f \, d\mathcal{L}^n \llcorner M) = \mathcal{L}^n \llcorner \nabla f(M), \qquad M = F_{2,\text{inv}} \cap \text{Leb}(\det \nabla^2 f). \tag{8.29}$$

Since ∇f transports the measure $\pi = \det \nabla^2 f \, d\mathcal{L}^n \llcorner M$ (indeed, $M \subset F_1$), we can apply Theorem 8.6-(iii) with $u = \det \nabla^2 f$, and setting $\omega = (\nabla f)_\# \pi$, we find

$$\lim_{r \to 0^+} \frac{\omega(B_r(\nabla f(x_0)))}{\omega_n r^n} = \frac{\det \nabla^2 f(x_0)}{\det \nabla^2 f(x_0)} = 1 \qquad \forall x_0 \in M. \tag{8.30}$$

In particular, $\omega \llcorner \nabla f(M) = \mathcal{L}^n \llcorner \nabla f(M)$. Since π is concentrated on M and $\omega = (\nabla f)_\# \pi$, it follows that ω is concentrated on $(\nabla f)(M)$, and thus (8.29) holds. $\qquad\square$

PART III

Applications to PDE and the Calculus of Variations and the Wasserstein Space

9

Isoperimetric and Sobolev Inequalities in Sharp Form

As detailed in the introduction, the third part of this book presents a series applications of the Brenier–McCann theorem to the analysis of various problems in PDE and the Calculus of Variations, and to the introduction of the Wasserstein space. We begin in this chapter by proving sharp forms of the Euclidean isoperimetric inequality and of the Sobolev inequality. By "sharp form" we mean that we are going to prove these inequalities with their sharp constants, although we will avoid the more technical discussions needed to characterize their equality cases.

9.1 A Jacobian–Laplacian Estimate

We begin our analysis by proving a proposition which lies at the heart of the method discussed in this chapter. The proposition is based on the **arithmetic-geometric mean inequality**: given *non-negative* numbers $\alpha_k \geq 0$, we have

$$\prod_{k=1}^{n} \alpha_k^{1/n} \leq \frac{1}{n} \sum_{k=1}^{n} \alpha_k, \tag{9.1}$$

with equality if and only if there is $\alpha \geq 0$ such that $\alpha_k = \alpha$ for every $k = 1, \ldots, n$.

Proposition 9.1 *If* $f : \mathbb{R}^n \to \mathbb{R} \cup \{+\infty\}$ *is convex with* $\Omega = \mathrm{IntDom}(f) \neq \emptyset$, *then, as Radon measures in* Ω,

$$\mathrm{Div}(\nabla f) \geq n \, (\det \nabla^2 f)^{1/n} \, d(\mathcal{L}^n \llcorner \Omega), \tag{9.2}$$

where Div *denotes the distributional divergence operator.*

Proof We have proved in Theorem 7.1 that $\nabla f \in BV_{\mathrm{loc}}(\Omega; \mathbb{R}^n)$ and that the distributional Hessian $D^2 f$ of f is an $\mathbb{R}^{n \times n}_{\mathrm{sym}}$-valued Radon measure in Ω such that, for every $v \in \mathbb{R}^n$, $D^2_{vv} f$ is a (standard, nonnegative) Radon measure

97

in Ω – where as usual we have set $D^2_{vw}f(\varphi) = v \cdot \int_{\mathbb{R}^n} f \nabla^2 \varphi(x)[w]\,dx$ for every $v,w \in \mathbb{R}^n$ and $\varphi \in C^\infty_c(\Omega)$. Moreover, we have the Radon–Nikodym decomposition of $D^2 f$ with respect to $\mathcal{L}^n \llcorner \Omega$,

$$D^2 f = \nabla^2 f \, d(\mathcal{L}^n \llcorner \Omega) + [D^2 f]^s, \qquad \text{where } [D^2 f]^s = D^2 f \llcorner Y,$$

for a suitable Borel set Y, see (A.14). In particular,

$$v \cdot \left(\int_{\mathbb{R}^n} \varphi \, d[D^2 f]^s \right)[v] \geq 0, \qquad \forall \varphi \in C^0_c(\Omega), \varphi \geq 0, \forall v \in \mathbb{R}^n,$$

so that for every $v \in \mathbb{R}^n$

$$D^2_{vv} f \geq v \cdot \nabla^2 f[v] \, d(\mathcal{L}^n \llcorner \Omega)$$

as Radon measures in Ω. Since $\nabla f \in L^1_{\text{loc}}(\Omega; \mathbb{R}^n)$ gives $D(\nabla f) = D^2 f$, we find that, as Radon measures in Ω,

$$\text{Div}(\nabla f) = \sum_{i=1}^n D^2_{e_i e_i} f \geq \text{trace}(\nabla^2 f) \, d\mathcal{L}^n \llcorner \Omega, \qquad (9.3)$$

where $\text{trace}(A) = \sum_{i=1}^n A_{ii}$ if $A \in \mathbb{R}^{n \times n}$. Now, for a.e. $x \in \Omega$, the absolutely continuous density $\nabla^2 f(x)$ of $D^2 f$ with respect to $\mathcal{L}^n$ is an $n \times n$ symmetric matrix (as the limit as $r \to 0^+$ of the $n \times n$-symmetric matrices $D^2 f(B_r(x))/\omega_n r^n$). Therefore $\nabla^2 f(x)$ can be represented, in an x-dependent orthonormal basis, as a diagonal matrix with nonnegative entries $\{\alpha_k(x)\}_{k=1}^n$, so that, by the arithmetic-geometric mean inequality (9.1) we find

$$\text{Trace}(\nabla^2 f(x)) = \sum_{k=1}^n \alpha_k(x) \geq n \prod_{k=1}^n \alpha_k(x)^{1/n} = n \, (\det \nabla^2 f(x))^{1/n}. \quad (9.4)$$

Combining (9.3) with (9.4) we conclude the proof. □

9.2 The Euclidean Isoperimetric Inequality

In Theorem 9.2 we present an OMT proof of the Euclidean isoperimetric inequality. Let us recall that if E is an open set with smooth boundary in $\mathbb{R}^n$ and outer unit normal ν_E, then the Gauss–Green formula

$$\int_E \nabla \varphi = \int_{\partial E} \varphi \, \nu_E \, d\mathcal{H}^{n-1} \qquad \forall \varphi \in C^\infty_c(\mathbb{R}^n), \qquad (9.5)$$

can be interpreted in distributional sense as saying that the characteristic function 1_E of E (which, trivially, is locally summable, and thus admits a distributional derivative) satisfies

$$D1_E = \nu_E \, \mathcal{H}^{n-1} \llcorner \partial E. \qquad (9.6)$$

In particular, $D1_E$ is an $\mathbb{R}^n$-valued Radon measure, since $|D1_E| = \mathcal{H}^{n-1} \llcorner \partial E$ and ∂E is locally $\mathcal{H}^{n-1}$-finite. Indeed, for every $x \in \partial E$ there exist an $(n-1)$-dimensional disk M centered at x, a cylinder K with section M, a smooth function $u : M \to \mathbb{R}$ with bounded gradient such that

$$K \cap \partial E = \text{graph of } u \text{ over } M,$$

which combined with (9.6) gives

$$|D1_E|(K) = \mathcal{H}^{n-1}(K \cap \partial E) = \int_M \sqrt{1 + |\nabla u|^2} < \infty.$$

Defining the **perimeter of** E by setting

$$P(E) = \mathcal{H}^{n-1}(\partial E) = |D1_E|(\mathbb{R}^n),$$

one has the **Euclidean isoperimetric inequality**, asserting that if $0 < |E| < \infty$, then

$$P(E) \geq P(B_{r(E)}) \qquad r(E) = \left(\frac{|E|}{\omega_n} \right)^{1/n},$$

where $r(E)$ is the radius such that $|B_{r(E)}| = |E|$, and where equality holds if and only if $E = x + B_{r(E)}$ for some $x \in \mathbb{R}^n$. Given that

$$P(B_{r(E)}) = n \, \omega_n^{1/n} \, |E|^{(n-1)/n},$$

the Euclidean isoperimetric inequality can be written as in (9.7).

Theorem 9.2   *If E is a bounded open set with smooth boundary in $\mathbb{R}^n$, then*

$$P(E) \geq n \, \omega_n^{1/n} \, |E|^{(n-1)/n}. \tag{9.7}$$

Proof   *Step one*: We prove that if $u \in C_c^\infty(\mathbb{R}^n)$ with $u \geq 0$ and $\int_{\mathbb{R}^n} u^{n/(n-1)} = 1$, then

$$\int_{\mathbb{R}^n} |\nabla u| \geq n \, \omega_n^{1/n}. \tag{9.8}$$

Indeed, let us consider the probability measures

$$\mu = u^{n/(n-1)} \, d\mathcal{L}^n, \qquad \nu = \frac{1_{B_1}}{\omega_n} \, d\mathcal{L}^n.$$

Since spt ν is compact, by Theorem 8.1 there exists a convex function $f : \mathbb{R}^n \to \mathbb{R}$ such that $(\nabla f)_\# \mu = \nu$, with $\nabla f(x) \in B_1$ for $\mathcal{L}^n$-a.e. $x \in E$, and

$$\det(\nabla^2 f(x)) = \omega_n \frac{u(x)^{n/(n-1)}}{1_{B_1}(\nabla f(x))}, \qquad \text{for } \mathcal{L}^n\text{-a.e. } x \in \{u > 0\}. \tag{9.9}$$

By integrating $h(y) = 1_{B_1}(y)^{-1/n}$ with respect to ν and using first $(\nabla f)_\# \mu = \nu$ and then Monge's transport condition (9.9), we find

$$1 = \int_{\mathbb{R}^n} 1_{B_1}(y)^{-1/n}\, d\nu(y) = \int_{\mathbb{R}^n} 1_{B_1}(\nabla f)^{-1/n}\, u^{n/(n-1)}$$

$$= \frac{1}{\omega_n^{1/n}} \int_{\mathbb{R}^n} \left(\frac{\det \nabla^2 f}{u^{n/(n-1)}} \right)^{1/n} u^{n/(n-1)}$$

$$= \frac{1}{\omega_n^{1/n}} \int_{\mathbb{R}^n} (\det \nabla^2 f)^{1/n}\, u \le \frac{1}{n\,\omega_n^{1/n}} \int_{\mathbb{R}^n} u\, d[\mathrm{Div}\,(\nabla f)].$$

where in the last inequality we have used (9.2) (which holds with $\Omega = \mathbb{R}^n$). Given that $u \in C_c^\infty(\mathbb{R}^n)$, by definition of distributional divergence we find that

$$n\omega_n^{1/n} \le \int_{\mathbb{R}^n} u\, d[\mathrm{Div}(\nabla f)] = - \int_{\mathbb{R}^n} \nabla f \cdot \nabla u \le \int_{\mathbb{R}^n} |\nabla u|$$

where in the last inequality we have used $\nabla u = 0$ on $\{u = 0\}$ and $\nabla f(x) \in B_1$ for $\mathcal{L}^n$-a.e. $x \in \{u > 0\}$. This proves (9.8).

Step two: Now let E be a bounded open set with smooth boundary in $\mathbb{R}^n$, and let $u_\varepsilon = 1_E \star \rho_\varepsilon$. Since E is bounded we have that $u_\varepsilon \in C_c^\infty(\mathbb{R}^n)$, and by (9.8) we have

$$\int_{\mathbb{R}^n} |\nabla u_\varepsilon| \ge n\omega_n^{1/n} \left(\int_{\mathbb{R}^n} u_\varepsilon^{n/(n-1)} \right)^{(n-1)/n}.$$

As $\varepsilon \to 0^+$ we have $u_\varepsilon(x) \to 1_E(x)$ at every Lebesgue point x of 1_E, and

$$\int_{\mathbb{R}^n} |\nabla u_\varepsilon| \to |D1_E|(\mathbb{R}^n) = P(E)$$

thanks to (A.22). Therefore (9.7) is proved.   □

9.3 The Sobolev Inequality on $\mathbb{R}^n$

The **sharp Sobolev inequality on** $\mathbb{R}^n$ states that if $1 < p < n$, $n \ge 2$, $u \in L^1_{\mathrm{loc}}(\mathbb{R}^n)$ is such that $\nabla u \in L^p(\mathbb{R}^n; \mathbb{R}^n)$, and $\{|u| > t\}$ has finite Lebesgue measure for every $t > 0$, then

$$\left(\int_{\mathbb{R}^n} |\nabla u|^p \right)^{1/p} \ge S(n,p) \left(\int_{\mathbb{R}^n} |u|^{p^\star} \right)^{1/p^\star}, \qquad p^\star = \frac{np}{n-p} \qquad (9.10)$$

with equality if and only if for some $a \in \mathbb{R}$, $x_0 \in \mathbb{R}^n$ and $r > 0$ one has $u = u_{a,x_0,r}$, where, setting $p' = p/(p-1)$,

$$u_{a,x_0,r}(x) = \frac{a}{(1 + r\,|x - x_0|^{p'})^{(n/p)-1}}, \qquad x \in \mathbb{R}^n. \qquad (9.11)$$

Evaluation of the $L^{p^\star}$-norm of $u_{a,x_0,r}$ and of the L^p-norm of $\nabla u_{a,x_0,r}$ allows one to give an explicit formula for $S(n,p)$ in terms of gamma functions. In Theorem 9.3 we prove (9.10) with the sharp constant $S(n,p)$ by using Brenier maps. The interest of establishing the Sobolev inequality in sharp form is paramount both from the physical and geometric viewpoint. In the latter direction, we may recall the geometric interpretation of (9.10) when $p = 2$ and $n \geq 3$. In this setting, the conformally flat Riemannian manifold $(M,g) = (\mathbb{R}^n, u^{4/(n-2)} \delta_{ij})$ defined by a smooth and positive function $u : \mathbb{R}^n \to (0,\infty)$ is such that

$$\int_{\mathbb{R}^n} |\nabla u|^2 = \frac{n-2}{4(n-1)} \int_M R_g, \qquad \int_{\mathbb{R}^n} u^{2^\star} = \mathrm{vol}_g(M),$$

where R_g denotes the scalar curvature of (M,g). In other words, $S(n,2)$ is the minimum of the total scalar curvature functional among conformally flat Riemannian manifolds with unit volume. Since (9.11) gives

$$u_{1,0,1}(x)^{4/(n-2)} = \frac{1}{(1 + |x|^2)^2} \qquad x \in \mathbb{R}^n,$$

and since $(1 + |x|^2)^{-2}$ is the conformal factor of the standard metric on the sphere $\mathbb{S}^n$ under stereographic projection onto $\mathbb{R}^n$, we conclude that the sharp Sobolev inequality with $p = 2$ is equivalent to say that *standard spheres minimize total scalar curvature among conformally flat manifolds with fixed volume.*

Theorem 9.3 (Sharp Sobolev inequality) *If* $n \geq 2$, $1 < p < n$, *and* $u \in L^1_{\mathrm{loc}}(\mathbb{R}^n)$ *with* $\nabla u \in L^p(\mathbb{R}^n; \mathbb{R}^n)$ *and* $\mathcal{L}^n(\{|u| > t\}) < \infty$ *for every* $t > 0$, *then*

$$\left(\int_{\mathbb{R}^n} |\nabla u|^p \right)^{1/p} \geq S(n,p) \left(\int_{\mathbb{R}^n} |u|^{p^\star} \right)^{1/p^\star}. \tag{9.12}$$

Moreover, equality holds if $u = u_{a,x_0,r}$ *for some* $a \in \mathbb{R}$, $x_0 \in \mathbb{R}^n$ *and* $r > 0$.

Proof *Step one*: We show that if $u,v \in C_c^\infty(\mathbb{R}^n)$ with $u,v \geq 0$ on $\mathbb{R}^n$ and

$$\int_{\mathbb{R}^n} u^{p^\star} = \int_{\mathbb{R}^n} v^{p^\star} = 1,$$

then

$$\frac{n}{p^\#} \frac{\int_{\mathbb{R}^n} v^{p^\#}}{\left(\int_{\mathbb{R}^n} |y|^{p'} v(y)^{p^\star} dy \right)^{1/p'}} \leq \left(\int_{\mathbb{R}^n} |\nabla u|^p \right)^{1/p}, \qquad p^\# = \frac{p(n-1)}{n-p}. \tag{9.13}$$

Let ∇f be the Brenier map from $\mu = u^{p^\star} d\mathcal{L}^n$ to $\nu = v^{p^\star} d\mathcal{L}^n$, so that

$$\int_{\mathbb{R}^n} h\, v^{p^\star} = \int_{\mathbb{R}^n} (h \circ \nabla f)\, u^{p^\star} \tag{9.14}$$

for every Borel function $h : \mathbb{R}^n \to [0, \infty]$. By Theorem 8.1, and since spt v is compact, we can assume that $\mathrm{Dom}(f) = \mathbb{R}^n$ and that $\nabla f(x) \in \{v > 0\}$ and

$$\det \nabla^2 f(x) = \frac{u(x)^{p^\star}}{v(\nabla f(x))^{p^\star}}, \qquad \text{for } \mathcal{L}^n\text{-a.e. } x \in \{u > 0\}. \tag{9.15}$$

By applying (9.14) to $h = v^{-p^\star/n}$, and noticing that $p^\# = p^\star - (p^\star/n)$, we find that

$$\int_{\mathbb{R}^n} v^{p^\#} = \int_{\mathbb{R}^n} v^{-p^\star/n} \, v^{p^\star} = \int_{\mathbb{R}^n} v(\nabla f)^{-p^\star/n} \, u^{p^\star},$$

and thus, by (9.15),

$$\int_{\mathbb{R}^n} v^{p^\#} = \int_{\mathbb{R}^n} \frac{(\det \nabla^2 f)^{1/n}}{u^{p^\star/n}} \, u^{p^\star} = \int_{\mathbb{R}^n} (\det \nabla^2 f)^{1/n} \, u^{p^\#} \le \frac{1}{n} \int_{\mathbb{R}^n} u^{p^\#} \, d[\mathrm{Div} \nabla f],$$

where in the last inequality we have used the key inequality (9.2) (with $\Omega = \mathbb{R}^n$). Since $u \in C_c^\infty(\mathbb{R}^n)$ we can integrate by parts and apply the Hölder inequality to find

$$n \int_{\mathbb{R}^n} v^{p^\#} \le \int_{\mathbb{R}^n} u^{p^\#} \, d[\mathrm{Div}(\nabla f)] = -p^\# \int_{\mathbb{R}^n} u^{p^\#-1} \nabla u \cdot \nabla f$$

$$\le p^\# \left(\int_{\mathbb{R}^n} u^{(p^\#-1)p'} |\nabla f|^{p'} \right)^{1/p'} \left(\int_{\mathbb{R}^n} |\nabla u|^p \right)^{1/p}.$$

Since $(p^\# - 1)p' = p^\star$ we can apply (9.14) again, this time with $h(y) = |y|^{p'}$, and find

$$n \int_{\mathbb{R}^n} v^{p^\#} \le p^\# \left(\int_{\mathbb{R}^n} |y|^{p'} v(y)^{p^\star} \, dy \right)^{1/p'} \left(\int_{\mathbb{R}^n} |\nabla u|^p \right)^{1/p}.$$

This proves (9.13).

Step two: By a non-trivial density argument that is omitted for the sake of brevity, (9.13) implies that if $u \in L^1_{\mathrm{loc}}(\mathbb{R}^n)$ with $\nabla u \in L^p(\mathbb{R}^n; \mathbb{R}^n)$ and $\mathcal{L}^n(\{|u| > t\}) < \infty$ for every $t > 0$, then

$$\left(\int_{\mathbb{R}^n} |\nabla u|^p \right)^{1/p} \ge S(n, p) \left(\int_{\mathbb{R}^n} |u|^{p^\star} \right)^{1/p^\star},$$

where $S(n, p)$ is defined as

$$S(n, p) = \sup \left\{ M(v) : v \in C_c^\infty(\mathbb{R}^n), v \ge 0, \int_{\mathbb{R}^n} v^{p^\star} = 1 \right\},$$

$$M(v) = \frac{n}{p^\#} \frac{\int_{\mathbb{R}^n} v^{p^\#}}{\left(\int_{\mathbb{R}^n} |y|^{p'} |v(y)|^{p^\star} \, dy \right)^{1/p'}}.$$

By letting $u(x) = v(x) = c(n,p)(1 + |x|^{p'})^{1-(n/p)}$ for a suitable choice of $c(n,p)$ that guarantees $\int_{\mathbb{R}^n} u^{p^\star} = \int_{\mathbb{R}^n} v^{p^\star} = 1$, a direct computation (or, better said, another omitted density argument) shows that

$$\frac{\|\nabla u\|_{L^p(\mathbb{R}^n)}}{\|u\|_{L^{p^\star}(\mathbb{R}^n)}} = M(v), \tag{9.16}$$

thus proving the optimality of (9.11) in (9.12). $\square$

Remark 9.4 The above proof leaves the impression that, without prior knowledge of the optimal functions in the Sobolev inequality, one could have not guessed their form from the mass transport argument. This is, however, a wrong impression, since the proof of Theorem 9.3 actually indicates a simple way to identify the specific profiles of the optimizers in the Sobolev inequality. Indeed, we can inspect the proof of Theorem 9.3 looking for functions u such that the choice $v = u$ saturates the duality inequality (9.13). Since in this case we can set $\nabla f(x) = x$ for every $x \in \mathbb{R}^n$, the only non-trivial inequality sign left in the argument is found in correspondence of the application of the Hölder inequality,

$$\int_{\mathbb{R}^n} u^{p^\#-1}(-\nabla u) \cdot x \leq \left(\int_{\mathbb{R}^n} |x|^{p'} u^{p^\star}(x)\, dx \right)^{1/p'} \left(\int_{\mathbb{R}^n} |\nabla u|^p \right)^{1/p}.$$

Equality holds in this Hölder inequality if and only if, for some $\lambda > 0$,

$$u^{p^\#-1} x = \lambda |\nabla u|^{p-2}(-\nabla u);$$

writing this condition for $u(x) = \zeta(|x|)$, with $\zeta = \zeta(r)$ a decreasing function, we find that

$$(-\zeta')^{p-1} = \frac{r}{\lambda} \zeta^{p^\#-1} \qquad \forall r > 0,$$

which is equivalent to $-\mu \zeta' \zeta^{-n/(n-p)} = r^{1/(p-1)}$, $\mu = \lambda^{1/(p-1)}$, for every $r \in \{\zeta > 0\}$. This is easily integrated to find that $\zeta(r)^{-p/(n-p)} = \alpha + \beta r^{p'}$ for positive constants α and β, and thus that $\zeta(r) = (\alpha + \beta r^{p'})^{1-(n/p)}$ for every $r > 0$.

10

Displacement Convexity and Equilibrium of Gases

Displacement convexity is one of the most important ideas stemming from the study of OMT problems. Following McCann's original presentation, in this chapter we introduce displacement convexity (and the related notion of displacement interpolation) in connection with an apparently simple problem in the Calculus of Variations, namely, establishing the uniqueness of equilibrium states in a basic variational model for self-interacting gases. A displacement convexity proof of the Brunn–Minkowski inequality closes the chapter.

10.1 A Variational Model for Self-interacting Gases

We consider a model where the state of a gas is entirely determined by its mass density ρ. We normalize the total mass of the gas to unit, and thus identify the state variable ρ with an absolutely continuous probability measure

$$\mu = \rho \, d\mathcal{L}^n \in \mathcal{P}_{ac}(\mathbb{R}^n).$$

The energy $\mathcal{E} = \mathcal{U} + \mathcal{V}$ of the state ρ is defined as the sum of an internal energy $\mathcal{U}$ and of a self-interaction energy $\mathcal{V}$: the **internal energy** $\mathcal{U}$ has the form

$$\mathcal{U}(\rho) = \int_{\mathbb{R}^n} U(\rho(x)) \, dx,$$

where $U : [0, \infty) \to \mathbb{R}$ is continuous, with $U(0) = 0$, while the **self-interaction energy** $\mathcal{V}$ has the form

$$\mathcal{V}(\rho) = \frac{1}{2} \int_{\mathbb{R}^n} \int_{\mathbb{R}^n} \rho(x) \, V(x - y) \, \rho(y) \, dx \, dy$$

corresponding[1] to a continuous function $V : \mathbb{R}^n \to \mathbb{R}$.

---

[1]   In the physical case $n = 3$, the dimensions of U are energy per unit volume, and those of V energy times squared unit mass.

Internal energies: The internal energy $\mathcal{U}$ models the behavior of the gas under compression and dilution. Compression and dilution corresponds, respectively, to the limits $\lambda \to 0^+$ and $\lambda \to \infty$ of the rescaled densities $\rho^\lambda(x) = \lambda^{-n} \rho(x/\lambda)$ ($\lambda > 0$). Since gases oppose compression and favor dilution, we want

$$\lambda \in (0,\infty) \mapsto \mathcal{U}(\rho^\lambda) = \int_{\mathbb{R}^n} U\left(\frac{\rho(x)}{\lambda^n}\right) \lambda^n \, dx$$

to be a decreasing function. By the arbitrariness of ρ, this means requiring that

$$\lambda \in (0,\infty) \mapsto \lambda^n U(\lambda^{-n}) \text{ is a decreasing function.} \tag{10.1}$$

Physically relevant examples of internal energy densities U satisfying (10.1) are given by

$$U(\rho) = \rho \log \rho, \tag{10.2}$$
$$U(\rho) = \rho^\gamma, \qquad\qquad \gamma \geq 1, \tag{10.3}$$
$$U(\rho) = -\rho^\gamma, \qquad\qquad 1 \geq \gamma \geq \frac{n-1}{n}. \tag{10.4}$$

With U as in (10.3) or (10.4), the functional $\mathcal{U}$ is well-defined on $\mathcal{P}_{\mathrm{ac}}(\mathbb{R}^n)$ with values, respectively in $[0,\infty]$ and in $[-\infty,0]$. The case when $U(\rho) = \rho \log \rho$, and thus $\mathcal{U}$ is the (negative) entropy of the gas, is of course more delicate, because both $U(\rho)^+ = \max\{U(\rho),0\}$ and $U(\rho)^- = \max\{-U(\rho),0\}$ may have infinite integrals. To avoid this difficulty one usually assumes a moment bound. For example, given $p \in [1,\infty)$, by taking into account that $\rho\,[\log(\rho) + |x|^p] = U(\rho\, e^{|x|^p})\, e^{-|x|^p}$ and by applying Jensen inequality to the convex function U and to the probability measure $c(p)\, e^{-|x|^p}\, dx$, it is easily seen that

$$\mathcal{U}(\rho) \geq -\int_{\mathbb{R}^n} |x|^p\, \rho(x)\, dx - \log\left(\int_{\mathbb{R}^n} e^{-|x|^p}\, dx\right) \quad \forall \mu = \rho\, d\mathcal{L}^n \in \mathcal{P}_{p,\mathrm{ac}}(\mathbb{R}^n). \tag{10.5}$$

Finally, as a remark that will play an important role in Theorem 10.5, we notice that in all the three examples listed above the function of λ defined in (10.1) is also *convex*, in addition to being decreasing.

Self-interaction energies: The energy term $\mathcal{V}$ takes into account the interaction between different particles in the gas mediated by the self-interaction potential V. The Newtonian potential $V(z) = 1/|z|$ for $z \in \mathbb{R}^3 \setminus \{0\}$ is, by and large, the physically most important example of a self-interaction potential: in this case, the attraction between particles *decreases* with an increase of their mutual distance. We shall rather focus on the case of interaction potentials V such that

$$V \text{ is strictly convex on } \mathbb{R}^n. \tag{10.6}$$

This assumption *excludes* the Newtonian potential, and refers more naturally to situations where the attraction between particles *increases* with their mutual distance.

We now turn to the analysis of the minimization problem

$$\inf \left\{ \mathcal{E}(\rho) = \mathcal{U}(\rho) + \mathcal{V}(\rho) : \mu = \rho \, d\mathcal{L}^n \in \mathcal{P}_{ac}(\mathbb{R}^n) \right\}, \qquad (10.7)$$

whose minimizers represent the equilibrium states of a gas modeled by U and V. By a classical technique in the Calculus of Variations, the concentration-compactness method, and under mild growth assumptions on U and V, one checks the existence of minimizers in (10.7): we do not provide the details of this existence argument, since they are not really related to OMT. We rather focus on the question of the *uniqueness* of minimizers, which is considerably trickier. The standard approach to uniqueness of minimizers, for example in the minimization problem $\min_K F$ of a real-valued function F defined over a convex subset K of a linear space, requires the **strictly convexity of** F: if for every $x_0, x_1 \in K$ and $t \in (0, 1)$ one has $F((1 - t)x_0 + t x_1) \leq (1 - t) F(x_0) + t F(x_1)$, with equality for some $t \in (0, 1)$ if and only if $x_0 = x_1$, then $\min_K F$ can be achieved only at one point; indeed $F(x_0) = F(x_1) = \min_K F = m$ and convexity imply that F is constantly equal to m along the segment $(1 - t)x_0 + t x_1, t \in (0, 1)$, thus implying, by strict convexity, that $x_0 = x_1$. As the reader may expect at this point of the discussion, the problem is that under the assumptions on U and V made above, the energy $\mathcal{E}$ will not be strictly convex with respect to the standard notion of convex interpolation on $\mathcal{P}_{ac}(\mathbb{R}^n)$.

By **standard interpolation** of $\mu_0 = \rho_0 \, d\mathcal{L}^n$ and $\mu_1 = \rho_1 \, d\mathcal{L}^n$ we mean the family of measures

$$\mu_t = ((1 - t) \rho_0 + t \rho_1) \, d\mathcal{L}^n \in \mathcal{P}_{ac}(\mathbb{R}^n), \qquad (10.8)$$

satisfying of course $\mu_{t=0} = \mu_0$ and $\mu_{t=1} = \mu_1$. The internal energy $\mathcal{U}$ is strictly convex along the standard interpolation (10.8) as soon as U is a strictly convex function on $(0, \infty)$, a condition which is met by all the examples in our list (10.2), (10.3) and (10.4). Thus, for these choices of the internal energy $\mathcal{U}$, the total energy $\mathcal{E}$ is strictly convex along (10.8) as soon as $\mathcal{V}$ is convex along (10.8). However, as we are going to see in a moment, under very mild assumptions on V (namely, $V(0) = 0$ and $V > 0$ on $\mathbb{R}^n \setminus \{0\}$), the self-interaction energy $\mathcal{V}$ may very well be concave along the interpolation (10.8)! Let us take, for example,

$$\rho_0^\varepsilon = \frac{1_{B_\varepsilon}(x_0)}{\omega_n \varepsilon^n}, \qquad \rho_1^\varepsilon = \frac{1_{B_\varepsilon}(x_1)}{\omega_n \varepsilon^n}, \qquad \varepsilon < |x_0 - x_1|, \qquad (10.9)$$

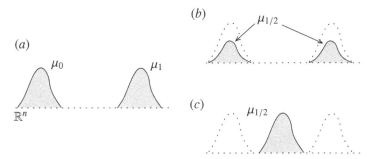

Figure 10.1 (a) Two probability densities in $\mathbb{R}^n$, obtained one as a translation of the other; (b) the standard convex interpolation with $t = 1/2$ is defined by taking the average of the two densities; (c) the new displacement interpolation is defined by continuously transporting μ_0 in μ_1 through a convex combination of the identity map and the Brenier map: in the depicted case, this amounts to a continuous translation.

so that $\rho_t^\varepsilon = (1 - t) \rho_0^\varepsilon + t \rho_1^\varepsilon$, and $\mu_t^\varepsilon = \rho_t^\varepsilon \, d\mathcal{L}^n \overset{*}{\rightharpoonup} \mu_t$ as $\varepsilon \to 0^+$, where

$$\mu_0 = \delta_{x_0}, \qquad \mu_1 = \delta_{x_1}, \qquad \mu_t = (1 - t) \delta_{x_0} + t \delta_{x_1} \qquad t \in (0, 1). \quad (10.10)$$

If, for example, $V(0) = 0$, then, letting $\varepsilon \to 0^+$, we find

$$\mathcal{V}(\rho_t^\varepsilon) \to \frac{1}{2} \int_{\mathbb{R}^n} d\mu_t(x) \int_{\mathbb{R}^n} V(x - y) \, d\mu_t(y)$$

$$= \frac{1}{2} \int_{\mathbb{R}^n} (1 - t) V(x - x_0) + t V(x - x_1) \, d\mu_t(x)$$

$$= \frac{1}{2} \{(1 - t) t V(x_1 - x_0) + t (1 - t) V(x_0 - x_1)\}$$

which is clearly a *concave* function of t as soon as $V > 0$ on $\mathbb{R}^n \setminus \{0\}$. Therefore we can essentially rule out the use of the standard convex interpolation (10.8) in addressing uniqueness of minimizers in (10.7).

The crucial remark to exit this impasse is that self-interaction energies $\mathcal{V}$ corresponding to convex potentials V are indeed convex if tested along an interpolation *between the supports* of the densities ρ_0 and ρ_1 (rather than between their values, as it is the case with (10.8)). This simple idea is illustrated in Figure 10.1, and can be preliminarily tested in the simple case when $\mu_0 = \sum_{i=1}^N \lambda_i \delta_{x_i}$ and $\mu_1 = \sum_{i=1}^N \lambda_i \delta_{y_i}$ for some $\lambda_i \in (0, 1)$ with $\sum_{i=1}^N \lambda_i = 1$. In this case, we are thinking about replacing the standard convex interpolation $\mu_t = (1 - t) \mu_0 + t \mu_1$, see (10.10), with the *displacement interpolation*[2]

---

[2] To be precise, (10.11) coincides with the formal notion of displacement interpolation (introduced in (10.12) for the case when $\mu \ll \mathcal{L}^n$, and in Proposition 11.1 for general μ and ν) if we take care of relabeling $\{y_i\}_i$ so that the map $x_i \mapsto y_i$ is optimal in $\mathbf{M}_2(\mu, \nu)$.

$$\mu_t = \sum_{i=1}^{N} \lambda_i \, \delta_{(1-t)\,x_i + t\,y_i}, \tag{10.11}$$

for which we easily see that

$$\mathcal{V}(\mu_t) = \frac{1}{2} \sum_{i,j=1}^{N} \lambda_i \, \lambda_j \, V\big((1-t)\,(x_j - x_i) + t\,(y_j - y_i)\big);$$

in particular, $t \mapsto \mathcal{V}(\mu_t)$ is convex as soon as V is. This simple promising picture apparently pertains just to the interpolation between discrete measures. To understand how to generalize (10.11) we notice that when two densities ρ_0 and ρ_1 are very concentrated, respectively, near two collections of points $\{x_i\}_{i=1}^{N}$ and $\{y_i\}_{i=1}^{N}$ in $\mathbb{R}^n$ as in the above example, then the Brenier map ∇f from μ_0 to μ_1 must take $\operatorname{spt}\mu_0 = \bigcup_{i=1}^{N} B_\varepsilon(x_i)$ into $\operatorname{spt}\mu_1 = \bigcup_{i=1}^{N} B_\varepsilon(y_i)$. If we are in a situation where the optimal transport plan from $\sum_{i=1}^{N} \lambda_i \, \delta_{x_i}$ to $\sum_{i=1}^{N} \lambda_i \, \delta_{y_i}$ is induced by a map taking x_i into (up to a reshuffling of indexes) y_i, then, provided ε is small enough, we expect $\nabla f(x) \approx y_i$ on $B_\varepsilon(x_i) \subset \operatorname{spt}\mu_0$. We thus generalize (10.11) by defining, for arbitrary $\mu_0 = \rho_0 \, d\mathcal{L}^n, \mu_1 = \rho_1 \, d\mathcal{L}^n \in \mathcal{P}_{\mathrm{ac}}(\mathbb{R}^n)$, the **displacement interpolation from** μ_0 **to** μ_1 as

$$\mu_t = ((1 - t)\,\mathbf{id} + t\,\nabla f)_{\#}\,\mu_0, \tag{10.12}$$

where ∇f is the Brenier map from μ_0 to μ_1. A non-trivial fact, proven in Proposition 10.2, is that $\mu_t \ll \mathcal{L}^n$ for every $t \in (0,1)$, i.e.,

$$\mu_t = ((1 - t)\,\mathbf{id} + t\,\nabla f)_{\#}\,\mu_0 = \rho_t \, d\mathcal{L}^n. \tag{10.13}$$

Assuming for a moment that (10.13) holds, we notice that if V is convex on $\mathbb{R}^n$, then $\mathcal{V}$ is convex along ρ_t: indeed, by applying first the push-forward identity (10.12), and then $(\nabla f)_{\#}\mu_0 = \mu_1$, we obtain

$$
\begin{aligned}
\mathcal{V}(\rho_t) &= \int_{\mathbb{R}^n} d\mu_t(x) \int_{\mathbb{R}^n} V(x - y) \, d\mu_t(y) \\
&= \int_{\mathbb{R}^n} \rho_0(x) \, dx \int_{\mathbb{R}^n} V\big((1-t)(x-y) + t\,(\nabla f(x) - \nabla f(y))\big) \rho_0(y) \, dy \\
&\leq (1 - t)\,\mathcal{V}(\rho_0) + t \int_{\mathbb{R}^n} \rho_0(x) \, dx \int_{\mathbb{R}^n} V(\nabla f(x) - \nabla f(y)) \rho_0(y) \, dy \\
&= (1 - t)\,\mathcal{V}(\rho_0) + t\,\mathcal{V}(\rho_1). \tag{10.14}
\end{aligned}
$$

Moreover, if V is strictly convex and equality holds in (10.14) for some $t \in (0,1)$, then equality holds in

$$V\big((1-t)(x-y) + t\,(\nabla f(x) - \nabla f(y))\big) \leq (1-t)\,V(x-y) + t\,V(\nabla f(x) - \nabla f(y))$$

for $\mu_0 \times \mu_0$-a.e. $(x,y) \in \mathbb{R}^n$, and thus $x - \nabla f(x)$ is constant on a set of full μ_0-measure on $\mathbb{R}^n$. This means that $\nabla f(x) = x - x_0$ for $\mathcal{L}^n$-a.e. $x \in \{\rho_0 > 0\}$, and thus, thanks to $(\nabla f)_\# \mu_0 = \mu_1$, that

$$\int_{\mathbb{R}^n} \varphi \, \rho_1 = \int_{\mathbb{R}^n} \varphi(\nabla f) \, \rho_0 = \int_{\mathbb{R}^n} \varphi(x - x_0) \, \rho_0(x) \, dx = \int_{\mathbb{R}^n} \varphi(y) \, \rho_0(y + x_0) \, dy$$

for every $\varphi \in C_c^0(\mathbb{R}^n)$. Therefore,

$$\rho_1(x) = \rho_0(x - x_0),$$

for $\mathcal{L}^n$-a.e. $x \in \{\rho_0 > 0\}$, and thus for $\mathcal{L}^n$-a.e. $x \in \mathbb{R}^n$. We thus have,

Proposition 10.1 (Displacement convexity of interaction energies) *If $V :$ $\mathbb{R}^n \to \mathbb{R}$ is convex, then the self-interaction energy $\mathcal{V}$ is convex along displacement interpolations in $\mathcal{P}_{\mathrm{ac}}(\mathbb{R}^n)$. Moreover, if V is strictly convex and $\mu_0 = \rho_0 \, d\mathcal{L}^n$ and $\mu_1 = \rho_1 \, d\mathcal{L}^n$ are such that $\mathcal{V}(\rho_t) = (1-t)\,\mathcal{V}(\rho_0) + t\,\mathcal{V}(\rho_1)$ for some $t \in (0,1)$, then there is $x_0 \in \mathbb{R}^n$ such that $\rho_0(x) = \rho_1(x - x_0)$ for $\mathcal{L}^n$-a.e. $x \in \mathbb{R}^n$.*

Proof By the above discussion and Proposition 10.2. □

This is of course very promising, but notice that, in fixing the convexity issues of $\mathcal{V}$, we have created a potential problem on the front of the internal energy $\mathcal{U}$. Indeed, if true, the convexity in $t \in (0,1)$ of

$$\mathcal{U}(\rho_t) = \int_{\mathbb{R}^n} U(\rho_t(x)) \, dx, \qquad \text{where} \quad \rho_t \, d\mathcal{L}^n = ((1-t)\,\mathbf{id} + t\,\nabla f)_\# \, \mu_0,$$

is definitely *not* an immediate consequence of a convexity assumption on U – as it was the case for the standard convex interpolation. Interestingly, the physically natural condition that $\lambda \mapsto \lambda^n \, U(\lambda^{-n})$ is convex and decreasing on $\lambda > 0$ is sufficient to establish the convexity of $\mathcal{U}$ along the displacement interpolation (10.12), and is tightly related to the theory of Brunn–Minkowski inequalities.

The rest of this chapter is organized as follows. We first discuss some basic properties of displacement interpolation, and then provide a general result about the displacement convexity of internal energies. Finally, we briefly discuss a proof of the Brunn–Minkowski inequality by means of displacement interpolation.

10.2 Displacement Interpolation

Whenever $\mu_0 \in \mathcal{P}(\mathbb{R}^n)$ is such that $\mu_0(S) = 0$ for every countably $(n-1)$-rectifiable set in $\mathbb{R}^n$ and $\mu_1 \in \mathcal{P}(\mathbb{R}^n)$, we can construct the Brenier map ∇f from μ_0 to μ_1 (Theorem 6.1) and define the **displacement interpolation** from μ_0 to μ_1 by setting

$$\mu_t = ((1-t)\,\mathbf{id} + t\,\nabla f)_\#\mu_0, \qquad t \in [0,1]. \tag{10.15}$$

At times the notation $[\mu_0, \mu_1]_t$ is used in place of μ_t, because it stresses the directionality of (10.15): we know that we can start from μ_0 and end-up on μ_1 by formula (10.15), but unless μ_1 also satisfies $\mu_1(S) = 0$ for every countably $(n-1)$-rectifiable set in $\mathbb{R}^n$, we do not have a Brenier map from μ_1 to μ_0 to define $[\mu_1, \mu_0]_t$. The following proposition is thus very useful, as it shows that displacement interpolation is closed in $\mathcal{P}_{\mathrm{ac}}(\mathbb{R}^n)$.

Proposition 10.2 *If $\mu_0 = \rho_0\,d\mathcal{L}^n \in \mathcal{P}_{\mathrm{ac}}(\mathbb{R}^n)$ and $f : \mathbb{R}^n \to \mathbb{R} \cup \{+\infty\}$ is a convex function such that ∇f transports[3] μ_0, then*

$$\mu_t = ((1-t)\mathbf{id} + t\,\nabla f)_\#\mu_0 \in \mathcal{P}_{\mathrm{ac}}(\mathbb{R}^n),$$

for every $t \in [0,1)$, i.e., $\mu_t = \rho_t\,d\mathcal{L}^n$ where $\rho_t \in L^1(\mathbb{R}^n)$, $\rho_t \geq 0$ and $\int_{\mathbb{R}^n} \rho_t = 1$.

Remark 10.3 Notice that in the setting of Proposition 10.2 we may have $\mu_1 = (\nabla f)_\#\mu_0 \notin \mathcal{P}_{\mathrm{ac}}(\mathbb{R}^n)$. The sole condition that $\mu_0 \in \mathcal{P}_{\mathrm{ac}}(\mathbb{R}^n)$ is enough to guarantee that $\mu_t \in \mathcal{P}_{\mathrm{ac}}(\mathbb{R}^n)$ whenever $t \in [0,1)$.

Proof of Proposition 10.2   Let $f_t(x) = (1-t)|x|^2/2 + t\,f(x)$ for $x \in \mathbb{R}^n$. Since $\nabla f_t = (1-t)\mathbf{id} + t\,\nabla f$ we clearly have that ∇f_t transports μ and that $\mu_t = (\nabla f_t)_\#\,\mu$. If F is the set of the differentiability points of f, then F is also the set of differentiability points of f_t. Moreover, if $x_1, x_2 \in F$, then

$$|\nabla f_t(x_1) - \nabla f_t(x_2)|\,|x_1 - x_2| \geq \big(\nabla f_t(x_1) - \nabla f_t(x_2)\big) \cdot (x_1 - x_2)$$

$$= (1-t)|x_1 - x_2|^2 + t\big(\nabla f(x_1) - \nabla f(x_2)\big) \cdot (x_1 - x_2)$$

$$\geq (1-t)|x_1 - x_2|^2,$$

where in the last inequality we have used the monotonicity of ∇f, namely $(\nabla f(x_1) - \nabla f(x_2)) \cdot (x_1 - x_2) \geq 0$ for all $x_1, x_2 \in \mathbb{R}^n$ (see (5.1)). We thus find,

$$|\nabla f_t(x_1) - \nabla f_t(x_2)| \geq (1-t)\,|x_1 - x_2| \qquad \forall x_1, x_2 \in F.$$

Setting $S_t = (\nabla f_t)(F)$, this means that $(\nabla f_t)^{-1}$ defines a map on S_t with values in $\mathbb{R}^n$ such that $\mathrm{Lip}((\nabla f_t)^{-1}; S_t) \leq (1-t)^{-1}$. In particular, for every Borel set $E \subset \mathbb{R}^n$ we have

$$\mathcal{L}^n\big((\nabla f_t)^{-1}(E)\big) = \mathcal{L}^n\big((\nabla f_t)^{-1}(E \cap S_t)\big) \leq (1-t)^{-n}\mathcal{L}^n(E \cap S_t) \leq (1-t)^{-n}\mathcal{L}^n(E),$$

so that $\mathcal{L}^n\big((\nabla f_t)^{-1}(E)\big) = 0$ if $\mathcal{L}^n(E) = 0$. In particular,

$$\mu_t(E) = \int_{(\nabla f_t)^{-1}(E)} \rho_0(x)\,dx = 0 \qquad \text{if } \mathcal{L}^n(E) = 0,$$

showing that $\mu_t \ll \mathcal{L}^n$ for every $t \in (0,1)$.   □

---

[3]   That is, ∇f is defined μ_0-a.e. on $\mathbb{R}^n$.

Example 10.4 We now look at some examples of displacement interpolation. Given $\mu \in \mathcal{P}_{ac}(\mathbb{R}^n)$ we study the displacement interpolations

$$[\mu, \tau_{x_0}\mu]_t, \qquad [\mu, Q_{\#}\mu]_t, \qquad [\mu, \mu^\lambda]_t,$$

corresponding to $x_0 \in \mathbb{R}^n$, $Q \in \mathbf{O}(n)$, $\lambda > 0$, and to the measures $\tau_{x_0}\mu(E) = \mu(E - x_0)$, $Q_{\#}\mu(E) = \mu(Q^{-1}(E))$ and $\mu^\lambda(E) = \mu(E/\lambda)$ (E a Borel set in $\mathbb{R}^n$). It is easily seen that $f(x) = |x|^2/2 + x_0 \cdot x$ is such that ∇f transports $\mu = \rho \, d\mathcal{L}^n$ into $\tau_{x_0}\mu$, so that $t\,x + (1 - t)\,\nabla f(x) = x + t\,x_0$, and thus

$$[\mu, \tau_{x_0}\mu]_t = \rho_t(x)\,dx \qquad \text{with} \quad \rho_t(x) = \rho(x - t\,x_0).$$

In particular, the displacement interpolation between μ and its translation $\tau_{x_0}\mu$ is obtained by continuously translating μ, i.e.,

$$[\mu, \tau_{x_0}\mu]_t = \tau_{t\,x_0}\,\mu.$$

Similarly, $f(x) = \lambda\,|x|^2/2$ is such that ∇f transports $\mu = \rho\,d\mathcal{L}^n$ into μ^λ, so that $t\,x + (1 - t)\,\nabla f(x) = [t + (1 - t)\lambda]x$ and

$$[\mu, \mu^\lambda]_t = \rho_t(x)\,dx \qquad \text{with} \quad \rho_t(x) = \frac{\rho(x/\lambda(t))}{\lambda(t)^n}, \qquad \lambda(t) = t + (1 - t)\lambda.$$

In particular, the displacement interpolation between μ and its rescaling μ^λ is obtained by continuously rescaling μ, i.e.,

$$[\mu, \mu^\lambda]_t = \mu^{\lambda(t)}.$$

However, the displacement interpolation between μ and $Q_{\#}\mu$ is not obtained by a continuous rotation process of μ! Indeed, being a (non-trivial) rotation and being the gradient of a convex function are incompatible conditions. More precisely, let ∇f be the Brenier map between $\mu = \rho\,d\mathcal{L}^n$ and $Q_{\#}\mu$, then we have

$$\int_{\mathbb{R}^n} \varphi(\nabla f)\,\rho = \int_{\mathbb{R}^n} \varphi(Q[x])\,\rho(x)\,dx \qquad \forall \varphi \in C_c^0(\mathbb{R}^n).$$

Of course if we could find a convex function f such that $\nabla f(x) = Q[x]$ for μ-a.e. $x \in \mathbb{R}^n$, then ∇f would be the Brenier map from μ to $Q_{\#}\mu$: however in that case the condition $\nabla^2 f = Q$ would imply that $Q \in \mathbb{R}_{\text{sym}}^{n \times n} \cap \mathbf{O}(n)$ with $Q \geq 0$, and thus that $Q = \mathrm{Id}$ (indeed, $Q^*Q = \mathrm{Id}$ and $Q = Q^*$ give $Q^2 = \mathrm{Id}$, so that $Q \geq 0$ implies $Q = \mathrm{Id}$). So unless we are in the trivial case when $Q = \mathrm{Id}$, the Brenier map ∇f from μ to $Q_{\#}\mu$ cannot be the rotation defined by Q.

10.3 Displacement Convexity of Internal Energies

The goal of this section is proving the following theorem, which provides a large class of displacement convex internal energies.

Theorem 10.5   *Let $U : [0,\infty) \to \mathbb{R}$ be continuous, with $U(0) = 0$ and*

$$\lambda \in (0,\infty) \mapsto \lambda^n U(\lambda^{-n}) \text{ is a convex decreasing function.} \qquad (10.16)$$

Let $\mathcal{K} \subset \mathcal{P}_{\mathrm{ac}}(\mathbb{R}^n)$ satisfy:

(i) *if $\mu_0, \mu_1 \in \mathcal{K}$, then $\mu_t \in \mathcal{K}$ for every $t \in (0,1)$, where $\mu_t = \rho_t \, d\mathcal{L}^n$ is the displacement interpolation from μ_0 to μ_1;*

(ii) *either "$\mathcal{U}(\rho)$ is well-defined in $\mathbb{R} \cup \{+\infty\}$ for every $\mu = \rho \, d\mathcal{L}^n \in \mathcal{K}$" or "$\mathcal{U}(\rho)$ is well-defined in $\mathbb{R} \cup \{-\infty\}$ for every $\mu = \rho \, d\mathcal{L}^n \in \mathcal{K}$."*

Then for every $\mu_0, \mu_1 \in \mathcal{K}$,

$$t \mapsto \mathcal{U}(\rho_t) \text{ is a convex function on } [0,1], \qquad (10.17)$$

where $\mu_t = \rho_t \, d\mathcal{L}^n$ is the displacement interpolation from μ_0 to μ_1.

Remark 10.6   When U is as in (10.3) and (10.4), then assumptions (i) and (ii) are satisfied by $\mathcal{K} = \mathcal{P}_{\mathrm{ac}}(\mathbb{R}^n)$. Indeed, (i) follows by Proposition 10.2, while (ii) is trivially true since $U \geq 0$ on $[0,\infty)$ if (10.3) holds, while $U \leq 0$ on $[0,\infty)$ if (10.4) holds. We claim that, when $U(\rho) = \rho \log \rho$, then $\mathcal{K} = \mathcal{P}_{p,\mathrm{ac}}(\mathbb{R}^n)$ (with $p \geq 1$) satisfies (i) and (ii). The validity of (ii) follows by (10.5), while the fact that $\rho_t \, d\mathcal{L}^n \in \mathcal{P}_{p,\mathrm{ac}}(\mathbb{R}^n)$ if $\mu_0, \mu_1 \in \mathcal{P}_{p,\mathrm{ac}}(\mathbb{R}^n)$ is proved by combining Proposition 10.2 with the following argument: if ∇f is the Brenier map from μ_0 to μ_1 and $f_t(x) = (1-t)|x|^2/2 + t f(x)$, then ∇f_t is the Brenier map from μ_0 to μ_t, so that

$$\int_{\mathbb{R}^n} |x|^p \, \rho_t(x) \, dx = \int_{\mathbb{R}^n} |\nabla f_t(x)|^p \, \rho_0(x) \, dx$$
$$\leq C(p) \left\{ \int_{\mathbb{R}^n} |x|^p \, \rho_0(x) \, dx + \int_{\mathbb{R}^n} |\nabla f(x)|^p \, \rho_0(x) \, dx \right\}$$
$$= C(p) \left\{ \int_{\mathbb{R}^n} |x|^p \, \rho_0(x) \, dx + \int_{\mathbb{R}^n} |x|^p \, \rho_1(x) \, dx \right\} < \infty,$$

thus proving that $\mathcal{P}_{p,\mathrm{ac}}(\mathbb{R}^n)$ is closed under displacement interpolation.

Proof of Theorem 10.5   Step one: We first show that if $A \in \mathbb{R}^{n \times n}_{\mathrm{sym}}$, $A \geq 0$, then

$$t \in [0,1] \mapsto \det\left((1-t)\,\mathrm{Id} + t\,A\right)^{1/n}$$

defines a concave function, which is strictly concave unless $A = \lambda\,\mathrm{Id}$ for some $\lambda \geq 0$. Indeed, $A = \mathrm{diag}(\lambda_i)$ with $\lambda_i \geq 0$ in a suitable orthonormal basis of $\mathbb{R}^n$. Therefore, setting $s = t/(1-t)$, we find that

$$\varphi(t) = \det\left((1-t)\,\mathrm{Id} + t\,A\right)^{1/n} = \prod_{i=1}^{n}\left((1-t) + t\,\lambda_i\right)^{1/n}.$$

$$= (1-t)\prod_{i=1}^{n}(1 + s\,\lambda_i)^{1/n}.$$

$$\geq (1-t)\left\{1 + \prod_{i=1}^{n}(s\,\lambda_i)^{1/n}\right\} = (1-t)\,\varphi(0) + t\,\varphi(1),$$

where in the inequality we have used

$$\prod_{i=1}^{n}(1 + a_i)^{1/n} \geq 1 + \prod_{i=1}^{n}a_i^{1/n} \qquad \forall a_i \geq 0. \tag{10.18}$$

Inequality (10.18) is obtained by applying the arithmetic–geometric mean inequality (9.1) to the sets of nonnegative numbers $\{1/(1+a_i)\}_{i=1}^{n}$ and $\{a_i/(1+a_i)\}_{i=1}^{n}$: indeed in this way one finds

$$\prod_{i=1}^{n}\frac{1}{(1+a_i)^{1/n}} + \prod_{i=1}^{n}\frac{a_i^{1/n}}{(1+a_i)^{1/n}} \leq \frac{1}{n}\sum_{i=1}^{n}\frac{1}{1+a_i} + \frac{1}{n}\sum_{i=1}^{n}\frac{a_i}{1+a_i} = 1.$$

Notice that equality holds in (10.18) if and only if there exists $a \geq 0$ such that $a_i = a$ for every i. In particular, if for some $t \in (0,1)$ we have $\varphi(t) = (1-t)\,\varphi(0) + t\,\varphi(1)$, then there exists $a \geq 0$ such that $a = t\,\lambda_i/(1-t)$ for every i, and thus $A = \lambda\,\mathrm{Id}$ for a constant $\lambda \geq 0$.

Step two: Now let $\mu_0 = \rho_0\,d\mathcal{L}^n$, $\mu_1 = \rho_1\,d\mathcal{L}^n$, let ∇f be the Brenier map from μ_0 to μ_1, and consider the convex function $f_t : \mathbb{R}^n \to \mathbb{R} \cup \{+\infty\}$ defined by

$$f_t(x) = (1-t)\frac{|x|^2}{2} + t\,f(x), \qquad x \in \mathbb{R}^n.$$

If F_1^t, F_2^t, $F_{2,\mathrm{inv}}^t$ denote the sets of differentiability, twice differentiability, and twice differentiability with positive Hessian of the convex function f_t, then for every $t \in (0,1)$ we have

$$F_1^t = F_1, \qquad F_{2,\mathrm{inv}}^t = F_2^t = F_2, \tag{10.19}$$

where as usual F_1 and F_2 are the corresponding sets for f. Since $|x|^2/2$ is smooth we trivially have $F_1^t = F_1$ and $F_2^t = F_2$. To complete the proof of (10.19) we notice that if $x \in F_2$ and $A = \nabla^2 f(x)$, then, in a suitable orthonormal basis, $A = \mathrm{diag}(\lambda_i)$ with $\lambda_i \geq 0$. Correspondingly, $(1-t)\mathrm{Id} + t\,A = \mathrm{diag}(1-t+t\,\lambda_i)$ with $1-t+t\,\lambda_i \geq 1-t > 0$, so that $(1-t)\mathrm{Id} + t\,A$ is invertible for every $t \in [0,1)$, thus showing $F_{2,\mathrm{inv}}^t = F_2^t$.

We prove (10.17). We notice that ∇f_t is the Brenier map from μ_0 to $\mu_t = (\nabla f_t)_\# \mu_0 = \mu_t$. We can thus apply Theorem 8.1 to μ_0 and μ_t and find that μ_0 is concentrated on a Borel set $Y_t \subset F_{2,\mathrm{inv}}^t = F_2$ such that

$$\rho_t(\nabla f_t) = \frac{\rho_0}{\det \nabla^2 f_t} \qquad \text{on } Y_t, \tag{10.20}$$

$$\int_{(\nabla f_t)(Y_t)} \varphi = \int_{Y_t} \varphi(\nabla f_t) \det \nabla^2 f_t, \tag{10.21}$$

for every Borel function $\varphi : \mathbb{R}^n \to [0, \infty]$. Thanks to assumptions (i) and (ii) we have that either $\varphi = U(\rho_t)^+$ or $\varphi = U(\rho_t)^-$ has finite integral, therefore we can first write (10.21) with $\varphi = U(\rho_t)$, and then apply (10.20), to find that

$$\int_{(\nabla f_t)(Y_t)} U(\rho_t) = \int_{Y_t} U\left(\frac{\rho_0}{\det \nabla^2 f_t}\right) \det \nabla^2 f_t \tag{10.22}$$

holds as an identity between extended real numbers. By $(\nabla f_t)_\# \mu_0 = \mu_t$ we have

$$\mu_t(\mathbb{R}^n \setminus (\nabla f_t)(Y_t)) \le \mu_0(\mathbb{R}^n \setminus Y_t) = 0,$$

i.e., $\rho_t = 0$ $\mathcal{L}^n$-a.e. on $\mathbb{R}^n \setminus (\nabla f_t)(Y_t)$, and since $U(0) = 0$ we conclude that

$$\int_{\mathbb{R}^n} U(\rho_t) = \int_{(\nabla f_t)(Y_t)} U(\rho_t). \tag{10.23}$$

Now, $Y_t \subset F_2$ and $\rho_0 = 0$ $\mathcal{L}^n$-a.e. on F_2, so that exploiting again $U(0) = 0$ we find that

$$\int_{Y_t} U\left(\frac{\rho_0}{\det \nabla^2 f_t}\right) \det \nabla^2 f_t = \int_{F_2} U\left(\frac{\rho_0}{\det \nabla^2 f_t}\right) \det \nabla^2 f_t, \tag{10.24}$$

where the integrand is well-defined also on F_2 since $F_{2,\text{inv}}^t = F_2$. By combining (10.22), (10.23) and (10.24) we conclude that

$$\mathcal{U}(\rho_t) = \int_{F_2} U\left(\frac{\rho_0}{\det \nabla^2 f_t}\right) \det \nabla^2 f_t,$$

as extended real numbers. For every $x \in F_2$ we now set

$$\lambda_x(t) = (\det \nabla^2 f_t(x))^{1/n}, \qquad t \in [0, 1],$$

$$\Phi_x(\lambda) = \lambda^n U\left(\frac{\rho_0(x)}{\lambda^n}\right), \qquad \lambda > 0,$$

and $\psi_x(t) = \Phi_x(\lambda_x(t))$. By step one, λ_x is concave on $[0, 1]$, and thus $\lambda_x(t) \ge (1 - t)\lambda_x(0) + t\lambda_x(1)$; therefore by exploiting first that Φ_x is decreasing and then that it is convex, we find that

$$\psi_x(t) = \Phi_x(\lambda_x(t))$$
$$\le \Phi_x((1 - t)\lambda_x(0) + t\lambda_x(1)) \tag{10.25}$$
$$\le (1 - t)\Phi_x(\lambda_x(0)) + t\Phi_x(\lambda_x(1)) \tag{10.26}$$
$$= (1 - t)\psi_x(0) + t\psi_x(1),$$

so that $\mathcal{U}(\rho_t) = \int_{F_2} \psi_x(t)\, dx$ is convex on $[0, 1]$, as claimed. $\qquad \square$

Remark 10.7 (Strict convexity and internal energy) It is interesting to see what happens if assumption (10.16) in Theorem 10.5 is strengthened to

$$\lambda \in (0, \infty) \mapsto \lambda^n U(\lambda^{-n}) \text{ is a } \textit{strictly} \text{ convex decreasing function.} \quad (10.27)$$

Under this assumption we have that if ρ_0 and ρ_1 are such that $\mathcal{U}(\rho_t) = (1 - t)\mathcal{U}(\rho_0) + t\mathcal{U}(\rho_1)$ for some $t \in [0, 1]$, then

$$\nabla^2 f(x) = \text{Id for } \mathcal{L}^n\text{-a.e. } x \in F_2. \quad (10.28)$$

Indeed, let $x \in F_2$ be such that $(1 - t)\psi_x(0) + t\psi_x(1) = \psi_x(t)$. Since Φ_x is strictly convex and decreasing, and thus also strictly decreasing, equality in (10.25) implies that $(1 - t)\lambda_x(0) + t\lambda_x(1) = \lambda_x(t)$, and thus by step one that $\nabla^2 f(x) = \sigma(x)\,\text{Id}$ for some $\sigma(x) \geq 0$. Equality in (10.26) implies that $\lambda_x(0) = \lambda_x(1)$, where

$$\lambda_x(0) = \det(\text{Id})^{1/n} = 1, \qquad \lambda_x(1) = \det(\nabla^2 f(x))^{1/n} = \sigma(x),$$

so that $\sigma(x) = 1$; this proves (10.28). Now F_2 is $\mathcal{L}^n$-equivalent to $\{\rho_0 > 0\}$. In a situation where $\{\rho_0 > 0\}$ is an open connected set, then (10.28) implies the existence of $x_0 \in \mathbb{R}^n$ such that $\nabla f(x) = x + x_0$ for every $x \in \{\rho_0 > 0\}$, and thus that $\rho_1 = \tau_{x_0}\rho_0$. If we just assume that $\{\rho_0 > 0\}$ is open, but possibly disconnected, then different connected components of $\{\rho_0 > 0\}$ may be compatible with different translations, and an additional argument (tailored on the specific problem under consideration) may be needed to conclude that $\rho_1 = \tau_{x_0}\rho_0$.

10.4 The Brunn–Minkowski Inequality

We close this chapter with an application of Theorem 10.5 to geometric inequalities, and give, in particular, a proof of the Brunn–Minkowski inequality. Let us recall that the **Minkowski sum of E_0 and E_1** subsets of $\mathbb{R}^n$ is defined as

$$E_0 + E_1 = \{x + y : x \in E_0, y \in E_1\}.$$

The **Brunn–Minkowski inequality** provides a lower bound on the volume of $E_0 + E_1$ in terms of the volumes of E_0 and E_1:

$$|E_0 + E_1|^{1/n} \geq |E_0|^{1/n} + |E_1|^{1/n}, \qquad \forall E_0, E_1 \in \mathcal{B}(\mathbb{R}^n). \quad (10.29)$$

This inequality, which at first sight may look a bit abstract, plays a central role in many mathematical problems. As an indication of its pervasiveness, we analyze its meaning in the special case when $E_1 = B_r(0)$. In this case, setting for brevity $E_0 = E$, we have that $E_0 + E_1 = I_r(E)$ is the open r-neighborhood of E, so that (10.29) gives a very general lower bound on the volume of $|I_r(E)|$.

$$|I_r(E)| \geq (|E|^{1/n} + r\,|B_1|^{1/n})^n \qquad \forall r > 0. \qquad (10.30)$$

Notice that equality holds in (10.30) if $E = B_R(x)$ for some $x \in \mathbb{R}^n$ and $R > 0$, a fact that suggest a link between (10.30) and Euclidean isoperimetry. Indeed, if E is a bounded open set with smooth boundary, then for r sufficiently small (in terms of the curvatures of ∂E) we have

$$|I_r(E)| = |E| + r\,P(E) + \mathrm{O}(r^2),$$

so that (10.30) implies

$$
\begin{aligned}
P(E) &= \lim_{r \to 0^+} \frac{|I_r(E)| - |E|}{r} \\
&\geq \lim_{r \to 0^+} \frac{(|E|^{1/n} + r\,|B_1|^{1/n})^n - |E|}{r} = n\,|B_1|^{1/n}\,|E|^{(n-1)/n},
\end{aligned}
$$

which is the sharp Euclidean isoperimetric inequality (9.7).

Proof of the Brunn–Minkowski inequality (10.29) Without loss of generality we can assume that $|E_0 + E_1| < +\infty$ and that $|E_0|, |E_1| > 0$. Since both E_0 and E_1 are non-empty, $E_0 + E_1$ contains both a translation of E_0 and of E_1, and thus we also have $|E_0|, |E_1| < \infty$. Therefore, the densities

$$\rho_0 = \frac{1_{E_0}}{|E_0|}, \qquad \rho_1 = \frac{1_{E_1}}{|E_1|},$$

define probability measures $\mu_0, \mu_1 \in \mathcal{P}_{\mathrm{ac}}(\mathbb{R}^n)$, and the internal energy

$$\mathcal{U}(\rho) = -\int_{\mathbb{R}^n} \rho(x)^{1-(1/n)}\,dx,$$

corresponding to $U(\rho) = -\rho^{1-(1/n)}$, $\rho > 0$, is such that

$$\mathcal{U}(\rho_0) = -|E_0|^{1/n}, \qquad \mathcal{U}(\rho_1) = -|E_1|^{1/n}.$$

Now let ∇f be the Brenier map from μ_0 to μ_1, so that the displacement interpolation μ_t from μ_0 to μ_1 satisfies $\mu_t = \rho_t\,d\mathcal{L}^n = (\nabla f_t)_\# \mu_0$ where $\nabla f_t = (1 - t)\,\mathbf{id} + t\,\nabla f$. By Theorem 8.1, and denoting by F the set of differentiability points of f, we have

$$\exists X_1 \subset E_0 \cap F \text{ s.t. } \mu_0 \text{ concentrated on } X_1 \text{ and } \nabla f(X_1) \subset E_1.$$

In particular, $\mu_t = (\nabla f_t)_\# \mu_0$ and the fact that F is also the set of differentiability points of f_t give

$$\mu_t \text{ is concentrated on } \nabla f_t(X_1).$$

By combining these two facts we find that for $\mathcal{L}^n$-a.e. $y \in \{\rho_t > 0\}$ there exists $x \in X_1$ such that $y = \nabla f_t(x) = (1-t)\, x + t\, \nabla f(x) \in (1-t)\, E_0 + t\, E_1$, which of course implies

$$|\{\rho_t > 0\}| \le |(1-t)\, E_0 + t\, E_1|. \tag{10.31}$$

We now take $t = 1/2$. By $|E_0 + E_1| < +\infty$ we have $|\{\rho_{1/2} > 0\}| < \infty$, while by displacement convexity of $\mathcal{U}$, recall Theorem 10.5, we find

$$
\begin{aligned}
-\frac{|E_0|^{1/n} + |E_1|^{1/n}}{2} &= \frac{\mathcal{U}(\rho_0) + \mathcal{U}(\rho_1)}{2} \\
&\ge \mathcal{U}(\rho_{1/2}) = \int_{\{\rho_{1/2} > 0\}} U(\rho_{1/2}(x))\, dx \\
&\ge |\{\rho_{1/2} > 0\}|\, U\!\left(\frac{1}{|\{\rho_{1/2} > 0\}|} \int_{\{\rho_{1/2}>0\}} \rho_{1/2}(x)\, dx\right) \\
&= -|\{\rho_{1/2} > 0\}|^{1/n},
\end{aligned}
\tag{10.32}
$$

where in the last inequality we have used Jensen's inequality, the convexity of $U(\rho)$ and the finiteness of $|\{\rho_{1/2} > 0\}|$. By combining (10.31) and (10.32) we obtain the Brunn–Minkowski inequality. $\square$

11

The Wasserstein Distance $\mathbf{W}_2$ on $\mathcal{P}_2(\mathbb{R}^n)$

In this chapter we introduce the $(L^2\text{-})$ **Wasserstein space** $(\mathcal{P}_2(\mathbb{R}^n), \mathbf{W}_2)$, which is the set of probability measures with finite second moments, $\mathcal{P}_2(\mathbb{R}^n)$, endowed with the $(L^2\text{-})$ **Wasserstein distance** $\mathbf{W}_2(\mu, \nu) = \mathbf{K}_2(\mu, \nu)^{1/2}$. We prove the completeness of $(\mathcal{P}_2(\mathbb{R}^n), \mathbf{W}_2)$, that the $\mathbf{W}_2$-convergence is equivalent to narrow convergence of Radon measures plus convergence of the second moments, and that displacement interpolation can be interpreted as "geodesic interpolation" in $(\mathcal{P}_2(\mathbb{R}^n), \mathbf{W}_2)$. The Wasserstein space is a central object in OMT, in all its applications to PDE, and in those to Riemannian and metric geometry.

11.1 Displacement Interpolation and Geodesics in $(\mathcal{P}_2(\mathbb{R}^n), \mathbf{W}_2)$

We begin by showing that every transport plan $\gamma \in \Gamma(\mu_0, \mu_1)$ induces a Lipschitz curve in $(\mathcal{P}_2(\mathbb{R}^n), \mathbf{W}_2)$ with end-points at μ_0 and μ_1, and that optimal transport plans saturate the corresponding Lipschitz bounds. As usual we set $\mathbf{p}(x, y) = x$ and $\mathbf{q}(x, y) = y$ for the standard projection operators on $\mathbb{R}^n \times \mathbb{R}^n$.

Proposition 11.1 *If $\mu_0, \mu_1 \in \mathcal{P}_2(\mathbb{R}^n)$ and $\gamma \in \Gamma(\mu_0, \mu_1)$, then*

$$\mu_t = ((1 - t)\mathbf{p} + t\mathbf{q})_{\#}\gamma, \qquad t \in [0, 1], \tag{11.1}$$

defines a Lipschitz curve in $(\mathcal{P}_2(\mathbb{R}^n), \mathbf{W}_2)$, in the sense that

$$\mathbf{W}_2(\mu_t, \mu_s) \le L|t - s|, \qquad \forall t, s \in [0, 1], \tag{11.2}$$

where L is the quadratic transport cost of γ, i.e.,

$$L = \left(\int_{\mathbb{R}^n \times \mathbb{R}^n} |x - y|^2 \, d\gamma(x, y) \right)^{1/2}.$$

If in addition γ is an optimal transport plan in $\mathbf{K}_2(\mu_0, \mu_1)$, then μ_t is called the **displacement interpolation from μ_0 to μ_1 induced by γ,** *and satisfies*

$$\mathbf{W}_2(\mu_t, \mu_s) = \mathbf{W}_2(\mu_0, \mu_1) \, |t - s| \qquad \forall t, s \in [0, 1]. \tag{11.3}$$

Remark 11.2 The terminology just introduced is consistent with the one introduced in (10.12). Indeed, if $\mu_0, \mu_1 \in \mathcal{P}_{2, \mathrm{ac}}(\mathbb{R}^n)$, then by Theorem 4.2 there exists a unique optimal transport plan γ in $\mathbf{K}_2(\mu_0, \mu_1)$, induced by the Brenier map ∇f from μ_0 to μ_1 through the formula $\gamma = (\mathbf{id} \times \nabla f)_{\#} \mu_0$, and with this choice of γ it is easily seen that

$$\left((1 - t)\, \mathbf{p} + t\, \mathbf{q}\right)_{\#} \gamma = \left((1 - t)\, \mathbf{id} + t\, \nabla f\right)_{\#} \mu_0.$$

The main difference between the case when $\mu_0, \mu_1 \in \mathcal{P}_{2, \mathrm{ac}}(\mathbb{R}^n)$ and the general case is simply that, in the general case, there could be more than one displacement interpolation between μ_0 and μ_1. This fact should be understood in analogy with the possible occurrence of multiple geodesics with same end-points in a Riemannian manifold (see also Remark 11.3).

Proof of Proposition 11.1   Let $\gamma \in \Gamma(\mu_0, \mu_1)$. If $0 \le t < s \le 1$ and $\mathbf{p}_t = (1 - t)$ $\mathbf{p} + t\, \mathbf{q} : \mathbb{R}^n \times \mathbb{R}^n \to \mathbb{R}^n$, then the map $(\mathbf{p}_t, \mathbf{p}_s) : \mathbb{R}^n \times \mathbb{R}^n \to \mathbb{R}^n \times \mathbb{R}^n$ can be used to define a plan

$$\gamma_{t,s} = (\mathbf{p}_t, \mathbf{p}_s)_{\#} \gamma \in \Gamma(\mu_t, \mu_s).$$

Therefore,

$$\mathbf{K}_2(\mu_t, \mu_s) \le \int_{\mathbb{R}^n \times \mathbb{R}^n} |x - y|^2 \, d\gamma_{t,s} = \int_{\mathbb{R}^n \times \mathbb{R}^n} |\mathbf{p}_t(x, y) - \mathbf{p}_s(x, y)|^2 \, d\gamma$$

$$= |t - s|^2 \int_{\mathbb{R}^n \times \mathbb{R}^n} |x - y|^2 \, d\gamma = |t - s|^2 \, L^2,$$

that is (11.2). If in addition γ is optimal in $\mathbf{K}_2(\mu_0, \mu_1)$, then we have proved that

$$\mathbf{W}_2(\mu_t, \mu_s) \le \mathbf{W}_2(\mu_0, \mu_1) \, |t - s|, \qquad \forall t, s \in [0, 1].$$

By applying this inequality on the intervals $[0, t]$, $[t, s]$ and $[s, 1]$, and by exploiting the triangular inequality in $(\mathcal{P}_2(\mathbb{R}^n), \mathbf{W}_2)$ (which is proved in Theorem 11.10 without making use of this proposition), we thus find that

$$\mathbf{W}_2(\mu_0, \mu_1) \le \mathbf{W}_2(\mu_0, \mu_t) + \mathbf{W}_2(\mu_t, \mu_s) + \mathbf{W}_2(\mu_s, \mu_1)$$

$$\le (t + (s - t) + (1 - s)) \, \mathbf{W}_2(\mu_0, \mu_1) = \mathbf{W}_2(\mu_0, \mu_1),$$

thus showing the validity of (11.3). $\qquad \square$

Remark 11.3 (Displacement interpolation as geodesic interpolation)   The notion of length of a smooth curve $\gamma : [0, 1] \to M$ on a Riemannian manifold

$$\ell(\gamma) = \int_0^1 \sqrt{g_{\gamma(t)}(\gamma'(t), \gamma'(t))} \, dt = \int_0^1 |\gamma'(t)|_{g_{\gamma(t)}} \, dt,$$

can be used, when M is path-connected, to define a metric d_g on M,

$$d_g(p,q) = \inf \{\ell(\gamma) : \gamma(0) = p, \gamma(1) = q\}. \tag{11.4}$$

Minimizers in $d_g(p,q)$ always exist, are possibly non-unique, and are called **minimizing geodesics**. More generally, if ∇ denotes the Levi-Civita connection of (M,g), then **geodesics** are defined as solutions of $\nabla_{\gamma'}\gamma' = 0$, i.e., as *critical points* of the length functional (among curves with fixed ends). The notion of minimizing geodesics is usually more restrictive than that of geodesic (as illustrated by a pair of complementary arcs with non-antipodal end-points inside an equatorial circle: both arcs are geodesics, but only one is a minimizing geodesic). A geodesic, as every other curve in M, can be reparametrized so to make its speed $|\gamma'|_{g_\gamma}$ constant, and constant speed geodesics satisfy

$$d_g(\gamma(t),\gamma(s)) = |t - s|\, d_g(\gamma(0),\gamma(1)) \qquad \forall t,s \in [0,1]. \tag{11.5}$$

Viceversa, if γ is a geodesic in M and (11.5) holds, then γ has constant speed. Generalizing on these considerations, it is customary to say that a constant speed geodesic in a metric space (X,d) is any curve $\gamma : [0,1] \to X$ such that $d(\gamma(t),\gamma(s)) = |t-s|\, d(\gamma(0),\gamma(1))$ for every $t,s \in [0,1]$. With this convention, in view of Proposition 11.1, displacement interpolation is the (constant speed) geodesic interpolation on $(\mathcal{P}_2(\mathbb{R}^n),\mathbf{W}_2)$.

11.2 Some Basic Remarks about $\mathbf{W}_2$

Remark 11.4 ($\mathbf{W}_2$ "extends" the Euclidean distance from $\mathbb{R}^n$ to $\mathcal{P}_2(\mathbb{R}^n)$)   Setting $d(x,y) = |x - y|$ for $x,y \in \mathbb{R}^n$, we see that $(\mathbb{R}^n,d)$ embeds isometrically into $(\mathcal{P}_2(\mathbb{R}^n),\mathbf{W}_2)$ through the map $x \in \mathbb{R}^n \mapsto \delta_x$. Indeed, if $x_0,x_1 \in \mathbb{R}^n$, then

$$\mathbf{W}_2(\delta_{x_0},\delta_{x_1}) = |x_0 - x_1|. \tag{11.6}$$

Remark 11.5 ($\mathbf{W}_2$ is lower semicontinuous under weak-star convergence) Indeed, let $\mu_j,\mu,\nu_j,\nu \in \mathcal{P}_2(\mathbb{R}^n)$ be such that $\mu_j \overset{*}{\rightharpoonup} \mu$ and $\nu_j \overset{*}{\rightharpoonup} \nu$ as $j \to \infty$ (and thus narrowly, since $\mu(\mathbb{R}^n) = \nu(\mathbb{R}^n) = 1$): we claim that

$$\mathbf{W}_2(\mu,\nu) \le \liminf_{j\to\infty} \mathbf{W}_2(\mu_j,\nu_j). \tag{11.7}$$

Up to extracting a subsequence we can assume the liminf is a limit. Moreover, if γ_j is optimal in $\mathbf{K}_2(\mu_j,\nu_j)$, then up to extracting a further subsequence $\gamma_j \overset{*}{\rightharpoonup} \gamma$ as $j \to \infty$ for some Radon measure γ on $\mathbb{R}^n \times \mathbb{R}^n$. As usual, given the narrow convergence of μ_j and ν_j, we have that $\gamma \in \Gamma(\mu,\nu)$ and that the convergence of γ_j to γ is narrow. Hence, given $\varphi \in C_c^0(\mathbb{R}^n \times \mathbb{R}^n)$ with $0 \le \varphi \le 1$ we find

$$\mathbf{K}_2(\mu,\nu) \le \int_{\mathbb{R}^n\times\mathbb{R}^n} |x-y|^2\, d\gamma = \sup_\varphi \int_{\mathbb{R}^n\times\mathbb{R}^n} |x-y|^2\, \varphi\, d\gamma$$

$$= \sup_\varphi \lim_{j\to\infty} \int_{\mathbb{R}^n\times\mathbb{R}^n} |x-y|^2\, \varphi\, d\gamma_j \le \lim_{j\to\infty} \mathbf{K}_2(\mu_j,\nu_j).$$

Notice that the inequality in (11.7) can be strict: for example, with $\mathbb{R}^n = \mathbb{R}$, if $\mu = \nu = \delta_0$ and $\mu_j = (1 - (1/j))\,\delta_0 + (1/j)\,\delta_j$, then $\mu_j \overset{*}{\rightharpoonup} \mu$ as $j \to \infty$, but

$$\mathbf{W}_2(\mu_j,\mu)^2 = \int_{\mathbb{R}} |x|^2\, d\mu_j(x) = j \to \infty.$$

Remark 11.6 (W$_2$ is a contraction under ε-regularization) Precisely, if $\mu,\nu \in \mathcal{P}_2(\mathbb{R}^n)$, $\mu_\varepsilon = \mu \star \rho_\varepsilon$ and $\nu_\varepsilon = \nu \star \rho_\varepsilon$ for a standard regularizing kernel ρ_ε, then

$$\mathbf{W}_2(\mu,\nu) \ge \mathbf{W}_2(\mu_\varepsilon\, d\mathcal{L}^n, \nu_\varepsilon\, d\mathcal{L}^n), \tag{11.8}$$

$$\mathbf{W}_2(\mu,\nu) = \lim_{\varepsilon\to 0^+} \mathbf{W}_2(\mu_\varepsilon\, d\mathcal{L}^n, \nu_\varepsilon\, d\mathcal{L}^n). \tag{11.9}$$

Indeed, if γ is an optimal plan for $\mathbf{W}_2(\mu,\nu)$ and $\gamma^{(\varepsilon)} \in \mathcal{P}(\mathbb{R}^n\times\mathbb{R}^n)$ is defined by

$$\int_{\mathbb{R}^n\times\mathbb{R}^n} \varphi\, d\gamma^{(\varepsilon)} = \int_{\mathbb{R}^n\times\mathbb{R}^n} d\gamma(x,y) \int_{\mathbb{R}^n} \varphi(x+z,y+z)\, \rho_\varepsilon(z)\, dz, \tag{11.10}$$

for every $\varphi \in C_c^0(\mathbb{R}^n\times\mathbb{R}^n)$, then $\gamma^{(\varepsilon)} \in \mathcal{P}(\mathbb{R}^n\times\mathbb{R}^n)$ and, for every $\varphi \in C_c^0(\mathbb{R}^n)$,

$$\int_{\mathbb{R}^n\times\mathbb{R}^n} \varphi(x)\, d\gamma^{(\varepsilon)}(x,y) = \int_{\mathbb{R}^n\times\mathbb{R}^n} d\gamma(x,y) \int_{\mathbb{R}^n} \varphi(x+z)\, \rho_\varepsilon(z)\, dz$$

$$= \int_{\mathbb{R}^n} d\mu(x) \int_{\mathbb{R}^n} \varphi(x+z)\, \rho_\varepsilon(z)\, dz$$

$$= \int_{\mathbb{R}^n} d\mu(x) \int_{\mathbb{R}^n} \varphi(w)\, \rho_\varepsilon(w-x)\, dw = \int_{\mathbb{R}^n} \varphi(\mu \star \rho_\varepsilon).$$

Thus $\gamma^{(\varepsilon)} \in \Gamma(\mu_\varepsilon\, d\mathcal{L}^n, \nu_\varepsilon\, d\mathcal{L}^n)$. Since $\mu_\varepsilon\, d\mathcal{L}^n$ and $\nu_\varepsilon\, d\mathcal{L}^n$ belong to $\mathcal{P}_2(\mathbb{R}^n)$, (11.10) holds with $\varphi(x,y) = |x-y|^2$, and thus

$$\mathbf{W}_2(\mu_\varepsilon\, d\mathcal{L}^n, \nu_\varepsilon\, d\mathcal{L}^n)^2 \le \int_{\mathbb{R}^n\times\mathbb{R}^n} |x-y|^2\, d\gamma^{(\varepsilon)}(x,y)$$

$$= \int_{\mathbb{R}^n\times\mathbb{R}^n} d\gamma(x,y) \int_{\mathbb{R}^n} |x+z-(y+z)|^2\, \rho_\varepsilon(z)\, dz$$

$$= \int_{\mathbb{R}^n\times\mathbb{R}^n} |x-y|^2\, d\gamma(x,y) = \mathbf{W}_2(\mu,\nu)^2.$$

This proves (11.8), and then (11.9) follows by (11.7) and (11.8).

Example 11.7 (Wasserstein distance and translations) This example should be helpful in developing some geometric intuition about the Wasserstein distance. Given two measures $\mu_0,\mu_1 \in \mathcal{P}_2(\mathbb{R}^n)$, let us consider the Wasserstein distance $\mathbf{W}_2(\mu_0, \tau_{x_0}\mu_1)$ between μ_0 and the translation of μ_1 by $x_0 \in \mathbb{R}^n$, namely, $\tau_{x_0}\mu_1(E) = \mu_1(E - x_0)$, $E \in \mathcal{B}(\mathbb{R}^n)$. Given (11.6), we would expect this distance to grow linearly in $|x_0|$, at least in certain directions. We have indeed that

$$\mathbf{W}_2(\mu_0, \tau_{x_0}\mu_1) \geq \sqrt{\mathbf{W}_2(\mu_0, \mu_1)^2 + |x_0|^2}, \quad \text{if} \quad x_0 \cdot (\mathrm{bar}\mu_1 - \mathrm{bar}\mu_0) \geq 0,$$
$$(11.11)$$

where $\mathrm{bar}\mu = \int_{\mathbb{R}^n} x \, d\mu(x)$ denotes the barycenter of μ. Indeed, let us first assume that $\mu_0, \mu_1 \ll \mathcal{L}^n$, and let ∇f be the Brenier map from μ_0 to μ_1. Since $x_0 + \nabla f$ is the gradient of a convex function and pushes forward μ_0 to $\tau_{x_0}\mu_1$, we have that $x_0 + \nabla f$ is the Brenier map from μ_0 to $\tau_{x_0}\mu_1$ and thus

$$\mathbf{W}_2(\mu_0, \tau_{x_0}\mu_1)^2 = \int_{\mathbb{R}^n} |x_0 + \nabla f(x) - x|^2 \, d\mu_0(x)$$

$$= \mathbf{W}_2(\mu_0, \mu_1)^2 + |x_0|^2 + 2x_0 \cdot \int_{\mathbb{R}^n} (\nabla f(x) - x) \, d\mu_0(x)$$

where $\int_{\mathbb{R}^n} \nabla f(x) \, d\mu_0(x) = \mathrm{bar}\mu_1$ by $(\nabla f)_{\#}\mu_0 = \mu_1$; this proves (11.11) in the absolutely continuous case. In the general case, since $x \in \mathbb{R}^n \mapsto \mathbf{W}_2(\mu_0, \tau_x \mu_1)$ is lower semicontinuous thanks to (11.7), it is enough to prove (11.11) for x_0 such that $x_0 \cdot (\mathrm{bar}\mu_1 - \mathrm{bar}\mu_0) > 0$. Since for every $\mu \in \mathcal{P}_2(\mathbb{R}^n)$ we have

$$\mathrm{bar}[(\mu \star \rho_\varepsilon) \, d\mathcal{L}^n] \to \mathrm{bar}\mu \quad \text{as } \varepsilon \to 0^+,$$

we find that $x_0 \cdot (\mathrm{bar}\mu_1^\varepsilon - \mathrm{bar}\mu_0^\varepsilon) > 0$ if $\mu_k^\varepsilon = (\mu_k \star \rho_\varepsilon) \, d\mathcal{L}^n$. Since $\mu_0^\varepsilon, \mu_1^\varepsilon \ll \mathcal{L}^n$ we find

$$\mathbf{W}_2(\mu_0^\varepsilon, \tau_{x_0}\mu_1^\varepsilon) \geq \sqrt{\mathbf{W}_2(\mu_0^\varepsilon, \mu_1^\varepsilon)^2 + |x_0|^2}$$

and then (11.11) follows by property (11.9) and by

$$\tau_{x_0}\mu_1^\varepsilon = (\tau_{x_0}\mu_1) \star \rho_\varepsilon \, d\mathcal{L}^n.$$

Remark 11.8 (Non-convexity of $\mathbf{W}_2$ with respect to displacement interpolation) The distance $|x - x_0|$ from a fixed point $x_0 \in \mathbb{R}^n$ defines a convex function in $\mathbb{R}^n$: in analogy, one may think that if $\nu \in \mathcal{P}_2(\mathbb{R}^n)$ and μ_t is the displacement interpolation defined by an optimal transport plan $\gamma \in \Gamma(\mu_0, \mu_1)$, then the function $t \mapsto \mathbf{W}_2(\mu_t, \nu)$ is convex on $t \in [0,1]$: but in general this is not true. The right analogy is with the situation seen in Riemannian manifolds, where the failing of this property is related to the non-uniqueness of minimizing geodesics. For example, on $\mathbb{S}^1 \subset \mathbb{R}^2$, set $\nu = (1,0)$ and, say, $\mu_t = (\cos(t\pi + \pi/2), \sin(t\pi + \pi))$, $0 < t < 1$. Notice that if d denotes the geodesic distance on $\mathbb{S}^1$, then $d(\mu_t, \nu) = t\pi + \pi/2$ for $t \in (0, 1/2]$ and $d(\mu_t, \nu) = (3/2)\pi - t\pi$ if $t \in [1/2, 1)$, so that $t \in (0,1) \mapsto d(\mu_t, \nu)$ is indeed

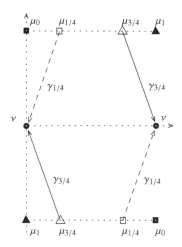

Figure 11.1 The measures μ_0, μ_1 and ν are depicted, respectively, by black squares, triangles, and disks. The displacement interpolations μ_t for $t = 1/4$ and $t = 3/4$ are depicted by empty squares and empty triangles, respectively. For every $t \neq 1/2$ there exists a unique optimal transport plan γ_t in $\mathbf{K}_2(\mu_t, \nu)$, but at $t = 1/2$ there are of course two. Since $t \mapsto \mathbf{W}_2(\mu_t, \nu)$ has a strict maximum at $t = 1/2$, and thus is not convex.

concave on $(0, 1)$. As similar example in $(\mathcal{P}_2(\mathbb{R}^2), \mathbf{W}_2)$ is obtained by setting, with $L > 1$,

$$\mu_0 = \frac{1}{2}\delta_{(0,L)} + \frac{1}{2}\delta_{(1,-L)}, \quad \mu_1 = \frac{1}{2}\delta_{(1,L)} + \frac{1}{2}\delta_{(0,-L)}, \quad \nu = \frac{1}{2}\delta_{(0,0)} + \frac{1}{2}\delta_{(1,0)}.$$
$$(11.12)$$

In this way

$$\mu_t = \frac{1}{2}\delta_{(t,L)} + \frac{1}{2}\delta_{(1-t,-L)},$$

see Figure 11.1, and the optimal plan γ_t in $\mathbf{K}_2(\mu_t, \nu)$ is clearly given by

$$\gamma_t = \frac{1}{2}\delta_{[(t,L),(0,0)]} + \frac{1}{2}\delta_{[(1-t,-L),(1,0)]} \qquad \forall t \in (0, 1/2],$$
$$\gamma_t = \frac{1}{2}\delta_{[(t,L),(1,0)]} + \frac{1}{2}\delta_{[(1-t,-L),(0,0)]} \qquad \forall t \in [1/2, 1).$$

In particular,

$$\mathbf{W}_2(\mu_t, \nu)^2 = t^2 + L^2, \qquad\qquad \forall t \in (0, 1/2],$$
$$\mathbf{W}_2(\mu_t, \nu)^2 = (t-1)^2 + L^2, \qquad \forall t \in [1/2, 1),$$

and $t \mapsto \mathbf{W}_2(\mu_t, \nu)$, having a strict maximum at $t = 1/2$, is not convex.

Remark 11.9 (Convexity of $\mathbf{W}_2^2$ with respect to standard interpolation)   We finally notice that, if $\mu_0, \mu_1, \nu \in \mathcal{P}_2(\mathbb{R}^n)$, then

$$\mathbf{W}_2((1-t)\,\mu_0 + t\,\mu_1, \nu)^2 \le (1-t)\,\mathbf{W}_2(\mu_0, \nu)^2 + t\,\mathbf{W}_2(\mu_1, \nu)^2. \qquad (11.13)$$

Indeed, if $\gamma_0 \in \Gamma(\mu_0, \nu)$ and $\gamma_1 \in \Gamma(\mu_1, \nu)$ are optimal transport plans for the quadratic cost, then $\gamma_t = (1-t)\,\gamma_0 + t\,\gamma_1 \in \Gamma((1-t)\,\mu_0 + t\,\mu_1, \nu)$ and thus

$$\mathbf{W}_2((1-t)\,\mu_0 + t\,\mu_1, \nu)^2 \le \int_{\mathbb{R}^n \times \mathbb{R}^n} |x - y|^2 d\gamma_t$$

$$= (1-t) \int_{\mathbb{R}^n \times \mathbb{R}^n} |x - y|^2 d\gamma_0 + t \int_{\mathbb{R}^n \times \mathbb{R}^n} |x - y|^2 d\gamma_1$$

which is (11.13).

11.3 The Wasserstein Space $(\mathcal{P}_2(\mathbb{R}^n), \mathbf{W}_2)$

We finally come to the main result of this chapter.

Theorem 11.10   *The Wasserstein space* $(\mathcal{P}_2(\mathbb{R}^n), \mathbf{W}_2)$ *is a complete metric space. Moreover, given* $\mu_j, \mu \in \mathcal{P}_2(\mathbb{R}^n)$, $j \in \mathbb{N}$, *we have that*

$$\mathbf{W}_2(\mu_j, \mu) \to 0, \qquad as \; j \to \infty, \qquad (11.14)$$

if and only if, as $j \to \infty$,

$$\mu_j \overset{*}{\rightharpoonup} \mu \qquad as \; Radon \; measures \; on \; \mathbb{R}^n, \qquad (11.15)$$

$$\int_{\mathbb{R}^n} |x|^2 \, d\mu_j(x) \to \int_{\mathbb{R}^n} |x|^2 \, d\mu(x). \qquad (11.16)$$

In particular, if (11.14) *holds, then* μ_j *narrowly converges to* μ *as* $j \to \infty$.

Proof   In step one we prove that $(\mathcal{P}_2(\mathbb{R}^n), \mathbf{W}_2)$ is a metric space, while step two and three discuss the equivalence between (11.14) and (11.15)–(11.16). Finally, in step four, we show the completeness of $(\mathcal{P}_2(\mathbb{R}^n), \mathbf{W}_2)$.

Step one: We prove that $\mathbf{W}_2$ defines a metric on $\mathcal{P}_2(\mathbb{R}^n)$. We have already noticed that $\mathbf{K}_2(\mu, \nu) = \mathbf{K}_2(\nu, \mu)$, hence $\mathbf{W}_2$ is symmetric. If $\mathbf{W}_2(\mu, \nu) = 0$ and γ is an optimal plan from μ to ν, then $0 = \int_{\mathbb{R}^n \times \mathbb{R}^n} |x - y|^2 \, d\gamma(x, y)$ implies that γ is concentrated on $\{(x, y) : x = y\}$: hence for every $E \in \mathcal{B}(\mathbb{R}^n)$ we have

$$\gamma(E \times (\mathbb{R}^n \setminus E)) = \gamma((\mathbb{R}^n \setminus E) \times E) = 0,$$

and thus

$$\mu(E) = \gamma(E \times \mathbb{R}^n) = \gamma(E \times E) = \gamma(\mathbb{R}^n \times E) = \nu(E),$$

that is, $\mu = \nu$. We are thus left to prove the triangular inequality

$$\mathbf{W}_2(\mu_1, \mu_3) \leq \mathbf{W}_2(\mu_1, \mu_2) + \mathbf{W}_2(\mu_2, \mu_3), \qquad \forall \mu_1, \mu_2, \mu_3 \in \mathcal{P}_2(\mathbb{R}^n). \quad (11.17)$$

Thanks to (11.9) it is sufficient to prove (11.17) with $\mu_k^\varepsilon = (\mu_k \star \rho_\varepsilon) \, d\mathcal{L}^n$ in place of μ_k. To this end, let ∇f_{12} and ∇f_{23} be the Brenier maps from μ_1^ε to μ_2^ε and from μ_2^ε to μ_3^ε respectively. Denoting by F_{12} and F_{23} the differentiability sets of f_{12} and f_{23}, we define $T : F_{12} \cap (\nabla f_{12})^{-1}(F_{23}) \to \mathbb{R}^n$ by setting

$$T(x) = \nabla f_{23}(\nabla f_{12}(x)) \qquad x \in F_{12} \cap (\nabla f_{12})^{-1}(F_{23}).$$

We immediately see that $T_\# \mu_1^\varepsilon = \mu_3^\varepsilon$ since $(\nabla f_{12})_\# \mu_1^\varepsilon = \mu_2^\varepsilon$ and $(\nabla f_{23})_\# \mu_2^\varepsilon = \mu_3^\varepsilon$: therefore,

$$\mathbf{W}_2(\mu_1^\varepsilon, \mu_3^\varepsilon) \leq \|T - \mathrm{Id}\|_{L^2(\mu_1^\varepsilon)} \leq \|T - \nabla f_{12}\|_{L^2(\mu_1^\varepsilon)} + \mathbf{W}_2(\mu_1^\varepsilon, \mu_2^\varepsilon).$$

We conclude since $(\nabla f_{12})_\# \mu_1^\varepsilon = \mu_2^\varepsilon$ gives

$$
\begin{aligned}
\|T - \nabla f_{12}\|_{L^2(\mu_1^\varepsilon)} &= \left(\int_{\mathbb{R}^n} |(\nabla f_{23}) \circ (\nabla f_{12}) - \nabla f_{12}|^2 \, d\mu_1^\varepsilon \right)^{1/2} \\
&= \left(\int_{\mathbb{R}^n} |\nabla f_{23} - \mathrm{Id}|^2 \, d\mu_2^\varepsilon \right)^{1/2} = \mathbf{W}_2(\mu_2^\varepsilon, \mu_3^\varepsilon).
\end{aligned}
$$

Step two: We prove that if $\mathbf{W}_2(\mu_j, \mu) \to 0$ as $j \to \infty$, then (11.15)–(11.16) holds. Of course it is enough to prove that (11.15) holds up to extracting subsequences. We start proving that there is a positive constant M such that

$$\sup_j \mu_j(\mathbb{R}^n \setminus B_R) \leq \frac{M}{R^2}, \qquad \forall R > 0. \quad (11.18)$$

Indeed if $\gamma_j \in \Gamma(\mu_j, \mu)$ is an optimal plan in $\mathbf{K}_2(\mu_j, \mu)$, then by first integrating the inequality

$$|x|^2 \leq (1 + \varepsilon) |y|^2 + \left(1 + \frac{1}{\varepsilon} \right) |x - y|^2 \qquad \forall x, y \in \mathbb{R}^n,$$

with respect to γ_j, and then by exploiting $\mathbf{p}_\# \gamma_j = \mu_j$ and $\mathbf{q}_\# \gamma_j = \mu$, we obtain

$$\int_{\mathbb{R}^n} |x|^2 \, d\mu_j(x) \leq (1 + \varepsilon) \int_{\mathbb{R}^n} |y|^2 \, d\mu(y) + C(\varepsilon) \, \mathbf{K}_2(\mu_j, \mu).$$

In particular, since $\mathbf{W}_2(\mu_j, \mu) \to 0$ as $j \to \infty$, we find

$$\limsup_{j \to \infty} \int_{\mathbb{R}^n} |x|^2 \, d\mu_j(x) \leq \int_{\mathbb{R}^n} |y|^2 \, d\mu(y), \quad (11.19)$$

which implies (11.18). Now, up to extracting a subsequence, $\gamma_j \overset{*}{\rightharpoonup} \gamma$ for some Radon measure γ on $\mathbb{R}^n \times \mathbb{R}^n$, and if $\varphi \in C_c^0(\mathbb{R}^n \times \mathbb{R}^n)$ with $0 \le \varphi \le 1$, then

$$\int_{\mathbb{R}^n \times \mathbb{R}^n} |x-y|^2 \, \varphi(x,y) \, d\gamma = \lim_{j \to \infty} \int_{\mathbb{R}^n \times \mathbb{R}^n} |x-y|^2 \, \varphi(x,y) \, d\gamma_j \le \lim_{j \to \infty} \mathbf{K}_2(\mu_j,\mu)=0,$$

so that

$$x = y \ \gamma\text{-a.e. on } \mathbb{R}^n \times \mathbb{R}^n. \tag{11.20}$$

Moreover (11.18) and $\gamma_j \in \Gamma(\mu_j,\mu)$ imply that γ_j converges *narrowly* to γ, so that we can pick any $\varphi \in C_c^0(\mathbb{R}^n)$ and find that

$$\lim_{j \to \infty} \int_{\mathbb{R}^n} \varphi \, d\mu_j = \lim_{j \to \infty} \int_{\mathbb{R}^n \times \mathbb{R}^n} \varphi(x) \, d\gamma_j(x,y) = \int_{\mathbb{R}^n \times \mathbb{R}^n} \varphi(x) \, d\gamma(x,y).$$

By (11.20) and again by narrow convergence of γ_j to γ, we find however that

$$\int_{\mathbb{R}^n \times \mathbb{R}^n} \varphi(x) \, d\gamma(x,y) = \int_{\mathbb{R}^n \times \mathbb{R}^n} \varphi(y) \, d\gamma(x,y) = \lim_{j \to \infty} \int_{\mathbb{R}^n \times \mathbb{R}^n} \varphi(y) \, d\gamma_j(x,y)$$

$$= \lim_{j \to \infty} \int_{\mathbb{R}^n} \varphi \, d\mu = \int_{\mathbb{R}^n} \varphi \, d\mu,$$

where in the penultimate identity we have used $\mathbf{q}_\# \gamma_j = \mu$. We have thus proved that $\mu_j \overset{*}{\rightharpoonup} \mu$ as $j \to \infty$. Finally (11.19) and $\mu_j \overset{*}{\rightharpoonup} \mu$ give

$$\int_{\mathbb{R}^n} |x|^2 \, d\mu = \sup_\varphi \int_{\mathbb{R}^n} \varphi \, |x|^2 \, d\mu = \sup_\varphi \lim_{j \to \infty} \int_{\mathbb{R}^n} \varphi \, |x|^2 \, d\mu_j$$

$$\le \liminf_{j \to \infty} \int_{\mathbb{R}^n} |x|^2 \, d\mu_j \le \int_{\mathbb{R}^n} |x|^2 \, d\mu,$$

where $\sup_\varphi$ runs among those $\varphi \in C_c^0(\mathbb{R}^n)$ with $0 \le \varphi \le 1$. This proves (11.16).

Step three: We prove that (11.15)–(11.16) imply $\mathbf{W}_2(\mu_j,\mu) \to 0$ as $j \to \infty$. Once again we only need to prove this up to the extraction of subsequences. We notice that (11.16) implies (11.19) and thus (11.18). In particular we find that (i):

$$\int_{\mathbb{R}^n} \varphi \, d\mu_j \to \int_{\mathbb{R}^n} \varphi \, d\mu, \qquad \forall \varphi \in C^0(\mathbb{R}^n) \text{ with quadratic growth,} \tag{11.21}$$

i.e., whenever $\varphi \in C^0(\mathbb{R}^n)$ is such that $|\varphi(x)| \le C \, (1 + |x|^2)$ for every $x \in \mathbb{R}^n$ and a suitable $C < \infty$; and (ii): for every $\varepsilon > 0$, we can find $\varphi_\varepsilon \in C_c^0(\mathbb{R}^n)$ with $0 \le \varphi_\varepsilon \le 1$, $\varphi_\varepsilon = 1$ on $B_{R(\varepsilon)/2}$ and $\varphi_\varepsilon = 0$ on $\mathbb{R}^n \setminus B_{R(\varepsilon)}$ for some $R(\varepsilon) \to +\infty$ as $\varepsilon \to 0^+$, so that

$$\sup_{\gamma \in \Gamma(\mu_j,\mu)} \int_{\mathbb{R}^n \times \mathbb{R}^n} |x - y|^2 \, [1 - \varphi_\varepsilon(x) \, \varphi_\varepsilon(y)] \, d\gamma < \varepsilon. \tag{11.22}$$

As a consequence, if we introduce the transport costs

$$c_p^\varepsilon(x,y) = |x - y|^p \, \varphi_\varepsilon(x) \, \varphi_\varepsilon(y) \qquad (x,y) \in \mathbb{R}^n \times \mathbb{R}^n, \qquad p = 1,2,$$

and notice that $c_2^\varepsilon \leq 2\,R(\varepsilon)\,c_1^\varepsilon$ and $c_1^\varepsilon \leq |x - y|$ on $\mathbb{R}^n \times \mathbb{R}^n$, then we find by (11.22)

$$\mathbf{K}_2(\mu_j, \mu) \leq \mathbf{K}_{c_2^\varepsilon}(\mu_j, \mu) + \varepsilon \leq 2\,R(\varepsilon)\,\mathbf{K}_{c_1^\varepsilon}(\mu_j, \mu) + \varepsilon \leq 2\,R(\varepsilon)\,\mathbf{K}_1(\mu_j, \mu) + \varepsilon,$$

for every j and $\varepsilon > 0$. We have thus reduced to prove that

$$\lim_{j \to \infty} \mathbf{K}_1(\mu_j, \mu) = 0, \tag{11.23}$$

and in turn, by the Kantorovich duality theorem with linear cost, Theorem 3.17, see in particular (3.38), it is enough to show that

$$\lim_{j \to \infty} \sup \left\{ \int_{\mathbb{R}^n} u \, d\mu_j - \int_{\mathbb{R}^n} u \, d\mu : u : \mathbb{R}^n \to \mathbb{R}, \mathrm{Lip}(u) \leq 1 \right\} = 0. \tag{11.24}$$

Let u_j be $1/j$-close to realize the $\sup_u$ in (11.24). Notice that we can add a constant to u_j without affecting the almost optimality property of u_j in $\sup_u$, so that we can also assume $u_j(0) = 0$. By the Ascoli-Arzelá theorem, up to extracting a further subsequence, we have that $u_j \to u$ locally uniformly on $\mathbb{R}^n$ for some u with $\mathrm{Lip}(u) \leq 1$ and $u(0) = 0$. Moreover, we clearly have $\max\{|u_j(x)|, |u(x)|\} \leq |x|$ for every $x \in \mathbb{R}^n$ and $j \in \mathbb{N}$. To conclude the proof we are left to show that

$$\lim_{j \to \infty} \int_{\mathbb{R}^n} u_j \, d\mu_j - \int_{\mathbb{R}^n} u_j \, d\mu = 0. \tag{11.25}$$

Indeed, given $R > 0$, we have that

$$\left| \int_{\mathbb{R}^n} u_j \, d\mu_j - \int_{\mathbb{R}^n} u_j \, d\mu \right| \leq \left| \int_{\mathbb{R}^n} u \, d\mu_j - \int_{\mathbb{R}^n} u \, d\mu \right| + 2 \, \|u_j - u\|_{C^0(B_R)}$$

$$+ \int_{\mathbb{R}^n \setminus B_R} |u_j| + |u| \, d\mu_j + \int_{\mathbb{R}^n \setminus B_R} |u_j| + |u| \, d\mu.$$

By $\max\{|u_j|, |u|\} \leq |x|$, we find that

$$\sup_j \int_{\mathbb{R}^n \setminus B_R} |u_j| + |u| \, d[\mu_j + \mu] \leq 2 \sup_j \left(\int_{\mathbb{R}^n \setminus B_R} |x|^2 \, d[\mu_j + \mu] \right)^{1/2} \to 0^+$$

as $R \to +\infty$, so that

$$\limsup_{j \to \infty} \left| \int_{\mathbb{R}^n} u_j \, d\mu_j - \int_{\mathbb{R}^n} u_j \, d\mu \right| \leq \limsup_{j \to \infty} \left| \int_{\mathbb{R}^n} u \, d\mu_j - \int_{\mathbb{R}^n} u \, d\mu \right| = 0,$$

where in the last identity we have used the fact that u has quadratic growth (since it has linear growth) in conjunction with (11.21).

Step four: Finally, let $\{\mu_j\}_j$ be a Cauchy sequence in $(\mathcal{P}_2(\mathbb{R}^n), \mathbf{W}_2)$. Given $\varepsilon > 0$ we can find $j(\varepsilon)$ such that if $j \geq j(\varepsilon)$, then $\mathbf{W}_2(\mu_j, \mu_{j(\varepsilon)}) \leq \varepsilon$. At the same time we can find $R(\varepsilon)$ such that $\mu_{j(\varepsilon)}(B_{R(\varepsilon)}) \geq 1 - \varepsilon$. In particular, if $u : \mathbb{R}^n \to [0,1]$ is such that $u = 1$ on $B_{R(\varepsilon)}$, $u = 0$ on $\mathbb{R}^n \setminus B_{R(\varepsilon)+1}$ and $\mathrm{Lip}(u) \leq 1$, then for every $k \geq j(\varepsilon)$ we find

$$\varepsilon \geq \mathbf{W}_2(\mu_k, \mu_{j(\varepsilon)}) \geq \mathbf{K}_1(\mu_k, \mu_{j(\varepsilon)}) \geq \int_{\mathbb{R}^n} u \, d\mu_{j(\varepsilon)} - \int_{\mathbb{R}^n} u \, d\mu_k,$$

$$\geq \mu_{j(\varepsilon)}(B_{R(\varepsilon)}) - \mu_k(B_{R(\varepsilon)+1}) \geq 1 - \varepsilon - \mu_k(B_{R(\varepsilon)+1}),$$

that is,

$$\mu_k(B_{R(\varepsilon)+1}) \geq 1 - 2\varepsilon, \qquad \forall k \geq j(\varepsilon).$$

In other words, $\sup_j \mu_j(\mathbb{R}^n \setminus B_R) \to 0$ as $R \to \infty$, so that we can find $\mu \in \mathcal{P}(\mathbb{R}^n)$ such that, up to extracting subsequences, μ_j converges to μ narrowly. Moreover,

$$\int_{\mathbb{R}^n} |x|^2 \, d\mu_j = \mathbf{W}_2(\mu_j, \delta_0)^2 \leq \left(\mathbf{W}_2(\mu_{j(\varepsilon)}, \delta_0) + \mathbf{W}_2(\mu_{j(\varepsilon)}, \mu_j) \right)^2$$

and $\mathbf{W}_2(\mu_j, \mu_{j(\varepsilon)}) \leq \varepsilon$ for $j \geq j(\varepsilon)$ imply that

$$\limsup_{j \to \infty} \int_{\mathbb{R}^n} |x|^2 \, d\mu_j(x) < \infty, \qquad (11.26)$$

so that $\mu \in \mathcal{P}_2(\mathbb{R}^n)$ by lower semicontinuity. By step four, $\mathbf{W}_2(\mu_j, \mu) \to 0$. $\square$

Remark 11.11 The proof of Theorem 11.10 is easily adapted to show that

$$(\mathcal{P}_1(\mathbb{R}^n), \mathbf{W}_1)$$

(where $\mathbf{W}_1(\mu, \nu) = \mathbf{K}_1(\mu, \nu)$) is a complete metric space, and to characterize the $\mathbf{W}_1$-convergence. The main difference concerns the proof of the triangular inequality, which in this case follows immediately from the Kantorovich duality formula (3.38).

12

Gradient Flows and the Minimizing Movements Scheme

One of the most interesting applications of OMT is the possibility of characterizing various parabolic PDE as **gradient flows** of physically meaningful energy functionals. The fundamental example is that of the heat equation, which, in addition to its classical characterization as the gradient flow of the *Dirichlet energy* in the Hilbert space L^2, can also be characterized, in a physically more compelling way, as the gradient flow of the *entropy functional* in the Wasserstein space. Given that *gradients* are defined by representing *differentials* via *inner products*, the most immediate framework for discussing gradient flows is definitely the Hilbertian one. For the same reason it is less clear how to define gradient flows in *metric spaces* which, like $(\mathcal{P}_2(\mathbb{R}^n), \mathbf{W}_2)$, do not obviously admit an Hilbertian structure. This point is addressed in this chapter with the introduction of the versatile and powerful algorithm known as the **minimizing movements scheme**. The implementation of this algorithm in the Wasserstein space is then discussed in Chapter 13, specifically, on the case study provided by the Fokker–Planck equation on $\mathbb{R}^n$.

In Section 12.1 we review the concepts of Lyapunov functional and of dissipation for gradient flows in finite dimensional Euclidean spaces, we present the general argument for convergence toward equilibrium of convex gradient flows (Theorem 12.1), and we illustrate the flexibility of these notions (see Remark 12.2). In Section 12.2 we heuristically exemplify these ideas in a proper PDE setting by looking at the case of the heat equation on $\mathbb{R}^n$. Finally, in Section 12.3, we present the minimizing movements scheme in the finite dimensional setting.

12.1 Gradient Flows in $\mathbb{R}^n$ and Convexity

Every smooth vector field $\mathbf{u} : \mathbb{R}^n \to \mathbb{R}^n$ defines a family of systems of ODE (depending on the initial condition $x_0 \in \mathbb{R}^n$)

$$\begin{cases} x'(t) = \mathbf{u}(x(t)), & t \geq 0, \\ x(0) = x_0. \end{cases} \quad (12.1)$$

There always exists a unique solution $x(t) = x(t; x_0)$ to (12.1) defined on a non-trivial interval $[0, t_0)$, with t_0 uniformly positive for x_0 is a compact set. We call the collection of these solutions the **flow generated by u**. When $t_0 = +\infty$ we can try to describe the long time behavior of these solutions by looking for quantities $H : \mathbb{R}^n \to \mathbb{R}$ which behave monotonically along the flow. It is customary to focus, for the sake of definiteness, on monotonically *decreasing* quantities, which are usually called either **entropies** (a terminology which creates confusion with the physical notion of entropy, which is actually *increasing* along diffusive processes) or **Lyapunov functionals**. In general, a flow admits many Lyapunov functionals: Indeed, a necessary and sufficient condition for being a Lyapunov functional is just that $\mathbf{u} \cdot \nabla H \leq 0$ on $\mathbb{R}^n$, as it is easily seen by using (12.1) to compute

$$\frac{d}{dt} H(x(t)) = x'(t) \cdot \nabla H(x(t)) = (\mathbf{u} \cdot \nabla H)(x(t)).$$

The **dissipation** $D : \mathbb{R}^n \to [0, \infty)$ of H along the flow defined by $\mathbf{u}$ is defined by

$$D = -\mathbf{u} \cdot \nabla H,$$

and of particular importance is the special case when (12.1) is a **gradient flow**, in the sense that

$$\mathbf{u} = -\nabla f$$

for an "energy functional" $f : \mathbb{R}^n \to \mathbb{R}$. In this case (12.1) takes the form

$$\begin{cases} x'(t) = -\nabla f(x(t)), & t > 0, \\ x(0) = x_0, \end{cases} \quad (12.2)$$

and $H = f$ is itself a Lyapunov functional of its own gradient flow, with the (somehow maximal) dissipation

$$D = -|\nabla f|^2.$$

Under (12.2), the flow $x(t)$ evolves according to an infinitesimal minimization principle, that is, the systems "sniffs around" for the nearest accessible states of lower energy by moving in the direction $-\nabla f(x(t))$. The dynamics of a gradient flow can be quite rich. Every critical point x_0 of f is stationary for the corresponding gradient flow, in the sense that $x(t) \equiv x_0$ for every $t > 0$ is the unique solution of (12.2) with initial datum x_0. Among critical points, local maxima will be unstable (under small perturbations of the initial datum, the solution will flow away); but all local minima will be possible asymptotic equilibria (limit points of $x(t)$ as $t \to \infty$). This last remark illustrate the great importance of understanding local minimizers, and not just global minimizers, in the study of variational problems. It also clarifies while *convexity* is such a powerful assumption in studying gradient flows. Indeed, as we had already reason to appreciate when discussing cyclical monotonicity, the general principle "critical points of convex functions are global minimizers" holds. As a reflection of this fact, and as proved in the following elementary theorem, convex gradient flows have particularly simple dynamics.

Theorem 12.1 (Convergence of convex gradient flows) *If $f : \mathbb{R}^n \to \mathbb{R}$ is a smooth convex function such that*

$$\lim_{|x| \to \infty} f(x) = +\infty, \tag{12.3}$$

then for every $x_0 \in \mathbb{R}^n$ there is a solution $x(t)$ of (12.2) defined for every $t \geq 0$. Moreover:

(a) *if f is strictly convex on $\mathbb{R}^n$, then f has a unique global minimizer x_{min} on $\mathbb{R}^n$ and for every $x_0 \in \mathbb{R}^n$ the solution of (12.2) is such that $x(t) \to x_{min}$ as $t \to \infty$;*

(b) *if f is uniformly convex on $\mathbb{R}^n$, in the sense that there exists $\lambda > 0$ such that*

$$v \cdot \nabla^2 f(x)[v] \geq \lambda |v|^2 \qquad \forall x, v \in \mathbb{R}^n, \tag{12.4}$$

then for every $x_0 \in \mathbb{R}^n$ the solution $x(t)$ of (12.2) converges to x_{min} exponentially fast, with

$$|x(t) - x_{min}| \leq e^{-\lambda t} |x_0 - x_{min}|, \qquad \forall t \geq 0. \tag{12.5}$$

Proof Thanks to (12.3) there is no loss of generality in assuming that $f \geq 0$. The flow exists for every time because, as already noticed, f itself is a Lyapunov functional of (12.2): Hence, $f(x(t)) \leq f(x_0)$ for every $t \in (0, t_0)$, and in particular $x(t)$ never leaves the closed, bounded (by (12.3)) set $\{f \leq f(x_0)\}$. Since ∇f is locally Lipschitz, a standard application of the Cauchy-Picard theorem proves that we can take $t_0 = +\infty$. By coercivity and strict convexity of f there exists a unique minimizer x_{min} of f on $\mathbb{R}^n$. We now prove that $x(t) \to x_{min}$ as $t \to \infty$. We first notice that

$$\frac{d}{dt} f(x(t)) = -|\nabla f(x(t))|^2,$$

implies the **dissipation inequality**

$$\int_0^\infty |\nabla f(x(t))|^2 dt \leq f(x_0). \tag{12.6}$$

Differentiating in turn $|\nabla f(x(t))|^2$, and using the second order characterization of convexity

$$v \cdot \nabla^2 f(x)[v] \geq 0 \qquad \forall u, v \in \mathbb{R}^n,$$

we deduce that

$$\frac{d}{dt} |\nabla f(x(t))|^2 = 2 \nabla f(x(t)) \cdot \nabla^2 f(x(t))[x'(t)] \tag{12.7}$$

$$= -2 \nabla f(x(t)) \cdot \nabla^2 f(x(t))[\nabla f(x(t))] \leq 0. \tag{12.8}$$

Combining (12.6) and (12.7) we see that $|\nabla f(x(t))| \to 0$ as $t \to \infty$. Since $f(x(t))$ is bounded and f is coercive, for every $t_j \to \infty$ there exists a subsequence z_j of $x(t_j)$ such that $z_j \to z_\infty$ as $j \to \infty$ for some $z_\infty \in \mathbb{R}^n$. By continuity of ∇f, we find $\nabla f(z_\infty) = 0$, and since f admits a unique global minimum and its convex, we find $z_\infty = x_{min}$. By the arbitrariness of t_j, we conclude that

$$\lim_{t \to \infty} x(t) = x_{min}. \tag{12.9}$$

To prove exponential convergence, we take the scalar product of

$$\nabla f(x(t)) - \nabla f(x_{min}) = \int_0^1 \nabla^2 f(sx(t) + (1-s)x_{min})[x(t) - x_{min}] \, ds$$

with $(x(t) - x_{min})$ and apply uniform convexity to find

$$\lambda \, |x(t) - x_{min}|^2 \leq (\nabla f(x(t)) - \nabla f(x_{min})) \cdot (x(t) - x_{min}).$$

But thanks to (12.2)

$$\frac{d}{dt} |x(t) - x_{min}|^2 = 2 \, (x(t) - x_{min}) \cdot x'(t)$$

$$= -2 \, (x(t) - x_{min}) \cdot (\nabla f(x(t)) - \nabla f(x_{min}))$$

$$\leq -2 \, \lambda \, |x(t) - x_{min}|^2$$

so that (12.5) follows. $\qquad \qquad \Box$

Remark 12.2 (Gradient flow identification)   Let $A \in \mathbb{R}^{n \times n}_{\text{sym}}$ be positive definite, and let $(x, y)_A = A[x] \cdot y$ denote the scalar product on $\mathbb{R}^n$ induced by A. Given a differentiable function $f : \mathbb{R}^n \to \mathbb{R}$, the **gradient of f at x defined by A** is the unique vector $\nabla^A f(x)$ such that $df_x[v] = (\nabla^A f(x), v)_A$. When $A = \text{Id}$ we are of course back to the usual definition of gradient, i.e., $\nabla^{\text{Id}} = \nabla$. Now, consider the system of ODE

$$x'(t) = -x(t), \tag{12.10}$$

and notice that (i) as soon as A is positive definite, $f_A(x) = (1/2) \, (x, x)_A$ is a Lyapunov functional of (12.10); (ii) clearly (12.10) is the gradient flow of

$f_{\mathrm{Id}}(x) = |x|^2/2$ (with respect to the usual definition of gradient); (iii) at the same time, (12.10) is also the gradient flow of $f_A(x) = (1/2)\,(x,x)_A$, if "gradient" means "gradient defined by A," since $\nabla^A f_A(x) = x$ for every $x \in \mathbb{R}^n$. In summary, *the same evolution equation can be seen as the gradient flow of different Lyapunov functionals by tailoring the notion of gradient on the considered functional* (where "tailoring" stands for "choosing an inner product"). This remark, which is a bit abstract and artificial in the Euclidean context, is actually quite substantial when working in the infinite dimensional setting of parabolic PDE. In the latter situation, one can typically guess different Lyapunov functionals (by differentiating a candidate functional, substituting the PDE in the resulting formula, and then seeking if the resulting expression can be shown to have a sign), and then can ask if the PDE under study is actually the gradient flow of any of these Lyapunov functionals (abstractly, by trying out different Hilbertian structures – but in the next section we will see more concretely how this works). Whenever it is possible to identify a PDE as the gradient flow of a convex functional, then the proof of Theorem 12.1 provides a sort of blueprint to address convergence to equilibrium.

12.2 Gradient Flow Interpretations of the Heat Equation

Following up on Remark 12.2, we now illustrate how to give different gradient flow interpretations of the most basic parabolic PDE,[1] namely, the **heat equation on $\mathbb{R}^n$**. Given an initial datum $\mu_0 = \rho_0\,d\mathcal{L}^n \in \mathcal{P}(\mathbb{R}^n)$, we look for $\rho = \rho(x,t) : \mathbb{R}^n \times [0,\infty) \to \mathbb{R}$ such that

$$\begin{cases} \partial_t \rho = \Delta \rho, & \text{in } \mathbb{R}^n \times (0,\infty), \\ \rho|_{t=0} = \rho_0. \end{cases} \tag{12.11}$$

One can easily find several Lyapunov functionals for (12.11). For example, setting

$$H(\rho) = \int_{\mathbb{R}^n} h(\rho), \qquad h : [0,\infty) \to \mathbb{R} \text{ smooth and convex},$$

---

[1] Clearly, we consider this example for purely illustrative purposes, because any reasonable question about the (unique) solution of (12.11) can be directly tackled by means of the resolution formula

$$\rho(x, t) = \int_{\mathbb{R}^n} \rho_0(y)\,\frac{e^{-|x-y|^2/4t}}{(4\pi t)^{n/2}}\,dy, \qquad (x, t) \in \mathbb{R}^n \times (0, \infty). \tag{12.12}$$

However, resolution formulas for PDE are definitely not the norm, hence the interest of exploring less direct methods of investigating their behavior.

we easily see that $H(\rho)$ decreases along the heat flow. Indeed, by informally applying the divergence theorem[2] to the vector field $h'(\rho) \nabla \rho$ we find that if ρ is a solution to (12.11), then $h'' \geq 0$ gives

$$\frac{d}{dt} H(\rho) = \int_{\mathbb{R}^n} h'(\rho) \partial_t \rho = \int_{\mathbb{R}^n} h'(\rho) \Delta \rho = -\int_{\mathbb{R}^n} h''(\rho) |\nabla \rho|^2 \leq 0.$$

Another natural[3] Lyapunov functional for the heat equation is the Dirichlet energy

$$H(\rho) = \frac{1}{2} \int_{\mathbb{R}^n} |\nabla \rho|^2.$$

Indeed, again thanks to an informal application of the divergence theorem, we see that if ρ solves (12.11), then

$$\frac{d}{dt} H(\rho) = \int_{\mathbb{R}^n} \nabla \rho \cdot \nabla(\partial_t \rho) = -\int_{\mathbb{R}^n} \partial_t \rho \, \Delta \rho = -\int_{\mathbb{R}^n} (\Delta \rho)^2 \leq 0.$$

Next we ask the question: Can we interpret the heat equation as the gradient flow of one of its many Lyapunov functionals? For example, it is easily seen that (12.11) can be interpreted as the gradient flow of the Dirichlet energy. On the Hilbert space $L^2(\mathbb{R}^n)$ endowed with the standard L^2-scalar product, we consider the functional H defined by

$$H(\rho) = \frac{1}{2} \int_{\mathbb{R}^n} |\nabla \rho|^2 \qquad \text{if } \rho \in L_0 = W^{1,2}(\mathbb{R}^n) \subset L^2(\mathbb{R}^n),$$

and by $H(\rho) = +\infty$ if, otherwise, $\rho \in L^2(\mathbb{R}^n) \setminus L_0$. Notice that H is convex on $L^2(\mathbb{R}^n)$, and that the differential of H at $\rho \in W^{1,2}(\mathbb{R}^n)$ in the direction $\varphi \in W^{1,2}(\mathbb{R}^n)$ is given by

$$dH_\rho[\varphi] = \lim_{t \to 0} \frac{H(\rho + t\varphi) - H(\rho)}{t} = \int_{\mathbb{R}^n} \nabla \rho \cdot \nabla \varphi, \qquad \forall \varphi \in W^{1,2}(\mathbb{R}^n).$$

In particular, if $\rho \in W^{2,2}(\mathbb{R}^n)$, then by

$$dH_\rho[\varphi] = -\int_{\mathbb{R}^n} \varphi \, \Delta \rho,$$

---

[2] Precisely, one applies the divergence theorem on a ball B_R to find

$$\int_{B_R} h'(\rho) \Delta \rho = -\int_{B_R} h''(\rho) |\nabla \rho|^2 + \int_{\partial B_R} h'(\rho) \frac{x}{|x|} \cdot \nabla \rho$$

and then needs to show the vanishing of the boundary integral in the limit $R \to \infty$. This requires suitable decays assumptions on the initial datum ρ_0 and checking their consequent validity on $\rho(\cdot, t)$ for every $t > 0$.

[3] Notice that the Dirichlet energy is (up to constants) the dissipation of the Lyapunov functional $\int_{\mathbb{R}^n} u^2$.

we can uniquely extend dH_ρ as a bounded linear functional on the whole $L^2(\mathbb{R}^n)$, with

$$\nabla^{L^2} H(\rho) = -\Delta\rho \qquad \forall \rho \in W^{2,2}(\mathbb{R}^n).$$

Therefore, setting $\rho(t) = \rho(x,t)$, the gradient flow equation

$$\frac{d}{dt}\rho(t) = -\nabla^{L^2}H(\rho(t)), \qquad \rho(0) = \rho_0,$$

is equivalent to (12.11).

What about Lyapunov functionals of the form $H(\rho) = \int_{\mathbb{R}^n} h(\rho)$? It can be seen, for example, that (12.11) is the gradient flow of the Lyapunov functional defined by $h(\rho) = \rho^2/2$ if gradients are computed in the scalar product of the dual space to $W^{1,2}(\mathbb{R}^n)$. The situation is however less clear in the case of a physically interesting Lyapunov functionals like the (negative) entropy

$$H(\rho) = \int_{\mathbb{R}^n} \rho \log \rho.$$

Informally (which here means, disregarding integrability issues, admissibility of the variations, etc.) we have

$$H(\rho + t\,\varphi) = H(\rho) + t\int_{\mathbb{R}^n}(1 + \log(\rho))\,\varphi + O(t^2) \qquad (12.13)$$

so that the L^2-gradient of H at ρ is $1 + \log\rho$ whenever $\log\rho \in L^2(\mathbb{R}^n)$, i.e.,

$$\nabla^{L^2} H(\rho) = 1 + \log\rho \quad \forall \rho \text{ such that } \log\rho \in L^2(\mathbb{R}^n);$$

but, of course, the corresponding gradient flow, $\partial_t\rho = -1 - \log(\rho)$, has nothing to do with the heat equation! *This is puzzling*, because one of the most natural interpretations of the heat equation is that of a model for the time evolution of the position probability density of a particle undergoing random molecular collisions, a process that is intuitively driven by the maximization of the physical entropy, i.e., by $-\int_{\mathbb{R}^n} \rho \log \rho$. Therefore, *there should be* one way to interpret $\rho_t = \Delta\rho$ as the gradient flow of $H(\rho) = \int_{\mathbb{R}^n} \rho \log \rho$, and this interpretation should be very significant from a physical viewpoint.

12.3 The Minimizing Movements Scheme

The minimizing movements scheme provides a perspective on the concept of gradient flow which is somehow more fundamental than the one based on the use of Hilbertian structures for representing differentials as gradients. The idea is looking at gradient flows as limits of discrete-in-time flows defined by sequences of minimization processes associated to the choice of a reference metric. We are now going to introduce this construction in the case of the finite dimensional example (12.2). Given a smooth, nonnegative function $f : \mathbb{R}^n \to [0, \infty)$ with bounded sublevel sets (i.e., $\{f \leq a\}$ is bounded for every $a > 0$), and given $h > 0$ and $x_0 \in \mathbb{R}^n$, we define a sequence

$$\{x_k^{(h)}\}_{k=1}^{\infty} \subset \mathbb{R}^n$$

by taking $x_0^{(h)} = x_0$ and, for $k \geq 1$,

$$x_k^{(h)} \quad \text{is the unique minimizer of} \quad f(x) + \frac{1}{2h} |x - x_{k-1}^{(h)}|^2. \qquad (12.14)$$

Correspondingly, we consider the **discrete flow of step** h, $x^{(h)} : [0, \infty) \to \mathbb{R}^n$, defined by

$$x^{(h)}(t) = x_{k-1}^{(h)} \quad \text{if} \quad (k-1)\,h \leq t < k\,h, \qquad k \geq 1, \qquad (12.15)$$

and claim that (i) as $h \to 0^+$, $x^{(h)}(t)$ has a locally uniform limit $x(t)$ on $[0, \infty)$; (ii) the pointwise limit $x(t)$ is the unique solution to the gradient flow (12.2), i.e., $x'(t) = -\nabla f(x(t))$ for $t > 0$, $x(0) = x_0$. We expect this to happen because the minimality property (12.14) implies the validity of the critical point condition

$$\frac{x_k^{(h)} - x_{k-1}^{(h)}}{h} = -\nabla f(x_k^{(h)}),$$

which, in turn, is a discrete-in-time approximation of $x'(t) = -\nabla f(x(t))$. The interest of this construction is that it is only the Euclidean *distance* that appears in the definition of the discrete flows $x^{(h)}$, and not the Euclidean scalar product. This feature makes possible to consider the same scheme starting from a function defined on a metric space: When the resulting scheme converges (as $h \to 0^+$) to a limit flow, we can say that such flow has the structure of a gradient flow (although no gradient was actually computed in the proper sense of representing a differential through an inner product), namely, of the (metric) gradient flow of the considered function with respect to the considered metric. The convergence of the minimizing movements scheme in the Euclidean setting is discussed in the next theorem. As in the case of Theorem 12.1, the proof is particularly interesting because it provides a blueprint for approaching the same problem in more general settings.

Theorem 12.3 (Convergence of the minimizing movements scheme) *Given a smooth $f : \mathbb{R}^n \to [0, \infty)$ with bounded sublevel sets, $x_0 \in \mathbb{R}^n$, and $h > 0$, let $x^{(h)}(t)$ be the discrete flow of step h defined by (12.14) and (12.15). Then, uniformly in h, $x^{(h)}(t)$ is locally bounded, and $(1/2)$-Hölder continuous above* *scale* $2h$, *on* $[0, \infty)$. *Moreover, its locally uniform limit* $x(t)$ *on* $[0, \infty)$ *is such that* $x(t) \in C^\infty([0, \infty); \mathbb{R}^n)$, $x'(t) = -\nabla f(x(t))$ *for every* $t > 0$ *and* $x(0) = x_0$.

Proof *Step one*: We claim that for some $M > 0$, depending on f and x_0 only,

$$\sup_{h>0} \sup_{t \geq 0} |x^{(h)}(t)| \leq M, \tag{12.16}$$

$$\sup_{h>0} \sup_{|t-s| \geq 2h} |x^{(h)}(t) - x^{(h)}(s)| \leq M \sqrt{|t-s|}. \tag{12.17}$$

Comparing the value of $x \mapsto f(x) + (1/2h)|x - x_{k-1}^{(h)}|^2$ at $x = x_{k-1}^{(h)}$ with the minimum value assumed at $x = x_k^{(h)}$, we find that

$$f(x_k^{(h)}) + (1/2h)|x_k^{(h)} - x_{k-1}^{(h)}|^2 \leq f(x_{k-1}^{(h)}) \qquad \forall k \geq 1. \tag{12.18}$$

Notice that (12.18) implies $f(x_k^{(h)}) \leq f(x_0^{(h)}) = f(x_0)$, i.e., f is decreasing along the discrete flow $x^{(h)}$. Since f has bounded sublevel sets, this proves (12.16). To prove (12.17), we start by adding up (12.18) on $k = 1, \ldots, N$ and canceling out the terms $f(x_k^{(h)})$, $k = 1, \ldots, N-1$, which appear on both sides, to find that

$$f(x_N^{(h)}) + \frac{1}{2h} \sum_{k=1}^{N} |x_k^{(h)} - x_{k-1}^{(h)}|^2 \leq f(x_0), \qquad \forall N \geq 1.$$

In particular, since $f \geq 0$, we find

$$\sum_{k=1}^{\infty} |x_k^{(h)} - x_{k-1}^{(h)}|^2 \leq 2h f(x_0). \tag{12.19}$$

If now $s > t \geq 0$ and $|t - s| \geq 2h$ then we can find integers $j > k \geq 1$ such that

$$(k-1)h \leq t < kh < (k+1)h \leq (j-1)h \leq s < jh, \qquad j - k - 1 \leq \frac{s-t}{h},$$

so that, by (12.19), we find

$$|x^{(h)}(s) - x^{(h)}(t)| = |x_{j-1}^{(h)} - x_{k-1}^{(h)}| \leq \sum_{\ell=k}^{j-1} |x_\ell^{(h)} - x_{\ell-1}^{(h)}|$$

$$\leq \sqrt{j-k} \left(\sum_{\ell=k}^{j-1} |x_\ell^{(h)} - x_{\ell-1}^{(h)}|^2 \right)^{1/2}$$

$$\leq \sqrt{1 + [(s-t)/h]} \sqrt{2h f(x_0)} \leq \sqrt{2f(x_0)} \sqrt{s-t+h} \leq M \sqrt{s-t},$$

since $s - t \geq 2h$. This proves (12.17).

Step two: We show that $x^{(h)}$ solves a "weak (integral) discrete form" of the gradient flow system of ODE $x(t)' = -\nabla f(x(t))$. More precisely, we prove that if $s > t \geq 0$, then

$$\left| x^{(h)}(s) - x^{(h)}(t) + \int_t^s \nabla f(x^{(h)}(r))\, dr \right| \le M\, h, \qquad (12.20)$$

for a constant M depending on f and x_0 only. Indeed, the critical point condition associated to the minimality property (12.14) gives

$$x_k^{(h)} - x_{k-1}^{(h)} = -h\, \nabla f\left(x_k^{(h)}\right), \qquad \forall k \ge 1, \qquad (12.21)$$

so that if $(k-1)\, h \le t < k\, h$ and $(j-1)\, h \le s < j\, h$ for integers $1 \le k \le j$, then

$$x^{(h)}(s) - x^{(h)}(t) = x_{j-1}^{(h)} - x_{k-1}^{(h)} = \sum_{\ell=k}^{j-1} x_\ell^{(h)} - x_{\ell-1}^{(h)} = -h \sum_{\ell=k}^{j-1} \nabla f\left(x_\ell^{(h)}\right)$$

$$= -\int_{kh}^{jh} \nabla f\left(x^{(h)}(r)\right)\, dr. \qquad (12.22)$$

Since $x^{(h)}(r) \in \{f \le f(x_0)\}$ for every $r \ge 0$ and ∇f is bounded by a constant $M/2$ depending only on f and x_0 on the bounded set $\{f \le f(x_0)\}$ we find that

$$\left| \int_{kh}^{jh} \nabla f(x^{(h)}(r))\, dr - \int_t^s \nabla f(x^{(h)}(r))\, dr \right| \le \frac{M}{2}\{|jh - s| + |kh - t|\} \le M\, h,$$

that is (12.20).

Step three: We claim that for each $h_j \to 0^+$ there exists a function $x(t) :$ $[0, \infty) \to \mathbb{R}^n$ such that, up to extract a not relabeled subsequence, $x^{(h_j)} \to x$ locally uniformly on $[0, \infty)$. If the claim holds, then $x(0) = x_0$ and, by letting $h = h_j$ in (12.20),

$$x(s) = x(t) - \int_t^s \nabla f(x(r))\, dr, \qquad \forall s > t \ge 0,$$

so that x is smooth on $[0, \infty)$, and it is actually the unique solution of $x'(t) = -\nabla f(x(t))$ such that $x(0) = x_0$. In particular, the limit of $x^{(h_j)}$ does not depend on the particular subsequence that had been extracted, and therefore $x^{(h)} \to x$ as $h \to 0^+$. We are left to prove the claim. The claim would follow from the Ascoli-Arzelá theorem, (12.16) and (12.17) if only (12.17) would hold without the restriction that $|t - s| \ge 2h$. However one can easily adapt the classical proof of the Ascoli-Arzelá theorem, and thus establish compactness, even under the (weaker than usual) condition (12.17) that the equicontinuity of $\{x^{(h)}\}_h$ only holds above the scale $2\, h$. The simple details are omitted.   □

13

The Fokker–Planck Equation in the Wasserstein Space

We continue the discourse started in Chapter 12 with an example of implementation of the minimizing movements scheme in the Wasserstein space. In Section 13.1 we introduce the Fokker–Planck equation and its associated free energy $\mathcal{F}$, which consists of a potential energy term and of the (negative) entropy term $\mathcal{S} = \int_{\mathbb{R}^n} \rho \log \rho$ already met in Chapter 10 (see (10.2)). We show that $\mathcal{F}$ has a unique critical point, which is also a stationary solution of the Fokker–Planck equation, and thus the natural candidate for describing the long time behavior of general solutions. In Section 13.2 we introduce the *inner variations* of functionals on $\mathcal{P}_{\mathrm{ac}}(\mathbb{R}^n)$, and (having in mind the importance of the critical point condition (12.21) in the proof of Theorem 12.3) derive the corresponding first variation formulae for the potential energy term in $\mathcal{F}$ and for $\mathbf{W}_2$. Taking inner variations of the entropy term $\mathcal{S}$ is more delicate, and is separately addressed in Section 13.3. Finally, in Section 13.4 we construct the minimizing movements scheme for $\mathcal{F}$ with respect to $\mathbf{W}_2$, and prove its convergence toward a solution to the Fokker–Planck equation, while in Section 13.5 we informally discuss how to prove convergence toward equilibrium in the Fokker–Planck equation when the free energy $\mathcal{F}$ is uniformly displacement convex (compare with the use of (12.4) in proving (12.5) in Theorem 12.1).

13.1 The Fokker–Planck Equation

The Fokker–Planck equation models the time evolution of the position probability density $\rho(t) = \rho(\cdot, t) : \mathbb{R}^n \to [0, \infty)$ of a particle moving under the action of a chemical potential $\Psi : \mathbb{R}^n \to [0, \infty)$ and of white noise forces due to molecular collisions. It takes the form

$$\frac{\partial \rho}{\partial t} = \mathrm{div}\,(\nabla \Psi(x)\rho) + \frac{1}{\beta}\Delta\rho, \qquad \text{on } \mathbb{R}^n \times [0, \infty), \qquad (13.1)$$

where β is a positive constant with the dimensions of inverse temperature. We impose the initial condition $\rho(x, 0) = \rho_0(x)$, for $\rho_0 : \mathbb{R}^n \to [0, \infty)$ such that

$$\int_{\mathbb{R}^n} \rho_0(x)\, dx = 1, \qquad \rho_0 \geq 0. \tag{13.2}$$

The conditions (13.2) are propagated in time by the Fokker–Planck equation: indeed, non-negativity is preserved by the maximum principle, while the total mass of $\rho(t)$ is constant in time since, at least informally,[1]

$$\frac{d}{dt} \int_{\mathbb{R}^n} \rho(t) = \int_{\mathbb{R}^n} \frac{\partial \rho}{\partial t} = \int_{\mathbb{R}^n} \operatorname{div}\left(\rho \nabla \Psi + \frac{\nabla \rho}{\beta}\right) = 0.$$

Therefore it makes sense to study the Fokker–Planck equation as a flow in $\mathcal{P}_{ac}(\mathbb{R}^n)$, under the identification between densities $\rho \in L^1(\mathbb{R}^n)$ with $\rho \geq 0$ and $\int_{\mathbb{R}^n} \rho = 1$ and probability measures $\mu = \rho\, d\mathcal{L}^n \in \mathcal{P}_{ac}(\mathbb{R}^n)$. In this direction, we notice that a stationary state for the Fokker–Planck equation, and thus a possible limit for the Fokker–Planck flow, exists in $\mathcal{P}_{ac}(\mathbb{R}^n)$ if the potential Ψ is such that $e^{-\beta\Psi} \in L^1(\mathbb{R}^n)$. Indeed, $\rho = e^{-\beta\Psi}$ is such that $\rho \nabla \Psi + \beta^{-1} \nabla \rho = 0$, so that, as soon as $e^{-\beta\Psi}$ has finite integral, the density ρ_* defined by

$$\rho_*(x) = \frac{e^{-\beta\Psi(x)}}{Z}, \qquad Z = \int_{\mathbb{R}^n} e^{-\beta\Psi(x)}\, dx, \tag{13.3}$$

is a stationary state for (13.1) and satisfies $\mu_* = \rho_*\, d\mathcal{L}^n \in \mathcal{P}_{ac}(\mathbb{R}^n)$.

A functional $\mathcal{F} : \mathcal{P}_{ac}(\mathbb{R}^n) \to \mathbb{R} \cup \{\pm\infty\}$ which is naturally associated with the Fokker–Planck equation is the free energy

$$\mathcal{F}(\rho) = \mathcal{E}(\rho) + \frac{1}{\beta} S(\rho) = \int_{\mathbb{R}^n} \Psi(x)\, \rho + \frac{1}{\beta} \int_{\mathbb{R}^n} \rho \log \rho \tag{13.4}$$

consisting of the sum of the chemical energy $\mathcal{E}$ defined by the potential Ψ and the (negative) entropy functional S. We now make two informal remarks pointing to the conclusion that it should be possible to interpret the Fokker–Planck equation as the gradient flow of the free energy $\mathcal{F}$.

Remark 13.1   The free energy $\mathcal{F}$ is a Lyapunov functional of (13.1). We can prove this informally (i.e., assuming that differentiations and integrations by parts can be carried over as expected) by noticing that if $\rho(t) = \rho(x,t)$ solves (13.1), then

$$\frac{d}{dt}\mathcal{F}(\rho(t)) = \int_{\mathbb{R}^n} \left(\Psi(x) + \frac{1 + \log \rho}{\beta}\right) \frac{\partial \rho}{\partial t}$$

$$= \int_{\mathbb{R}^n} \left(\Psi(x) + \frac{1 + \log \rho}{\beta}\right) \left(\operatorname{div}\left(\nabla\Psi(x)\rho\right) + \frac{1}{\beta}\Delta\rho\right)$$

---

[1]   A formal derivation would require establishing a suitably strong decay at infinity for ρ.

$$= -\int_{\mathbb{R}^n} \left(\nabla\Psi(x) + \frac{1}{\beta}\frac{\nabla\rho}{\rho}\right) \cdot \left(\nabla\Psi(x)\rho + \frac{\nabla\rho}{\beta}\right)$$

$$= -\int_{\mathbb{R}^n} \left|\nabla\left(\Psi + \frac{\log\rho}{\beta}\right)\right|^2 \rho \le 0.$$

In particular,

$$\mathcal{D}(\rho) = \int_{\mathbb{R}^n} \left|\nabla\left(\Psi + \frac{\log\rho}{\beta}\right)\right|^2 \rho, \tag{13.5}$$

is the dissipation functional of the free energy $\mathcal{F}$ along the Fokker–Planck flow.

Remark 13.2 The free energy $\mathcal{F}$ admits a unique critical point, which is the stationary state ρ_* described in (13.3). More precisely, if $\mu = \rho\,d\mathcal{L}^n \in \mathcal{P}_{ac}(\mathbb{R}^n)$ is such that ρ is continuous with $\{\rho > 0\} = \mathbb{R}^n$, and if ρ is a critical point of $\mathcal{F}$ in the sense that

$$\frac{d}{ds}\bigg|_{s=0} \mathcal{F}(\rho + s\,\varphi) = 0, \qquad \forall\varphi \in C_c^\infty(\mathbb{R}^n), \quad \int_{\mathbb{R}^n} \varphi = 0, \tag{13.6}$$

then $\rho = \rho_*$ with ρ_* as in (13.3). Indeed, $\{\rho > 0\} = \mathbb{R}^n$ implies that for every φ as above one has $(\rho + s\,\varphi)\,d\mathcal{L}^n \in \mathcal{P}_{ac}(\mathbb{R}^n)$ whenever s is sufficiently small, and then (13.6) boils down to

$$\int_{\mathbb{R}^n} \left(\Psi + \frac{1 + \log\rho}{\beta}\right)\varphi = 0, \qquad \forall\varphi \in C_c^\infty(\mathbb{R}^n), \quad \int_{\mathbb{R}^n} \varphi = 0.$$

This implies that $\Psi + \beta^{-1}(1 + \log\rho) = $ constant on $\mathbb{R}^n$, i.e., $\rho = \rho_*$.

We close this section presenting an additional motivation for studying the Fokker–Planck equation, namely, we prove that the Fokker–Planck equation with quadratic potential is the **blow-up flow of the heat equation**. (This remark plays no role in the analysis of the minimizing movements scheme done in the subsequent sections.)

Remark 13.3 Let us consider the solution $\rho = \rho(x,t)$ of the heat equation on the whole space $\mathbb{R}^n$, that is $\partial_t\rho = \Delta\rho$ on $\mathbb{R}^n$, with initial datum $\rho_0 \in L^1(\mathbb{R}^n)$, $\rho_0 \ge 0$. As it is easily deduced from (12.12), we know that $\rho(t) \to 0^+$ uniformly on $\mathbb{R}^n$, with $0 \le \rho \le t^{-n/2}\|\rho_0\|_{L^1(\mathbb{R}^n)}$. It is thus interesting to "blow-up" the heat flow near an arbitrary point $x_0 \in \mathbb{R}^n$, trying to capture more precisely the nature of this exponential decay toward 0. To this end is we consider L^1-norm preserving rescalings of ρ of the form

$$\sigma(y,\tau) = \phi(\tau)^n \rho(\phi(\tau)(y - x_0), \psi(\tau)), \qquad y \in \mathbb{R}^n, \tau \ge 0,$$

determined by functions $\phi,\psi : [0,\infty) \to (0,\infty)$ with $\phi(0) = 1$ and $\psi(0) = 0$. Notice that, whatever the choice of ϕ and ψ we make, the *Ansatz* made on σ is such that

$$\int_{\mathbb{R}^n} \sigma(\tau) = \int_{\mathbb{R}^n} \rho(\psi(\tau)) = \int_{\mathbb{R}^n} \rho_0 \qquad \forall \tau > 0.$$

We now want to choose ϕ and ψ wisely, so that the dynamics of σ is also described by a parabolic PDE: to this end, we compute

$$\nabla_y^2 \sigma = \phi^{n+2} \nabla_x^2 \rho,$$

$$\partial_\tau \sigma = n \phi^{n-1} \phi' \rho + \phi^n \phi' (y - x_0) \cdot \nabla_x \rho + \phi^n \psi' \partial_t \rho$$

$$= \frac{\phi'}{\phi} \operatorname{div}_y ((y - x_0) \sigma) + \phi^n \psi' \partial_t \rho.$$

In particular, if $\phi' = \phi$ and $\phi^n \psi' = \phi^{n+2}$, that is (taking into account the initial conditions $\phi(0) = 1$ and $\psi(0) = 0$), if

$$\phi(\tau) = e^\tau, \qquad \psi(\tau) = \frac{e^{2\tau} - 1}{2},$$

then σ solves the Fokker–Planck equation with $\Psi(y) = |y - x_0|^2/2$ and $\beta = 1$:

$$\partial_\tau \sigma = \operatorname{div} ((y - x_0) \sigma) + \Delta \sigma. \qquad (13.7)$$

As much as the heat equation describes the dynamics of the position probability density ρ of a particle undergoing random molecular collisions, the above scaling analysis shows that, near a point x_0 in space, the heat equation dynamics is the superposition of random molecular collisions with the action of the force field $y - x_0$ pointing away from x_0. Based on our previous discussion on the Fokker–Planck equation, we expect this refined local dynamics to have the asymptotic equilibrium state $\sigma(y) = e^{-|y-x_0|^2/2}$. This result, read in terms of the heat equation by inverting $\tau = \psi^{-1}(t) = \log \sqrt{1 + 2t}$, means that

$$(1 + 2t)^{n/2} \rho\left(\sqrt{1 + 2t} (y - x_0), t \right) \approx e^{-|y-x_0|^2/2} \qquad \text{as } t \to +\infty,$$

a prediction that is easily confirmed by working with the resolution formula (12.12).

13.2 First Variation Formulae for Inner Variations

Implementing the minimizing movements scheme for the free energy $\mathcal{F}$ with respect to the Wasserstein distance $\mathbf{W}_2$ involves the minimization in ρ of functionals of the form

$$\int_{\mathbb{R}^n} \Psi \rho + \int_{\mathbb{R}^n} \rho \log \rho + \frac{1}{2h} \mathbf{W}_2(\rho \, d\mathcal{L}^n, \rho_0 \, d\mathcal{L}^n)^2, \qquad (13.8)$$

associated to a given $\rho_0 \, d\mathcal{L}^n \in \mathcal{P}_{2,\mathrm{ac}}(\mathbb{R}^n)$. As in the finite dimensional case (see Theorem 12.3, and (12.21) in particular), the critical point condition (or

Euler–Lagrange equation) of the functional (13.8) plays a crucial role in proving the convergence of the discrete-in-time flow as $h \to 0^+$. Therefore, in this and in the next section, we introduce a notion of variation for the functional (13.8), and then characterize the corresponding notion of critical point.

Given $\mu = \rho \, d\mathcal{L}^n \in \mathcal{P}_{ac}(\mathbb{R}^n)$ we can take variations μ_t of μ in the form

$$\mu_t = (\rho + t \, \varphi) \, d\mathcal{L}^n, \tag{13.9}$$

whenever $\varphi \in C_c^0(\mathbb{R}^n)$ is such that $\mathrm{spt}\,\varphi \subset\subset \{\rho > \delta\}$ for some $\delta > 0$, t is sufficiently small with respect to δ and $\|\varphi\|_{C^0(\mathbb{R}^n)}$, and provided $\int_{\mathbb{R}^n} \varphi = 0$. Indeed, under these conditions, we trivially have $\mu_t \in \mathcal{P}_{ac}(\mathbb{R}^n)$, and therefore we can try to differentiate at $t = 0$ any functional defined on $\mathcal{P}_{ac}(\mathbb{R}^n)$ along the curve $t \mapsto \mu_t$. However, given the discussion contained in Section 10.1 (see, in particular (12.13)), we do not expect the Fokker–Planck equation and the free energy $\mathcal{F}$ to be linked through the "outer" variations defined in (13.9). We shall rather resort to a kind of variation that looks more natural from the point of view of displacement convexity (see Section 10.1), that is the notion of "inner" variation.

Given $\varepsilon > 0$, a vector field $\mathbf{u} \in C_c^\infty(\mathbb{R}^n; \mathbb{R}^n)$, and $\Phi \in C^\infty(\mathbb{R}^n \times (-\varepsilon, \varepsilon); \mathbb{R}^n)$, we say that $\{\Phi_t\}_{|t|<\varepsilon}$ is an **inner variation with initial velocity u** if, for each $|t| < \varepsilon$, $\Phi_t(x) = \Phi(x, t)$ defines a smooth diffeomorphism $\Phi_t : \mathbb{R}^n \to \mathbb{R}^n$ with $\{\Phi_t \neq \mathbf{id}\} \subset\subset \mathbb{R}^n$ and such that, uniformly on $x \in \mathbb{R}^n$, we have

$$\begin{aligned} \Phi_t(x) &= x + t\,\mathbf{u}(x) + \mathrm{O}(t^2), \\ \nabla\Phi_t(x) &= \mathrm{Id} + t\,\nabla\mathbf{u}(x) + \mathrm{O}(t^2) \end{aligned} \qquad \text{as } t \to 0. \tag{13.10}$$

Notice that given $\mathbf{u} \in C_c^\infty(\mathbb{R}^n; \mathbb{R}^n)$, there are several ways to define inner variations having $\mathbf{u}$ has their initial velocity. The simplest one is taking $\Phi_t(x) = x + t\,\mathbf{u}(x)$: for a suitable small ε all the desired properties will follow. Another possibility is considering the ODE $x'(t) = \mathbf{u}(x(t))$ with initial condition $x(0) = x_0$; denoting by $x(t; x_0)$ the solution of such problem, $\Phi(x_0, t) = x(t; x_0)$ defines an inner variation with initial velocity $\mathbf{u}$. In any case, as soon as (13.10) holds, then the first order Taylor expansion of the determinant at the identity matrix, namely $\det(\mathrm{Id} + t\,A) = 1 + t\,\mathrm{trace}(A) + \mathrm{O}(t^2)$ for $t \to 0$ (see, e.g., [Mag12, Lemma 17.4]), allows one to deduce that

$$\det \nabla\Phi_t = 1 + t\,\mathrm{div}\,\mathbf{u} + \mathrm{O}(t^2) \qquad \text{uniformly on } \mathbb{R}^n \text{ as } t \to 0, \tag{13.11}$$

an important formula which relates the Jacobian of Φ_t to the divergence of $\mathbf{u}$. Inner variations are well-suited to work in $\mathcal{P}_{ac}(\mathbb{R}^n)$, as stated in the following proposition.

Proposition 13.4 (Inner variations in $\mathcal{P}_{ac}(\mathbb{R}^n)$) *If $\{\Phi_t\}_{|t|<\varepsilon}$ is an inner variation and $\mu = \rho\, d\mathcal{L}^n \in \mathcal{P}_{ac}(\mathbb{R}^n)$, then*

$$\mu_t = (\Phi_t)_\# \mu \in \mathcal{P}_{ac}(\mathbb{R}^n) \qquad \forall |t| < \varepsilon, \tag{13.12}$$

and if we write $\mu_t = \rho_t\, d\mathcal{L}^n$, then ρ_t satisfies

$$\rho(x) = \rho_t(\Phi_t(x))\, \det \nabla\Phi_t(x), \qquad \forall x \in \mathbb{R}^n, |t| < \varepsilon. \tag{13.13}$$

Finally, if $\mu \in \mathcal{P}_{2,ac}(\mathbb{R}^n)$, then $\mu_t \in \mathcal{P}_{2,ac}(\mathbb{R}^n)$ for every $|t| < \varepsilon$.

Proof We have already proved similar conclusions in the non-smooth case where the role of Φ_t was played by $(1-t)\,\mathbf{id} + t\,\nabla f$, for ∇f the gradient of a convex map which transports μ – so we are giving the details just for the ease of the reader. First, we trivially have $\mu_t(\mathbb{R}^n) = \mu(\Phi_t^{-1}(\mathbb{R}^n)) = \mu(\mathbb{R}^n) = 1$. Second, compare with Proposition 10.2, one has $(\Phi_t)_\#\mu \ll \mathcal{L}^n$ because Φ_t^{-1} is a Lipschitz map on $\mathbb{R}^n$ and $\mu \ll \mathcal{L}^n$: therefore, if $E \in \mathcal{B}(\mathbb{R}^n)$ and $\mathcal{L}^n(E) = 0$, then $\mathcal{L}^n(\Phi_t^{-1}(E)) \le (\mathrm{Lip}\Phi_t^{-1})^n\,\mathcal{L}^n(E) = 0$ and thus

$$(\Phi_t)_\#\mu(E) = \int_{\Phi_t^{-1}(E)} \rho(x)\, dx = 0,$$

proving (13.12). Third, if $\mu_t = \rho_t\, d\mathcal{L}^n$ and $\varphi \in C_c^0(\mathbb{R}^n)$, then by $\mu_t = (\Phi_t)_\#\mu$ and by applying the area formula to the map Φ_t we find

$$\int_{\mathbb{R}^n} \varphi(\Phi_t)\, \rho = \int_{\mathbb{R}^n} \varphi\, d[(\Phi_t)_\#\mu]$$
$$= \int_{\mathbb{R}^n} \varphi\, \rho_t = \int_{\mathbb{R}^n} \varphi(\Phi_t)\, \rho_t(\Phi_t)\, \det\nabla\Phi_t,$$

which is (13.13). Finally, by (13.10) and since $\mathbf{u}$ is bounded on $\mathbb{R}^n$, there exists $C > 0$ such that $|\Phi_t(x)| \le C(1 + |x|)$ for every $x \in \mathbb{R}^n$ and $|t| < \varepsilon$, therefore

$$\int_{\mathbb{R}^n} |x|^2\, d\mu_t(x) = \int_{\mathbb{R}^n} |\Phi_t(x)|^2\, d\mu(x) \le C\left\{1 + \int_{\mathbb{R}^n} |x|^2\, d\mu(x)\right\},$$

and $\mu_t \in \mathcal{P}_{2,ac}(\mathbb{R}^n)$ if $\mu \in \mathcal{P}_{2,ac}(\mathbb{R}^n)$. $\qquad\square$

We now discuss first variation formulae for the various terms appearing in 13.8. We start with the chemical energy $\mathcal{E}$ induced by the potential Ψ, on which we make the following assumptions,

$$\begin{cases} \Psi \in C^\infty(\mathbb{R}^n), \\ \Psi(x) \ge 0, \\ |\nabla\Psi(x)| \le C\,(1 + |x|), \qquad \forall x \in \mathbb{R}^n. \end{cases} \tag{13.14}$$

As a first consequence, $\mathcal{E}(\rho) = \int_{\mathbb{R}^n} \Psi\,\rho \in [0,\infty)$ for every $\mu = \rho\, d\mathcal{L}^n \in \mathcal{P}_{2,ac}(\mathbb{R}^n)$.

Proposition 13.5 *If* Ψ *satisfies* (13.14), $\{\Phi_t\}_{|t|<\varepsilon}$ *is an inner variation with initial velocity* $\mathbf{u} \in C_c^\infty(\mathbb{R}^n; \mathbb{R}^n)$, $\mu = \rho \, d\mathcal{L}^n \in \mathcal{P}_{2,\mathrm{ac}}(\mathbb{R}^n)$, *and we set* $\rho_t \, d\mathcal{L}^n = (\Phi_t)_\# \mu$, *then*

$$\int_{\mathbb{R}^n} \Psi \, \rho_t = \int_{\mathbb{R}^n} \Psi \, \rho + t \int_{\mathbb{R}^n} (\mathbf{u} \cdot \nabla \Psi) \, \rho + O(t^2). \tag{13.15}$$

Proof By $(\Phi_t)_\# \mu = \mu_t$ and since Ψ is nonnegative we can use the push-forward formula to find that

$$\int_{\mathbb{R}^n} \Psi \, \rho_t = \int_{\mathbb{R}^n} \Psi(\Phi_t) \, \rho.$$

By smoothness of Ψ and by the properties of inner variations, we have

$$\Psi(\Phi_t(x)) = \Psi(x) + t \, \mathbf{u}(x) \cdot \nabla \Psi(x) + O(t^2) \qquad \text{uniformly on } x \in \mathbb{R}^n \text{ as } t \to 0$$

so that the conclusion follows. □

Next, we obtain a first variation formula for the Wasserstein distance from a reference measure. In general this is a first variation inequality (rather than identity) because $\mathbf{W}_2$ itself is obtained from a minimization process. When combined with a minimality condition (like in the typical applications to the minimizing movements scheme in $(\mathcal{P}_2(\mathbb{R}^n), \mathbf{W}_2)$, see e.g., step three in the proof of Theorem 13.10), the inequality becomes of course an identity.

Proposition 13.6 *If* $\nu \in \mathcal{P}_2(\mathbb{R}^n)$, $\{\Phi_t\}_{|t|<\varepsilon}$ *is an inner variation with initial velocity* $\mathbf{u} \in C_c^\infty(\mathbb{R}^n; \mathbb{R}^n)$, $\mu = \rho \, d\mathcal{L}^n \in \mathcal{P}_{2,\mathrm{ac}}(\mathbb{R}^n)$ *and we set* $\mu_t = (\Phi_t)_\# \mu$, *then*

$$\mathbf{W}_2(\mu_t, \nu)^2 \le \mathbf{W}_2(\mu, \nu)^2 + 2t \int_{\mathbb{R}^n} \mathbf{u} \cdot (x - T) \, \rho + O(t^2), \tag{13.16}$$

where T is any transport map from μ to ν (the Brenier map ∇f from μ to ν being a possible choice).

Proof Since $T \circ \Phi_t^{-1}$ transports μ_t into ν, by $(\Phi_t)_\# \mu = \mu_t$ we find

$$\mathbf{W}_2(\mu_t, \nu)^2 \le \int_{\mathbb{R}^n} |T(\Phi_t^{-1}(y)) - y|^2 \, d\mu_t(y) = \int_{\mathbb{R}^n} |T - \Phi_t|^2 \, d\mu.$$

Since $\Phi_t(x) = x + t \, \mathbf{u}(x) + O(t^2)$ uniformly on $x \in \mathbb{R}^n$ as $t \to 0$, we find that

$$
\begin{aligned}
|T(x) - \Phi_t(x)|^2 &= |(T(x) - x) - t \, \mathbf{u}(x) + O(t^2)|^2 \\
&= |T(x) - x|^2 + 2t \, \mathbf{u}(x) \cdot (x - T(x)) + O(t^2)
\end{aligned}
$$

uniformly on $x \in \mathbb{R}^n$ as $t \to 0$. An integration with respect to μ, combined with $\mathbf{W}_2(\mu, \nu)^2 \le \int_{\mathbb{R}^n} |T(x) - x|^2 \, d\mu(x)$, gives (13.16). □

We now discuss informally (i.e., disregarding integrability issues) the problem of finding the first variation formula for an internal energy of the form

$$\mathcal{U}(\rho) = \int_{\mathbb{R}^n} U(\rho(x)) \, dx,$$

corresponding, say, to a continuous $U : [0, \infty) \to \mathbb{R}$ with $U(0) = 0$. (The particular case of the entropy functional $\mathcal{S}(\rho) = \int_{\mathbb{R}^n} \rho \log \rho$ will be rigorously addressed in the next section.) Our informal claim is that, when all the technical details fall in place, the first variation formula along an inner variation $\{\Phi_t\}_{|t|<\varepsilon}$ with initial velocity $\mathbf{u}$ takes the form

$$\frac{d}{dt}\Big|_{t=0} \mathcal{U}(\rho_t) = \int_{\{\rho>0\}} (U(\rho) - \rho \, U'(\rho)) \operatorname{div} \mathbf{u}, \qquad (13.17)$$

where, as usual, $\rho_t \, d\mathcal{L}^n = (\Phi_t)_\# \mu$ and $\mu = \rho \, d\mathcal{L}^n$. Indeed, by $U(0) = 0$ we have

$$\mathcal{U}(\rho_t) = \int_{\{\rho_t>0\}} \frac{U(\rho_t)}{\rho_t} \rho_t = \int_{\{\rho>0\}} \frac{U(\rho_t(\Phi_t))}{\rho_t(\Phi_t)} \rho.$$

By plugging (13.13) into this formula we find that

$$\mathcal{U}(\rho_t) = \int_{\{\rho>0\}} U\left(\frac{\rho(x)}{\det \nabla \Phi_t(x)}\right) \det \nabla \Phi_t(x) \, dx.$$

By (13.11), and assuming that we can exchange integration and differentiation, we find

$$\begin{aligned}
\mathcal{U}(\rho_t) &= \int_{\{\rho>0\}} U\Big(\rho - t\,\rho\operatorname{div}\mathbf{u} + O(t^2)\Big)\,(1 + t\operatorname{div}\mathbf{u} + O(t^2)) \\
&= \int_{\{\rho>0\}} \Big(U(\rho) - t\,U'(\rho)\,\rho\operatorname{div}\mathbf{u} + O(t^2)\Big)\,(1 + t\operatorname{div}\mathbf{u} + O(t^2)) \\
&= \mathcal{U}(\rho) + t \int_{\{\rho>0\}} (U(\rho) - \rho\,U'(\rho))\operatorname{div}\mathbf{u} + O(t^2),
\end{aligned}$$

and (13.17) follows. Looking back at the examples of internal energies listed in Chapter 10, we find that

$$\frac{d}{dt}\Big|_{t=0} \mathcal{U}(\rho_t) = -\int_{\{\rho>0\}} \rho \operatorname{div} \mathbf{u}, \qquad \text{if } U(\rho) = \rho \log \rho,$$

$$\frac{d}{dt}\Big|_{t=0} \mathcal{U}(\rho_t) = (1 - \gamma) \int_{\{\rho>0\}} \rho^\gamma \operatorname{div} \mathbf{u}, \qquad \text{if } U(\rho) = \rho^\gamma.$$

In particular, when $\gamma = 1$ and $U(\rho) = \rho$, the first variation is always identically 0 (as expected, since we have $\int_{\mathbb{R}^n} \rho = \int_{\mathbb{R}^n} \rho_t = 1$ for every t).

13.3 Analysis of the Entropy Functional

We now take a closer look at the entropy functional

$$S(\rho) = \int_{\mathbb{R}^n} \rho \log \rho,$$

where some care has to be paid toward integrability issues. We shall work in the class

$$\mathcal{A} = \left\{ \rho : \rho \, d\mathcal{L}^n \in \mathcal{P}_{2,\mathrm{ac}}(\mathbb{R}^n) \right\} = \left\{ \rho : \int_{\mathbb{R}^n} \rho = 1, \rho \geq 0, M(\rho) < \infty \right\} \quad (13.18)$$

$$M(\rho) = \int_{\mathbb{R}^n} |x|^2 \, \rho(x) \, dx. \quad (13.19)$$

Thanks to (10.5) with $p = 2$, we have

$$S(\rho) \geq -M(\rho) - \log \left(\int_{\mathbb{R}^n} e^{-|x|^2} \right), \qquad \forall \rho \in \mathcal{A}, \quad (13.20)$$

so that S is well-defined, with values in $\mathbb{R} \cup \{+\infty\}$, on $\mathcal{A}$. However, this is pretty much all that is implied by (13.20), as exemplified in the following remarks.

Remark 13.7 (*S* is unbounded from below on $\mathcal{A}$)   In the limit $R \to +\infty$, the densities

$$\rho_R = \frac{1_{B_R}}{\omega_n \, R^n}, \qquad R > 0,$$

correspond to a "cloud of gas" that rarefies indefinitely over increasingly larger portions of space. We thus expect the physical entropy of such configurations to diverge to $+\infty$ as $R \to +\infty$, and indeed

$$S(\rho_R) = \log(1/\omega_n) + n \, \log(1/R) \to -\infty \qquad \text{as } R \to +\infty.$$

Of course in this case we have $M(\rho_R) = O(R^2)$ as $R \to +\infty$.

Remark 13.8 (*S* can take the value $+\infty$ on $\mathcal{A}$)   Consider, for example,

$$\rho(x) = c_0 \frac{1_{B_{1/2} \setminus \{0\}}(x)}{|x| \, \log^2(|x|)}, \qquad x \in \mathbb{R}.$$

By the first identity in

$$\int \frac{dr}{r \log^2(r)} = -\frac{1}{\log(r)}, \qquad \int \frac{dr}{r \, (-\log(r))} = -\log(-\log(r)), \quad (13.21)$$

we have $\rho \in L^1(\mathbb{R})$, so that we can get $\int_{\mathbb{R}} \rho = 1$ by suitably choosing c_0. Moreover, $M(\rho) < \infty$ since $\{\rho > 0\}$ is bounded so that $\rho \in \mathcal{A}$. At the same time $\rho \log \rho \geq 1/|x| \, |\log |x||$ for $0 < |x| < 1/2$, so that the second identity in (13.21) gives $S(\rho) = +\infty$.

In the next theorem we collect various properties of the entropy functional S over $\mathcal{A}$. Notice (13.22), which improves the lower bound in (13.20) with the appearance of a sublinear power of $M(\rho)$: this will be useful in subsequent compactness arguments.

Theorem 13.9 (Properties of the entropy functional S)   *The following properties hold:*

(i) Entropy-moment bound: *If $\rho \in \mathcal{A}$ and $\alpha \in (n/(n+2), 1)$, then*

$$0 \geq \int_{\{\rho < 1\}} \rho \log \rho \geq -C(n, \alpha)\, (1 + M(\rho))^{\alpha}. \tag{13.22}$$

(ii) Lower semicontinuity: *If $\rho_j, \rho \in \mathcal{A}$, $\rho_j\, d\mathcal{L}^n \overset{*}{\rightharpoonup} \rho\, d\mathcal{L}^n$ as $j \to \infty$, and*

$$\sup_j M(\rho_j) < \infty,$$

then

$$S(\rho) \leq \liminf_{j \to \infty} S(\rho_j). \tag{13.23}$$

(iii) Differentiability of S: *If $\{\Phi_t\}_{|t| < \varepsilon}$ is an inner variation with initial velocity $\mathbf{u} \in C_c^\infty(\mathbb{R}^n; \mathbb{R}^n)$, $\rho \in \mathcal{A}$ is such that $S(\rho) < \infty$, and $\rho_t\, d\mathcal{L}^n = (\Phi_t)_{\#}(\rho\, d\mathcal{L}^n)$, then $\rho_t \in \mathcal{A}$, $S(\rho_t) < \infty$ for every $|t| < \varepsilon$, and*

$$S(\rho_t) = S(\rho) - t \int_{\mathbb{R}^n} \rho \operatorname{div} \mathbf{u} + O(t^2) \qquad as\ t \to 0. \tag{13.24}$$

Proof   *Step one*: We prove that for every $R \geq 0$

$$0 \geq \int_{\{\rho < 1\} \setminus B_R} \rho \log \rho \geq -\frac{C(n, \alpha)}{1 + R^{2\alpha - n(1-\alpha)}}\, (1 + M(\rho))^{\alpha}. \tag{13.25}$$

In particular, (13.22) follows from (13.25) by taking $R = 0$. To prove (13.25): if $C(\alpha)$ is such that $r\,|\log r| \leq C(\alpha) r^\alpha$ for every $r \in (0, 1)$, then for every $R \geq 0$,

$$\int_{\{\rho < 1\} \setminus B_R} \rho\,|\log \rho| \leq C(\alpha) \int_{\{\rho < 1\} \setminus B_R} \rho^\alpha$$

$$\leq C(\alpha) \left(\int_{\mathbb{R}^n} (1 + |x|^2)\rho \right)^\alpha \left(\int_{\mathbb{R}^n \setminus B_R} \frac{dx}{(1 + |x|^2)^{\alpha/(1-\alpha)}} \right)^{1-\alpha},$$

which indeed gives (13.25).

Step two: We prove (13.23) (notice that this is an inequality between elements of $\mathbb{R} \cup \{+\infty\}$). Let $\rho_j^\varepsilon = \rho_j \star \eta_\varepsilon$ and $\rho^\varepsilon = \rho \star \eta_\varepsilon$ where η_ε is a standard regularizing kernel. Since $\rho_j\, d\mathcal{L}^n \overset{*}{\rightharpoonup} \rho\, d\mathcal{L}^n$ implies $\rho_j^\varepsilon(x) \to \rho^\varepsilon(x)$ for every

$x \in \mathbb{R}^n$, since $\mathcal{L}^n \llcorner B_{R-\varepsilon}$ is a finite measure, and since $r \, \log r$ is bounded from below, we can apply Fatou's lemma to deduce that

$$\int_{B_{R-\varepsilon}} \rho^\varepsilon \, \log \rho^\varepsilon \leq \liminf_{j \to \infty} \int_{B_{R-\varepsilon}} \rho_j^\varepsilon \, \log \rho_j^\varepsilon$$

$$\leq \liminf_{j \to \infty} \int_{B_{R-\varepsilon}} dx \int_{\mathbb{R}^n} \rho_j(y) \, \log(\rho_j(y)) \, \eta_\varepsilon(x - y) \, dy$$

where the second inequality follows by Jensen inequality applied to the convex function $r \, \log r$ and to the probability measure $\eta_\varepsilon(x - y) \, dy$. Setting $\lambda_\varepsilon(y) = \int_{B_{R-\varepsilon} \cap B_\varepsilon(y)} \eta_\varepsilon(y - x) \, dx$, so that $0 \leq \lambda_\varepsilon \leq 1$ on $\mathbb{R}^n$ with $\lambda_\varepsilon = 1$ on $B_{R-2\varepsilon}$, by Fubini's theorem we find

$$\int_{B_{R-\varepsilon}} dx \int_{\mathbb{R}^n} \rho_j(y) \, \log(\rho_j(y)) \, \eta_\varepsilon(x - y) \, dy = \int_{B_R} \lambda_\varepsilon \, \rho_j \, \log(\rho_j)$$

$$\leq \int_{B_R \cap \{\rho_j > 1\}} \rho_j \, \log(\rho_j) + \int_{B_R \cap \{\rho_j < 1\}} \lambda_\varepsilon \, \rho_j \, \log(\rho_j)$$

$$= \int_{B_R} \rho_j \, \log(\rho_j) + \int_{(B_R \setminus B_{R-2\varepsilon}) \cap \{\rho_j < 1\}} (1 - \lambda_\varepsilon) \, |\rho_j \, \log(\rho_j)|,$$

and thus, by $\mathcal{L}^n(B_R \setminus B_{R-2\varepsilon}) \leq C(n) \, \varepsilon$ and $|r \, \log(r)| \leq 1/e$ for $r \in (0, 1)$,

$$\int_{B_{R-\varepsilon}} \rho^\varepsilon \, \log \rho^\varepsilon \leq \liminf_{j \to \infty} \int_{B_R} \rho_j \, \log \rho_j + C(n) \, \varepsilon.$$

Since $1_{B_{R-\varepsilon}} \, \rho^\varepsilon \to 1_{B_R} \, \rho \, \mathcal{L}^n$-a.e. on $\mathbb{R}^n$ as $\varepsilon \to 0^+$, $\mathcal{L}^n \llcorner B_R$ is a finite measure, and $r \, \log r$ is continuous and bounded from below, we can apply again Fatou's lemma to find

$$\int_{B_R} \rho \, \log \rho \leq \liminf_{\varepsilon \to 0^+} \int_{B_{R-\varepsilon}} \rho^\varepsilon \, \log \rho^\varepsilon \leq \liminf_{j \to \infty} \int_{B_R} \rho_j \, \log \rho_j. \quad (13.26)$$

Let $\log^+(r) = \max\{0, \log r\}$. Since $r \, \log^+(r)$ is convex and nonnegative we can apply Fatou's lemma on the unbounded domains $\mathbb{R}^n \setminus B_R$ and $\mathbb{R}^n \setminus B_{R+\varepsilon}$, and exploit again Jensen's inequality, to find that

$$\int_{\mathbb{R}^n \setminus B_R} \rho \, \log^+(\rho) \leq \liminf_{\varepsilon \to 0^+} \int_{\mathbb{R}^n \setminus B_{R+\varepsilon}} \rho^\varepsilon \, \log^+(\rho^\varepsilon)$$

$$\int_{\mathbb{R}^n \setminus B_{R+\varepsilon}} \rho^\varepsilon \, \log^+(\rho^\varepsilon) \leq \liminf_{j \to \infty} \int_{\mathbb{R}^n \setminus B_R} \rho_j \, \log^+(\rho_j) \qquad \forall \varepsilon > 0,$$

thus concluding that

$$\int_{\mathbb{R}^n \setminus B_R} \rho \, \log^+(\rho) \leq \liminf_{j \to \infty} \int_{\mathbb{R}^n \setminus B_R} \rho_j \, \log^+(\rho_j). \quad (13.27)$$

By (13.25) and since $M(\rho_j) \leq m$ for every j, we see that for a given $\alpha \in (n/(n+2), 1)$ we have

$$0 \geq \int_{\{\rho_j < 1\} \backslash B_R} \rho_j \log \rho_j \geq -\frac{C(n, \alpha)(1+m)^\alpha}{1 + R^{2\alpha - n(1-\alpha)}},$$

and thus

$$\limsup_{R \to \infty} \left| \int_{\{\rho_j < 1\} \backslash B_R} \rho_j \log \rho_j \right| = 0,$$

which combined with (13.27) and $\log(r) = \log^+(r) + 1_{(0,1)}(r) \log(r)$ gives

$$\int_{\mathbb{R}^n \backslash B_R} \rho \log^+(\rho) \leq \liminf_{j \to \infty} \int_{\mathbb{R}^n \backslash B_R} \rho_j \log \rho_j. \qquad (13.28)$$

Finally, adding up (13.26) and (13.28) and noticing that

$$S(\rho) \leq \int_{B_R} \rho \log \rho + \int_{\mathbb{R}^n \backslash B_R} \rho \log^+(\rho),$$

we deduce the validity of (13.23).

Step three: We finally prove (13.24). The fact that $\rho_t \in \mathcal{A}$ was proved in Proposition 13.6, and it shows $S(\rho_t) > -\infty$. By the push-forward property $(\Phi_t)_\#(\rho \, d\mathcal{L}^n) = \rho_t \, d\mathcal{L}^n$ we have that

$$\int_{\mathbb{R}^n} \rho_t \log \rho_t = \int_{\mathbb{R}^n} \rho \log(\rho_t(\Phi_t)) \qquad (13.29)$$

as an identity between elements of $\mathbb{R} \cup \{+\infty\}$. By (13.13),

$$\log(\rho_t(\Phi_t)) = \log \rho - \log(\det \nabla \Phi_t) \qquad (13.30)$$

where we can assume that $|\log(\det \nabla \Phi_t)| \leq 1$ on $\mathbb{R}^n$ for every $|t| < \varepsilon$. In particular, (13.29) gives $S(\rho_t) \leq S(\rho) + 1 < \infty$. Moreover, by (13.30) and (13.11),

$$\log(\rho_t(\Phi_t)) = \log \rho - t \operatorname{div} \mathbf{u} + O(t^2)$$

uniformly on $\mathbb{R}^n$ as $t \to 0$, which, combined with (13.29), gives (13.24). $\qquad \square$

13.4 Implementation of the Minimizing Movements Scheme

We are now ready to implement the minimizing movements scheme on the free energy $\mathcal{F}$. For the sake of simplicity, in the following we shall set

$$\mathcal{W}_2(\rho, \sigma) = \mathbf{W}_2(\rho \, d\mathcal{L}^n, \sigma \, d\mathcal{L}^n),$$

whenever ρ and σ are densities such that $\rho \, d\mathcal{L}^n, \sigma \, d\mathcal{L}^n \in \mathcal{P}_2(\mathbb{R}^n)$. We start by discussing the basic step in the construction of the discrete flow.

Theorem 13.10 (Basic step in the construction of the discrete flow) *Let $\Psi \in C^\infty(\mathbb{R}^n)$, with $\Psi \geq 0$ and $|\nabla\Psi(x)| \leq C(1 + |x|)$ for every $x \in \mathbb{R}^n$, and let $\mathcal{F} : \mathcal{A} \to \mathbb{R} \cup \{+\infty\}$ be defined as*

$$\mathcal{F}(\rho) = \mathcal{E}(\rho) + \mathcal{S}(\rho) = \int_{\mathbb{R}^n} \Psi\,\rho + \int_{\mathbb{R}^n} \rho\,\log\rho,$$

where $\mathcal{A} = \{\rho : \rho\,d\mathcal{L}^n \in \mathcal{P}_{2,\mathrm{ac}}(\mathbb{R}^n), M(\rho) < \infty\}$. If $\rho_0 \in \mathcal{A}$ is such that

$$\mathcal{F}(\rho_0) < \infty, \tag{13.31}$$

then the variational problem

$$\inf\left\{\mathcal{F}(\rho) + \frac{1}{2h}\,W_2(\rho,\rho_0)^2 : \rho \in \mathcal{A}\right\}, \tag{13.32}$$

admits a unique minimizer $\rho_0^{(h)}$ which satisfies the energy estimate

$$\mathcal{F}(\rho_0^{(h)}) + \frac{1}{2h}\,W_2(\rho_0^{(h)},\rho_0)^2 \leq \mathcal{F}(\rho_0). \tag{13.33}$$

Moreover, if ∇f is the Brenier map from $\mu_0^{(h)} = \rho_0^{(h)}\,d\mathcal{L}^n$ to $\mu_0 = \rho_0\,d\mathcal{L}^n$, then the following Euler–Lagrange equation holds: for every $\mathbf{u} \in C_c^\infty(\mathbb{R}^n;\mathbb{R}^n)$,

$$\int_{\mathbb{R}^n} [(\mathbf{u}\cdot\nabla\Psi) - \mathrm{div}\,\mathbf{u}]\,\rho_0^{(h)} + \frac{1}{h}\int_{\mathbb{R}^n} \mathbf{u}\cdot(x - \nabla f)\,\rho_0^{(h)} = 0. \tag{13.34}$$

Proof Step one: We set $M(\rho) = \int_{\mathbb{R}^n} |x|^2\rho(x)\,dx$ and set for brevity

$$\mathcal{F}^{(h)}(\rho;\rho_0) = \mathcal{F}(\rho) + \frac{1}{2h}\,W_2(\rho,\rho_0)^2.$$

Clearly $\mathcal{F}^{(h)}(\rho_0;\rho_0) = \mathcal{F}(\rho_0) < \infty$ so that the infimum in (13.32) is finite and, if a minimizer exists, it trivially satisfies (13.33). Now let us consider a minimizing sequence $\{\rho_j\}_j$ in (13.32), and set $\mu_j = \rho_j\,d\mathcal{L}^n$. The triangular inequality for $\mathbf{W}_2$ gives

$$M(\rho_j) = \mathbf{W}_2(\mu_j,\delta_0)^2 \leq (\mathbf{W}_2(\mu_0,\delta_0) + \mathbf{W}_2(\mu_0,\mu_j))^2$$
$$= (M(\rho_0)^{1/2} + W_2(\rho_0,\rho_j))^2 \leq 2\,M(\rho_0) + 2\,W_2(\rho_0,\rho_j)^2, \tag{13.35}$$

where $M(\rho_0) < \infty$ by $\rho_0 \in \mathcal{A}$. Then, by (13.22) and for $\alpha \in (n/(n+2),1)$ fixed,

$$\mathcal{F}^{(h)}(\rho_j;\rho_0) = \mathcal{F}(\rho_j) + \frac{1}{2h}\,W_2(\rho_j,\rho_0)^2$$
$$\geq \int_{\{\rho_j<1\}} \rho_j\,\log\rho_j + \frac{1}{2h}\left\{\frac{M(\rho_j)}{2} - M(\rho_0)\right\}$$
$$\geq \frac{M(\rho_j)}{4h} - C(n,\alpha)\left(1 + M(\rho_j)\right)^\alpha - \frac{M(\rho_0)}{2h}. \tag{13.36}$$

Since we can assume

$$\mathcal{F}^{(h)}(\rho_j; \rho_0) \le \mathcal{F}(\rho_0) + (1/j) < \infty, \qquad \forall j,$$

by (13.36) and thanks to $\alpha \in (0, 1)$, we find that $\{M(\rho_j)\}_j$ is bounded. In particular, again by (13.22), we have

$$\int_{\{\rho_j < 1\}} \rho_j \log \rho_j \ge -C_*,$$

for a constant C_* depending on n, ρ_0, h and a fixed value of $\alpha \in (n/(n+2), 1)$. In particular, $\Psi \ge 0$ gives

$$\mathcal{F}(\rho_0) + (1/j) \ge S(\rho_j) \ge \int_{\{\rho_j > 1\}} \rho_j \log \rho_j - C_*,$$

and thus $\{\int_{\mathbb{R}^n} \rho_j \,|\log(\rho_j)|\}_j$ is bounded. Since $r > 0 \mapsto r\,|\log(r)|$ is super-linear at infinity, $\{\rho_j\}_j$ is equiintegrable, and by the Dunford–Pettis criterion[2] $\{\rho_j\}_j$ admits a not-relabeled subsequence such that $\rho_j \rightharpoonup \rho_\infty$ in the weak L^1-convergence, where $\rho_\infty \in L^1(\mathbb{R}^n)$. We claim that ρ_∞ is a minimizer in (13.32). Notice that the weak L^1-convergence of ρ_j to ρ_∞ implies immediately the narrow convergence of μ_j to $\mu_\infty = \rho_\infty \, d\mathcal{L}^n$ (in particular, $\mu_\infty \in \mathcal{P}(\mathbb{R}^n)$). Moreover, if φ ranges over $C_c^0(\mathbb{R}^n; [0, 1])$, then

$$M(\rho_\infty) = \sup_\varphi \int_{\mathbb{R}^n} |x|^2 \varphi \, \rho_\infty = \sup_\varphi \lim_{j \to \infty} \int_{\mathbb{R}^n} |x|^2 \varphi \, \rho_j \le \lim_{j \to \infty} M(\rho_j) < \infty,$$

so that $\rho_\infty \in \mathcal{A}$, i.e., ρ_∞ belongs to the minimization class of (13.32). The potential energy term $\int_{\mathbb{R}^n} \Psi(x)\, \rho$ is lower semicontinuous by a similar argument, while $(2h)^{-1}\, W_2(\rho, \rho_0)^2$ is lower semicontinuous by (11.7). Since $\{M(\rho_j)\}_j$ is bounded, the entropy S is lower semicontinuous by Theorem 13.9-(ii). Therefore $\mathcal{F}^{(h)}(\cdot; \rho_0)$ is lower semicontinuous along the sequence ρ_j, and this proves that ρ_∞ is a minimizer of $\mathcal{F}^{(h)}(\cdot; \rho_0)$ in $\mathcal{A}$.

Step two: We prove that $\mathcal{F}^{(h)}(\cdot; \rho_0)$ has a unique minimizer in $\mathcal{A}$. To see this it suffices to use standard convex interpolation. Indeed, if $\sigma, \tau \in \mathcal{A}$ are minimizers of $\mathcal{F}^{(h)}(\cdot; \rho_0)$ on $\mathcal{A}$, then $\rho = (\sigma + \tau)/2 \in \mathcal{A}$. The potential energy $\mathcal{E}$ is linear under standard convex interpolation, i.e., $\mathcal{E}((\sigma + \tau)/2) = (\mathcal{E}(\sigma) + \mathcal{E}(\tau))/2$, while $\rho \mapsto W_2(\rho \, d\mathcal{L}^n, \mu_0)^2$ is convex thanks to Remark 11.9. Finally, the entropy is strictly convex by the strict convexity of $U(\rho) = \rho \log \rho$, therefore $\mathcal{F}^{(h)}(\cdot; \rho_0)$ is strictly convex on $\mathcal{A}$ with respect to standard convex interpolation.

Step three: We are left to show that the unique minimizer $\rho_0^{(h)}$ of $\mathcal{F}^{(h)}(\cdot; \rho_0)$ satisfies (13.34). Indeed by combining minimality with Proposition 13.5, Proposition 13.6 and of Theorem 13.9-(iii) we find that, setting

---

[2]   See, e.g., [AFP00, Theorem 1.38].

$$(\Phi_t)_\#(\rho_0^{(h)}\, d\mathcal{L}^n) = \rho_t^{(h)}\, d\mathcal{L}^n$$

we have, in the limit as $t \to 0^+$,

$$F^{(h)}(\rho_0^{(h)}; \rho_0) \leq \mathcal{F}^{(h)}(\rho_t^{(h)}; \rho_0)$$

$$\leq \mathcal{F}^{(h)}(\rho_0^{(h)}; \rho_0) + t\left\{ \int_{\mathbb{R}^n} [(\mathbf{u} \cdot \nabla\Psi) - \operatorname{div} \mathbf{u}]\, \rho_0^{(h)} + \frac{1}{h} \int_{\mathbb{R}^n} \mathbf{u} \cdot (x - \nabla f)\, \rho_0^{(h)} \right\}$$

$$+ O(t^2),$$

Hence $t \mapsto \mathcal{F}^{(h)}(\rho_t^{(h)}; \rho_0)$ is differentiable at $t = 0$ (which was not obvious since Proposition 13.6 does not prove the differentiability of $\mathcal{W}_2(\rho_t^{(h)}, \rho_0)$ at $t = 0$!), with zero derivative. This proves (13.34), and completes the proof of the theorem. □

Under the assumptions on Ψ and ρ_0 in Theorem 13.10, for every $h > 0$ we define the **discrete Fokker–Planck flow of step** h,

$$\rho^{(h)} : [0, \infty) \to \mathcal{A}$$

by setting

$$\rho^{(h)}(t) = \rho_{k-1}^{(h)} \qquad \text{if } t \in [(k-1)\,h, k\,h), k \geq 1,$$

where the sequence $\{\rho_k^{(h)}\}_{k=0}^\infty$ is iteratively defined by setting $\rho_0^{(h)} = \rho_0$ and, for $k \geq 1$, by letting $\rho_k^{(h)}$ be the unique minimizer in the variational problem

$$\inf \left\{ \mathcal{F}(\rho) + \frac{1}{2h}\, \mathcal{W}_2(\rho, \rho_{k-1}^{(h)})^2 : \rho \in \mathcal{A} \right\}.$$

This definition makes sense thanks to Theorem 13.10, which also gives

$$\mathcal{F}(\rho_k^{(h)}) + \frac{1}{2h}\, \mathcal{W}_2(\rho_k^{(h)}, \rho_{k-1}^{(h)})^2 \leq \mathcal{F}(\rho_{k-1}^{(h)}), \tag{13.37}$$

and that, if ∇f_k is the Brenier map from $\mu_k^{(h)} = \rho_k^{(h)}\, d\mathcal{L}^n$ to $\mu_{k-1}^{(h)}$, then

$$\int_{\mathbb{R}^n} [(\mathbf{u} \cdot \nabla\Psi) - \operatorname{div} \mathbf{u}]\, \rho_k^{(h)} + \frac{1}{h} \int_{\mathbb{R}^n} \mathbf{u} \cdot (x - \nabla f_k)\, \rho_k^{(h)} = 0. \tag{13.38}$$

for every $\mathbf{u} \in C_c^\infty(\mathbb{R}^n; \mathbb{R}^n)$. We now prove the convergence as $h \to 0^+$ of these discrete flows to a solution of the Fokker–Planck equation (13.1), a key result due to Jordan, Kinderlehrer and Otto. Notice the importance of proving convergence as $h \to 0^+$, and not just along some sequence $h_j \to 0^+$, given the numerical interest of these discrete-in-time approximation scheme.

Theorem 13.11 (Jordan–Kinderleherer–Otto theorem) *Let $\rho^{(h)}$ be the discrete Fokker–Planck flow of step h corresponding to Ψ and ρ_0 as in Theorem*

13.10. Then $\rho^{(h)}$ is an approximate weak solution to the Fokker–Planck equation, in the sense that for every $\varphi \in C_c^\infty(\mathbb{R}^n)$ and $\zeta \in C_c^\infty([0,T))$ one has

$$\left| \zeta(0) \int_{\mathbb{R}^n} \varphi \, \rho_0 + \int_{\mathbb{R}^n \times (0,\infty)} \rho^{(h)} \left\{ \varphi \, \zeta' - \zeta \, (\nabla \Psi \cdot \nabla \varphi - \Delta \varphi) \right\} \right| \le C_T \, h, \quad (13.39)$$

where C_T depends on Ψ, ζ and φ and T. Moreover, there exists

$$\rho \in C^\infty(\mathbb{R}^n \times (0,\infty))$$

such that, for every $T > 0$,

$$\rho^{(h)} \rightharpoonup \rho \quad \text{weakly in } L^1((0,T) \times \mathbb{R}^n) \text{ as } h \to 0^+, \quad (13.40)$$

and ρ is the unique solution of

$$\partial_t \rho = \operatorname{div}(\rho \, \nabla \Psi) + \Delta \rho \qquad \text{on } \mathbb{R}^n,$$

such that $\rho(t) \to \rho_0$ in $L^1(\mathbb{R}^n)$ as $t \to 0^+$.

Proof The structure of the proof is the same as in Theorem 12.3. In step one we obtain bounds on the discrete flows $\rho^{(h)}$ that are uniform in h and in $t \in [0,T]$ for $T < \infty$, but are not uniform in $t \in [0,\infty)$, as are those obtained in (12.16) and (12.17): the matter here is made more complicated, with respect to the situation considered in Theorem 12.3, by the fact that $\mathcal{F}$ is not bounded from below. In step two, we use the results of Sections 13.2 and 13.3 to deduce that discrete flow solves the weak discrete formulation (13.39) of the Fokker–Planck equation (which is analogous to (12.20)). In step three we exploit the bounds of step one to obtain sequential compactness of $\rho^{(h)}$ in the weak-L^1 convergence, that we then use to pass to the limit in the weak discrete formulation (13.39) of the Fokker–Planck equation, showing that weak-L^1 subsequential limits of the discrete flow must solve a weak formulation of (13.1), see (13.56). Finally, in step four, we show that there is a unique solution of the weak equation (13.56) under suitable momentum and energy bounds (that are inherited from the estimates of step one). This shows that there is only one possible weak-L^1 subsequential limit for the discrete flows, thus proving convergence as $h \to 0^+$, and completing the proof.

Step one: We obtain some uniform in h bounds on the discrete flow: if $T > 0$, $N h \le T$, then for every $k = 1, \ldots, N$ we have

$$M(\rho_k^{(h)}) \le C_T, \quad (13.41)$$

$$\int_{\{\rho_k^{(h)} > 1\}} \rho_k^{(h)} \log(\rho_k^{(h)}) \le C_T, \quad (13.42)$$

$$\int_{\mathbb{R}^n} \rho_k^{(h)} \Psi \le C_T, \tag{13.43}$$

$$\sum_{k=0}^{N} \mathcal{W}_2(\rho_k^{(h)}, \rho_{k-1}^{(h)})^2 \le C_T h, \tag{13.44}$$

where $C_T = C(n, \Psi, \rho_0, T)$. Indeed, by adding up (13.37) over $j = 1, \ldots, k$, and canceling the common term $\sum_{j=1}^{k-1} \mathcal{F}(\rho_j^{(h)})$ on both sides of the resulting inequality, we find that

$$\mathcal{F}(\rho_k^{(h)}) + \frac{1}{2h} \sum_{j=1}^{k} \mathcal{W}_2(\rho_j^{(h)}, \rho_{j-1}^{(h)})^2 \le \mathcal{F}(\rho_0). \tag{13.45}$$

By $\Psi \ge 0$ and (13.22) we find that, for every $\alpha \in (n/(n+2), 1)$,

$$\mathcal{F}(\rho_k^{(h)}) \ge \int_{\{\rho_k^{(h)} < 1\}} \rho_k^{(h)} \log(\rho_k^{(h)}) \ge -C(n, \alpha)\left(1 + M(\rho_k^{(h)})\right)^\alpha, \tag{13.46}$$

so that (13.45) implies

$$\int_{\mathbb{R}^n} \rho_k^{(h)} \Psi + \int_{\{\rho_k^{(h)} > 1\}} \rho_k^{(h)} \log(\rho_k^{(h)}) + \frac{1}{2h} \sum_{j=1}^{k} \mathcal{W}_2(\rho_j^{(h)}, \rho_{j-1}^{(h)})^2$$
$$\le \mathcal{F}(\rho_0) + C(n, \alpha)\left(1 + M(\rho_k^{(h)})\right)^\alpha. \tag{13.47}$$

Now, by arguing as in (13.35), we find that, if $1 \le k \le N$ and $Nh \le T$, then

$$M(\rho_k^{(h)}) \le 2\mathcal{W}_2(\rho_k^{(h)}, \rho_0)^2 + 2M(\rho_0)$$
$$\le 2k \sum_{j=1}^{k} \mathcal{W}_2(\rho_j^{(h)}, \rho_{j-1}^{(h)})^2 + 2M(\rho_0)$$
$$\le 4kh\left\{\mathcal{F}(\rho_0) + C(n, \alpha)(1 + M(\rho_k^{(h)}))^\alpha\right\} + 2M(\rho_0). \tag{13.48}$$

where in the last inequality we have used (13.47). Since $\alpha < 1$ and $kh \le Nh \le T$, one easily deduces (13.41) from (13.48); and in turn, by combining (13.41) with (13.47) we deduce (13.42), (13.43) and (13.44).

Step two: We prove the validity of (13.39), that is, we prove that if $T > 0$, then for every $\varphi \in C_c^\infty(\mathbb{R}^n)$ and $\zeta \in C_c^\infty([0, T))$ one has

$$\left|\zeta(0) \int_{\mathbb{R}^n} \varphi \rho_0 + \int_{\mathbb{R}^n \times (0, \infty)} \rho^{(h)} \left\{\varphi \zeta' - \zeta\left(\nabla\Psi \cdot \nabla\varphi - \Delta\varphi\right)\right\}\right| \le C_T h, \tag{13.49}$$

where C_T depends on Ψ, ζ and φ and T. Indeed, setting for the sake of brevity

$$\omega[\varphi] = (\nabla\varphi \cdot \nabla\Psi) - \Delta\varphi, \tag{13.50}$$

we notice that, by taking $\mathbf{u} = \nabla\varphi$ in (13.38), we find that, for every $k \geq 1$,

$$\int_{\mathbb{R}^n} \omega[\varphi]\,\rho_k^{(h)} + \frac{1}{h}\int_{\mathbb{R}^n} \nabla\varphi \cdot (x - \nabla f_k)\,\rho_k^{(h)} = 0. \tag{13.51}$$

We can eliminate the Brenier map ∇f_k, and make the discrete time derivative of $\rho_k^{(h)}$ show up, by noticing that, thanks to Taylor's formula,

$$\left|\varphi(\nabla f_k(x)) - \varphi(x) - (\nabla f_k(x) - x) \cdot \nabla\varphi(x)\right| \leq \frac{\|\nabla^2\varphi\|_{C^0(\mathbb{R}^n)}}{2}\,|x - \nabla f_k(x)|^2,$$

so that, taking into account that $\int_{\mathbb{R}^n} \varphi(\nabla f_k)\,\rho_k^{(h)} = \int_{\mathbb{R}^n} \varphi\,\rho_{k-1}^{(h)}$, we obtain

$$\left|\int_{\mathbb{R}^n} \varphi\,\rho_{k-1}^{(h)} - \int_{\mathbb{R}^n} \varphi\,\rho_k^{(h)} - \int_{\mathbb{R}^n} \nabla\varphi \cdot (\nabla f_k - x)\,\rho_k^{(h)}\right|$$

$$\leq \frac{\|\nabla^2\varphi\|_{C^0(\mathbb{R}^n)}}{2}\int_{\mathbb{R}^n} |x - \nabla f_k|^2\,\rho_k^{(h)} = \frac{\|\nabla^2\varphi\|_{C^0(\mathbb{R}^n)}}{2}\,W_2(\rho_k^{(h)},\rho_{k-1}^{(h)})^2.$$

This last inequality, combined with (13.51), gives

$$\left|\int_{\mathbb{R}^n} \omega[\varphi]\,\rho_k^{(h)} + \varphi\,\frac{\rho_k^{(h)} - \rho_{k-1}^{(h)}}{h}\right| \leq \frac{\|\nabla^2\varphi\|_{C^0(\mathbb{R}^n)}}{2h}\,W_2(\rho_k^{(h)},\rho_{k-1}^{(h)})^2, \tag{13.52}$$

for every $k \geq 1$. We now do the discrete analogous of "multiplying by a time test function and integrating in time," that is, we multiply (13.52) by $\zeta_k = \zeta(k\,h)$ and add up over $k = 1,\ldots,N$, for N large enough to have $\operatorname{spt}\zeta \subset\subset [0, N\,h)$: in this way we get

$$\left|\sum_{k=1}^{N} \zeta_k\left\{\int_{\mathbb{R}^n} h\,\omega[\varphi]\,\rho_k^{(h)} + \varphi\,(\rho_k^{(h)} - \rho_{k-1}^{(h)})\right\}\right| \leq C\sum_{k=1}^{N} W_2(\rho_k^{(h)},\rho_{k-1}^{(h)})^2 \leq C_T\,h, \tag{13.53}$$

where we have used (13.44) and C_T depends on T (as well as on n, Ψ, ρ_0, $\|\zeta\|_{C^0(\mathbb{R})}$ and $\|\nabla^2\varphi\|_{C^0(\mathbb{R}^n)}$). To find continuous analogues to terms on the left-hand side of (13.53) we start noticing that, thanks to $\zeta_N = 0$, we have

$$\sum_{k=1}^{N} \zeta_k \int_{\mathbb{R}^n} \varphi\,(\rho_{k-1}^{(h)} - \rho_k^{(h)}) = \zeta(h)\int_{\mathbb{R}^n} \varphi\,\rho_0 + \sum_{k=1}^{N-1} (\zeta_{k+1} - \zeta_k)\int_{\mathbb{R}^n} \varphi\,\rho_k^{(h)}$$

$$= \zeta(h)\int_{\mathbb{R}^n} \varphi\,\rho_0 + \int_h^\infty \zeta'\,dt\int_{\mathbb{R}^n} \varphi\,\rho^{(h)}(t),$$

so that

$$\left|\sum_{k=1}^{N} \zeta_k \int_{\mathbb{R}^n} \varphi\,(\rho_{k-1}^{(h)} - \rho_k^{(h)}) - \left\{\zeta(0)\int_{\mathbb{R}^n} \varphi\,\rho_0 + \int_0^\infty \zeta'\,dt\int_{\mathbb{R}^n} \varphi\,\rho^{(h)}(t)\right\}\right| \leq C\,h, \tag{13.54}$$

with C depending on $\|\zeta'\|_{C^0(\mathbb{R})}$ and $\|\varphi\|_{C^0(\mathbb{R}^n)}$; analogously, taking into account that by definition (13.50) of $\omega[\varphi]$ and by $|\nabla\Psi(x)| \le C\,(1+|x|)$ for every $x \in \mathbb{R}^n$ we have

$$\|\omega[\varphi]\|_{C^0(\mathbb{R}^n)} \le \|\nabla^2\varphi\|_{C^0(\mathbb{R}^n)} + C\,\|\nabla\varphi\|_{C^0(\mathbb{R}^n)}\,\{1 + \mathrm{diam}(\mathrm{spt}\,\varphi)\},$$

we estimate that

$$\left| h\sum_{k=1}^{N} \zeta_k \int_{\mathbb{R}^n} \omega[\varphi]\,\rho_k^{(h)} - \int_0^\infty \zeta\,dt \int_{\mathbb{R}^n} \omega[\varphi]\,\rho^{(h)}(t) \right| \tag{13.55}$$

$$\le \sum_{k=1}^{N} \int_{kh}^{(k+1)\,h} |\zeta_k - \zeta| \int_{\mathbb{R}^n} |\omega[\varphi]|\,\rho^{(h)}(t) + \int_0^h |\zeta|\,dt \int_{\mathbb{R}^n} |\omega[\varphi]|\,\rho^{(h)}(t) \le C\,h,$$

where C depends on $\|\varphi\|_{C^2(\mathbb{R}^n)}$ and $\|\zeta\|_{C^1(\mathbb{R})}$. By combining (13.53), (13.54) and (13.55) we find that

$$\left| \zeta(0) \int_{\mathbb{R}^n} \varphi\,\rho_0 + \int_{\mathbb{R}^n \times (0,\infty)} \{\zeta'\,\varphi - \zeta\,\omega[\varphi]\}\rho^{(h)}(t) \right| \le C_T\,h,$$

that is (13.49).

Step three: We prove that for every $h_j \to 0^+$ there exists a not-relabeled subsequence of h_j and a measurable function $\rho : (0,\infty) \times \mathbb{R}^n \to [0,\infty)$ such that $\int_{\mathbb{R}^n} \rho(t) = 1$ for a.e. $t > 0$,

$$\zeta(0) \int_{\mathbb{R}^n} \varphi\,\rho_0 + \int_{\mathbb{R}^n \times (0,\infty)} \rho\,\big\{\zeta\,(\nabla\Psi \cdot \nabla\varphi - \Delta\varphi) + \varphi\,\zeta'\big\} = 0, \tag{13.56}$$

for every $\zeta \in C_c^\infty([0,\infty))$ and $\varphi \in C_c^\infty(\mathbb{R}^n)$, and such, that for every $T > 0$,

$$\|M(\rho(\cdot))\|_{L^\infty(0,T)} \le C_T \qquad \text{for every } T > 0, \tag{13.57}$$

$$\left\| \int_{\mathbb{R}^n} \rho(x,\cdot)\,\Psi(x)\,dx \right\|_{L^\infty(0,T)} \le C_T. \tag{13.58}$$

Indeed, if we set $U(r) = r\,\max\{0,\log(r)\}$, then (13.42) implies that

$$\sup_{h>0} \sup_{0 \le t \le T} \int_{\mathbb{R}^n} U(\rho^{(h)}(t))\,dx \le C_T.$$

Since $U(r)/r \to +\infty$ as $r \to +\infty$, by the Dunford–Pettis criterion[3] we find that for every $h_j \to 0^+$ there exists a not-relabeled subsequence of h_j and a measurable function $\rho : (0,\infty) \times \mathbb{R}^n \to [0,\infty)$ such that $\rho \in L^1((0,T) \times \mathbb{R}^n)$ for every $T < \infty$ and

$$\rho^{(h_j)} \rightharpoonup \rho \qquad \text{weakly in } L^1((0,T) \times \mathbb{R}^n) \text{ for every } T < \infty. \tag{13.59}$$

---

[3] See, e.g., [AFP00, Theorem 1.38].

We notice that (13.59) implies (13.56) by setting $h = h_j$ and letting $j \to \infty$ in (13.47). Moreover, testing (13.59) on $1_{\mathbb{R}^n \times I}(x,t)$ for a bounded Borel set $I \subset (0,\infty)$ shows that

$$\int_{\mathbb{R}^n \times I} \rho(x,t)\, dx\, dt = \lim_{j \to \infty} \int_{\mathbb{R}^n \times I} \rho^{(h_j)}(x,t)\, dx\, dy = \mathcal{L}^1(I).$$

Should $\mathcal{L}^1(\{t : \int_{\mathbb{R}^n} \rho(t) \neq 1\})$ be of positive measure, we could choose I so to reach a contradiction in the above identity. Therefore

$$\int_{\mathbb{R}^n} \rho(t) = 1 \qquad \text{for a.e. } t > 0. \tag{13.60}$$

Given $t_0 \in (0,T)$, we can similarly test (13.59) with $1_{B_R \times (t_0-\varepsilon, t_0+\varepsilon)}(x,t)$ to find

$$\int_{t_0-\varepsilon}^{t_0+\varepsilon} M(\rho(t))\, dt = \lim_{R \to +\infty} \int_{B_R \times (t_0-\varepsilon, t_0+\varepsilon)} |x|^2\, \rho(x,t)\, dx\, dt$$

$$\leq \lim_{R \to +\infty} \limsup_{j \to \infty} \int_{B_R \times (t_0-\varepsilon, t_0+\varepsilon)} |x|^2\, \rho^{(h_j)}(x,t)\, dx\, dt \leq C_T\, 2\,\varepsilon,$$

thanks to (13.41): in particular, by the Lebesgue's points theorem, we find (13.57). A similar argument, based on (13.43), proves (13.58).

Step four: We conclude the proof by showing that there exists a unique function $\rho : \mathbb{R}^n \times (0,\infty) \to \mathbb{R}$ such that $\rho \in L^1((0,T) \times \mathbb{R}^n)$ for every $T > 0$, $\int_{\mathbb{R}^n} \rho(t) = 1$ for a.e. $t > 0$, and (13.56), (13.57) and (13.58) hold. We first notice that if ρ is a smooth solution of the Fokker–Planck equation (13.1), that is, if $\partial_t \rho = \operatorname{div}(\rho \nabla \Psi) + \Delta \rho$ on $\mathbb{R}^n$ with $\rho(0) = \rho_0$, then multiplication of (13.1) by $\varphi(x)\, \zeta(t)$ with $\varphi \in C_c^0(\mathbb{R}^n)$ and $\zeta \in C_c^0([0,\infty))$ leads to (13.56). Viceversa, one can show that any solution ρ of (13.56) which satisfies (13.57) and (13.58) is actually a classical smooth solution of (13.1) such that $\rho(t) \to \rho_0$ strongly in $L^1(\mathbb{R}^n)$ as $t \to 0^+$. Since this argument really pertains the theory of linear parabolic PDEs more than OMT, we omit the details.[4] We rather explain in detail why there is a unique smooth solution of (13.1) such that $\rho(t) \to \rho_0$ strongly in $L^1(\mathbb{R}^n)$ as $t \to 0^+$ and (13.57) and (13.58) hold: indeed, this step if crucial for proving the convergence of the whole scheme as $h \to 0^+$! To this end, let σ denote the *difference* of two such solutions, so that $\sigma \in C^\infty(\mathbb{R}^n \times (0,\infty))$, and

$$\partial_t \sigma = \operatorname{div}(\sigma \nabla \Psi) + \Delta \sigma, \tag{13.61}$$

$$\sigma(t) \to 0 \text{ in } L^1(\mathbb{R}^n) \text{ as } t \to 0^+, \tag{13.62}$$

$$\sup_{0 < t < T} \int_{\mathbb{R}^n} |\sigma(t)|\, (1 + |x|^2) \leq C_T. \tag{13.63}$$

---

[4]   Such details are found in [JKO98, p. 14–16].

Now let $U_\delta : \mathbb{R} \to [0,\infty)$ be defined by $U_\delta(r) = \sqrt{r^2 + \delta^2}$, so that U_δ is a smooth convex function and $U_\delta(r) \to |r|$ as $\delta \to 0^+$. We notice that $\sigma_\delta = U_\delta(\sigma)$ is such that $\partial_t \sigma_\delta = U_\delta'(\sigma) \partial_t \sigma$, $\nabla \sigma_\delta = U_\delta'(\sigma) \nabla \sigma$ and $\Delta \sigma_\delta = U_\delta''(\sigma) |\nabla \sigma|^2 + U_\delta'(\sigma) \Delta \sigma$, so that

$$
\begin{aligned}
\partial_t \sigma_\delta - \mathrm{div}\,(\sigma_\delta \nabla \Psi + \nabla \sigma_\delta) &= \partial_t \sigma_\delta - \Delta \sigma_\delta - \nabla \sigma_\delta \nabla \Psi - \sigma_\delta \Delta \Psi \\
&= U_\delta'(\sigma) \partial_t \sigma - U_\delta'(\sigma) \Delta \sigma - U_\delta'(\sigma) \nabla \sigma \cdot \nabla \Psi - \sigma_\delta \Delta \Psi - U_\delta''(\sigma) |\nabla \sigma|^2 \\
&= U_\delta'(\sigma)(\partial_t \sigma - \mathrm{div}\,(\sigma \nabla \Psi + \nabla \sigma)) + U_\delta'(\sigma)\sigma \Delta \Psi - \sigma_\delta \Delta \Psi - U_\delta''(\sigma)|\nabla \sigma|^2 \\
&\leq (U_\delta'(\sigma) \sigma - \sigma_\delta) \Delta \Psi,
\end{aligned}
$$

where we have used (13.61) and $U_\delta'' \geq 0$. Multiplying by $\varphi \in C_c^0(\mathbb{R}^n)$ with $\varphi \geq 0$ we find

$$
\frac{d}{dt} \int_{\mathbb{R}^n} \varphi \, \sigma_\delta + \int_{\mathbb{R}^n} \sigma_\delta \, (\nabla \Psi \cdot \nabla \varphi - \Delta \varphi) \leq \int_{\mathbb{R}^n} \varphi \, (U_\delta'(\sigma) \sigma - \sigma_\delta) \, \Delta \Psi,
$$

and thus, after integration over $(0,t)$ and thanks to (13.62)

$$
\begin{aligned}
\int_{\mathbb{R}^n} \varphi \, \sigma_\delta(t) - U_\delta(0) \int_{\mathbb{R}^n} \varphi &+ \int_{\mathbb{R}^n \times (0,t)} \sigma_\delta \, (\nabla \Psi \cdot \nabla \varphi - \Delta \varphi) \\
&\leq \int_{\mathbb{R}^n \times (0,t)} \varphi \, (U_\delta'(\sigma) \sigma - \sigma_\delta) \, \Delta \Psi.
\end{aligned}
$$

Since $U_\delta'(r) \to \mathrm{sign}(r)$ and $U_\delta(r) \to |r|$ uniformly on $r \in \mathbb{R}$ as $\delta \to 0^+$, by (13.63) and by dominated convergence we deduce that

$$
\int_{\mathbb{R}^n} \varphi \, |\sigma(t)| + \int_{\mathbb{R}^n \times (0,t)} |\sigma| \, (\nabla \Psi \cdot \nabla \varphi - \Delta \varphi) \leq 0. \tag{13.64}
$$

Finally, we exploit the lack of homogeneity in the test function: testing (13.64) with $\varphi(x) = \varphi_0(\varepsilon x)$ for some fixed $\varphi_0 \in C_c^0(\mathbb{R}^n)$ with $\varphi_0 \geq 0$ and $\varepsilon > 0$, we find that

$$
\int_{\mathbb{R}^n} (\varphi_0 \circ \varepsilon \mathrm{id}) \, |\sigma(t)| + \int_{\mathbb{R}^n \times (0,t)} |\sigma| \, (\varepsilon \nabla \Psi \cdot (\nabla \varphi_0 \circ \varepsilon \mathrm{id}) - \varepsilon^2 (\Delta \varphi_0 \circ \varepsilon \, \mathrm{id})) \leq 0.
$$

By the boundedness of $\nabla \varphi_0$ and $\Delta \varphi_0$, and since $\varphi_0(\varepsilon x) \to \varphi_0(0)$ locally uniformly on $\mathbb{R}^n$ as $\varepsilon \to 0^+$, we conclude that $\sigma(t) = 0$ a.e. on $\mathbb{R}^n$ for every $t > 0$ thanks to the arbitrariness of $\varphi_0(0)$. □

Remark 13.12 One can prove, in analogy with (12.17), that the discrete flow $\rho^{(h)}$ is 1/2-Hölder continuous in the Wesserstein distance (above time scale h), in the sense that

$$
\sup_{0 \leq t, s \leq T} \mathcal{W}_2(\rho^{(h)}(t), \rho^{(h)}(s)) \leq C_T \sqrt{|s - t| + h}, \tag{13.65}
$$

compare with (12.17). Indeed, if $0 \leq t < s \leq T$, then there exists integers $k \leq j$ such that

$$(k-1)\,h \leq t < k\,h \leq (j-1)\,h \leq s < j\,h \leq T.$$

If $j = k$, then the left-hand side of (13.65) is 0. If $j = k + 1$, then by (13.44)

$$W_2(\rho^{(h)}(t), \rho^{(h)}(s)) = W_2(\rho^{(h)}_{k-1}, \rho^{(h)}_k) \leq C_T\,\sqrt{h},$$

and (13.65) holds. Finally, if $j \geq k + 2$, then $s - t \geq (j - k - 1)\,h$ and (13.44) give

$$W_2\big(\rho^{(h)}(t), \rho^{(h)}(s)\big) = W_2\big(\rho^{(h)}_{k-1}, \rho^{(h)}_{j-1}\big) \leq \sum_{\ell=k}^{j-2} W_2\big(\rho^{(h)}_{\ell+1}, \rho^{(h)}_\ell\big)$$

$$\leq \sqrt{j-k-1}\left(\sum_{\ell=k}^{j-1} W_2\big(\rho^{(h)}_{\ell+1}, \rho^{(h)}_\ell\big)^2\right)^{1/2}$$

$$\leq \sqrt{j-k-1}\left(\sum_{\ell=k}^{j-1} W_2\big(\rho^{(h)}_{\ell+1}, \rho^{(h)}_\ell\big)^2\right)^{1/2} \leq \sqrt{\frac{s-t}{h}}\,\sqrt{C_T\,h},$$

which implies (13.65) with $C_T\,\sqrt{s-t}$ on the right-hand side.

13.5 Displacement Convexity and Convergence to Equilibrium

In Theorem 12.1 we have proved that if $f : \mathbb{R}^n \to \mathbb{R}$ is uniformly convex, in the sense that for some $\lambda > 0$ we have

$$v \cdot \nabla^2 f(x)[v] \geq \lambda\,|v|^2 \qquad \forall x, v \in \mathbb{R}^n,$$

then the unique solution of $x'(t) = -\nabla f(x(t))$ converges exponentially to the unique critical point (and global minimum) $x_{\min}$ of f, in the sense that

$$|x(t) - x_{min}| \leq e^{-\lambda t}\,|x(0) - x_{min}|, \qquad \forall t \geq 0.$$

Given the identification (Theorem 13.11) of the Fokker–Planck equation (13.1) with the gradient flow of the free energy

$$\mathcal{F}(\rho) = \mathcal{E}(\rho) + \mathcal{S}(\rho) = \int_{\mathbb{R}^n} \Psi(x)\,\rho + \int_{\mathbb{R}^n} \rho \log \rho, \tag{13.66}$$

on the metric space $(\mathcal{A}, W_2)$, $\mathcal{A} = \{\rho : \rho\,d\mathcal{L}^n \in \mathcal{P}_{2,\mathrm{ac}}(\mathbb{R}^n)\}$, we may thus ask if the Fokker–Planck flow $\rho(t)$ converges to the equilibrium state

$\rho_\infty = e^{-\Psi}/Z$ of $\mathcal{F}$ ($Z = \int_{\mathbb{R}^n} e^{-\Psi}$). The answer is affirmative, and one can obtain the following type of statement:

> if Ψ is λ-uniformly convex on $\mathbb{R}^n$,
>
> and if the initial datum ρ_0 is comparable to ρ_∞,
>
> then $\rho(t)$ converges to ρ_∞ in $\mathbf{W}_2$ as $t \to \infty$,
>
> at an explicit exponential rate which depends on λ. $\qquad$ (13.67)

The assumption on Ψ is the existence of $\lambda > 0$ such that

$$\tau \cdot \nabla^2 \Psi(x)[\tau] \geq \lambda \, |\tau|^2 \qquad \forall x, \tau \in \mathbb{R}^n, \tag{13.68}$$

while ρ_0 is comparable to ρ_∞ if there are $a \in (0, 1)$ and $b > 1$ such that

$$a \, \rho_\infty \leq \rho_0 \leq b \, \rho_\infty \qquad \text{on } \mathbb{R}^n. \tag{13.69}$$

The main goal of this section is providing an informal justification of (13.67), highlighting in the process some key ideas that, more generally, are useful in addressing the same kind of questions in other contexts.

Following the blueprint of the proof of Theorem 12.1, convergence to equilibrium is deduced from two basic inequalities: **(i):** a control on $\mathcal{W}_2(\rho, \rho_\infty)^2 = \mathbf{W}_2(\rho \, d\mathcal{L}^n, \rho_\infty \, d\mathcal{L}^n)^2$ in terms of the (nonnegative) energy gap $\mathcal{F}(\rho) - \mathcal{F}(\rho_\infty)$; **(ii):** a control on the energy gap $\mathcal{F}(\rho) - \mathcal{F}(\rho_\infty)$ in terms of the dissipation $\mathcal{D}(\rho)$ of $\mathcal{F}$ along the Fokker–Planck flow, see (13.5). We now sketch the arguments needed for obtaining these two results.

Sketch of proof of (i): We show that

$$\mathcal{F}(\rho) - \mathcal{F}(\rho_\infty) \geq \frac{\lambda}{2} \mathcal{W}_2(\rho, \rho_\infty)^2, \qquad \forall \rho \in \mathcal{A}. \tag{13.70}$$

Indeed, let $\rho_t \, d\mathcal{L}^n = (\nabla f_t)_\# \mu$, where ∇f is the transport map from $\mu = \rho \, d\mathcal{L}^n$ to $\mu_\infty = \rho_\infty \, d\mathcal{L}^n$, and where $\nabla f_t = (1 - t) \, \mathbf{id} + t \, \nabla f$. By (13.68) we obtain a quantification of the gap in the basic convexity inequality for Ψ, namely

$$(1 - t) \, \Psi(x) + t \, \Psi(y) \geq \Psi((1 - t) \, x + t \, y) + t \, (1 - t) \frac{\lambda}{2} |x - y|^2, \tag{13.71}$$

for every $x, y \in \mathbb{R}^n$ and $t \in [0, 1]$. Setting $y = \nabla f(x)$ in (13.71), multiplying by ρ the resulting inequality, and integrating over $\mathbb{R}^n$ we obtain

$$(1 - t) \, \mathcal{E}(\rho) + t \, \mathcal{E}(\rho_\infty) = (1 - t) \int_{\mathbb{R}^n} \Psi \, \rho + t \int_{\mathbb{R}^n} \Psi(\nabla f) \, \rho \tag{13.72}$$

$$\geq \int_{\mathbb{R}^n} \Psi(\nabla f_t) \, \rho + t(1 - t) \frac{\lambda}{2} \int_{\mathbb{R}^n} |x - \nabla f|^2 \rho = \mathcal{E}(\rho_t) + t(1 - t) \frac{\lambda}{2} \mathcal{W}_2(\rho, \rho_\infty)^2.$$

Recalling that S is displacement convex thanks to Theorem 10.5, we have

$$(1-t)\,S(\rho) + t\,S(\rho_\infty) \geq S(\rho_t). \tag{13.73}$$

Since ρ_∞ is the absolute minimizer of $\mathcal{F}$ over $\mathcal{P}_{2,\mathrm{ac}}(\mathbb{R}^n)$, we can add up (13.72) and (13.73) to obtain

$$(1-t)\,\mathcal{F}(\rho) + t\,\mathcal{F}(\rho_\infty) \geq \mathcal{F}(\rho_t) + \frac{\lambda}{2}t(1-t)\,\mathcal{W}_2(\rho,\rho_\infty)^2$$

$$\geq \mathcal{F}(\rho_\infty) + \frac{\lambda}{2}t(1-t)\,\mathcal{W}_2(\rho,\rho_\infty)^2.$$

Rearranging terms, simplifying $1-t$, and then letting $t \to 1^-$, we find (13.70).

Sketch of proof of (ii): We prove the "energy–dissipation inequality"

$$\int_{\mathbb{R}^n} \left| \nabla\Psi + \frac{\nabla\rho}{\rho} \right|^2 \rho \geq 2\lambda\,(\mathcal{F}(\rho) - \mathcal{F}(\rho_\infty)), \tag{13.74}$$

(compare with (13.5)). Let us consider the quantification of the supporting hyperplane inequality for Ψ obtained from (13.68), namely,

$$\Psi(y) \geq \Psi(x) + \nabla\Psi(x)\cdot(y-x) + \frac{\lambda}{2}|x-y|^2, \qquad \forall x,y \in \mathbb{R}^n. \tag{13.75}$$

Taking $y = \nabla f(x)$ in (13.75), multiplying by ρ, and integrating over $\mathbb{R}^n$ we obtain

$$\mathcal{E}(\rho_\infty) \geq \mathcal{E}(\rho) + \int_{\mathbb{R}^n} \nabla\Psi\cdot(\nabla f - x)\,\rho + \frac{\lambda}{2}\,\mathcal{W}_2(\rho,\rho_\infty)^2.$$

At the same time since $\rho = \rho_\infty(\nabla f)\det\nabla^2 f\ \mathcal{L}^n$-a.e. on $\{\rho > 0\}$, we have that

$$S(\rho_\infty) = \int_{\mathbb{R}^n} \rho_\infty\,\log(\rho_\infty) = \int_{\mathbb{R}^n} \rho\,\log(\rho_\infty(\nabla f))$$

$$= \int_{\mathbb{R}^n} \rho\,\log\left(\frac{\rho}{\det\nabla^2 f}\right) = S(\rho) - \int_{\mathbb{R}^n} \rho\,\log(\det\nabla^2 f).$$

Adding up the two inequalities we have proved that

$$\mathcal{F}(\rho) - \mathcal{F}(\rho_\infty) + \frac{\lambda}{2}\,\mathcal{W}_2(\rho,\rho_\infty)^2 \leq \int_{\mathbb{R}^n} \nabla\Psi\cdot(x-\nabla f)\,\rho + \int_{\mathbb{R}^n} \rho\,\log(\det\nabla^2 f). \tag{13.76}$$

Now, for $\mathcal{L}^n$-a.e. $x \in \{\rho > 0\}$ we have that $\nabla^2 f(x)$ is a diagonal matrix in a suitable orthonormal basis, therefore, by the elementary inequality

$$\log\left(\prod_{k=1}^n (1+\lambda_k)\right) \leq \sum_{k=1}^n \lambda_k,$$

we deduce that, for $\mathcal{L}^n$-a.e. $x \in \{\rho > 0\}$,

$$\log(\det \nabla^2 f(x)) = \log (\det[\text{Id} + (\nabla^2 f(x) - \text{Id})]) \leq \text{trace} (\nabla^2 f(x) - \text{Id}),$$

and thus, by the non-negativity of $D^2 f$, that $\log(\det \nabla^2 f) \, d\mathcal{L}^n \leq \text{div} (\nabla f - x)$ in the sense of distributions on $\mathbb{R}^n$. Now, by (13.69) and by the maximum principle, we have

$$a \, \rho_\infty \leq \rho(t) \leq b \, \rho_\infty \quad \text{on } \mathbb{R}^n, \text{ for every } t > 0. \tag{13.77}$$

By (13.77), $0 \leq \Psi \leq C(1 + |x|^2)$ on $\mathbb{R}^n$, and the fact that ∇f is the Brenier map from $\mu = \rho \, d\mathcal{L}^n$ to $\mu_\infty = \rho_\infty \, d\mathcal{L}^n$, we can obtain bounds on $|\nabla f - x|$ such that the following integration by parts

$$\int_{\mathbb{R}^n} \rho \log (\det \nabla^2 f) \leq \int_{\mathbb{R}^n} \rho \, d[\text{div} (\nabla f - x)] = \int_{\mathbb{R}^n} \nabla \rho \cdot (x - \nabla f),$$

can be justified, and then can be combined with (13.76) to obtain (13.74),

$$\mathcal{F}(\rho) - \mathcal{F}(\rho_\infty) + \frac{\lambda}{2} W_2(\rho, \rho_\infty)^2 \leq \int_{\mathbb{R}^n} \left[\nabla\Psi + \frac{\nabla\rho}{\rho} \right] \cdot (x - \nabla f) \, \rho$$

$$\leq \left(\int_{\mathbb{R}^n} \left| \nabla\Psi + \frac{\nabla\rho}{\rho} \right|^2 \rho \right)^{1/2} \left(\int_{\mathbb{R}^n} |\nabla f - x|^2 \, \rho \right)^{1/2}$$

$$\leq \frac{1}{2\lambda} \int_{\mathbb{R}^n} \left| \nabla\Psi + \frac{\nabla\rho}{\rho} \right|^2 \rho + \frac{\lambda}{2} W_2(\rho, \rho_\infty)^2,$$

where in the last inequality we have used $ab \leq (a^2/2\lambda) + (b^2 \, \lambda/2)$.

Convergence to equilibrium: Finally we combine (13.70) and (12.6). Indeed, differentiating $\mathcal{F}(\rho(t))$ along (13.1), as done in Section 13.1, we find that

$$\frac{d}{dt} \mathcal{F}(\rho(t)) = - \int_{\mathbb{R}^n} \left| \nabla\Psi + \frac{\nabla\rho(t)}{\rho(t)} \right|^2 \rho(t).$$

Therefore (12.6) gives

$$\frac{d}{dt} \left(\mathcal{F}(\rho(t)) - \mathcal{F}(\rho_\infty) \right) \leq -2\lambda \, (\mathcal{F}(\rho(t)) - \mathcal{F}(\rho_\infty))$$

from which in turn we deduce, using also (13.70),

$$(\mathcal{F}(\rho_0) - \mathcal{F}(\rho_\infty)) \, e^{-2\lambda t} \geq \mathcal{F}(\rho(t)) - \mathcal{F}(\rho_\infty) \geq \frac{\lambda}{2} W_2(\rho(t), \rho_\infty)^2.$$

In summary, we have proved

$$W_2(\rho(t), \rho_\infty) \leq e^{-\lambda t} \sqrt{\frac{2}{\lambda} (\mathcal{F}(\rho_0) - \mathcal{F}(\rho_\infty))},$$

that is exponential convergence to equilibrium in Wasserstein distance.

14

The Euler Equations and Isochoric Projections

This chapter contains one more physically intriguing application of the Brenier theorem *and* the prelude to a crucial insight on the geometry of the Wasserstein space. Both developments originate in the study of the Euler equations for an incompressible fluid. This is one of the most fascinating and challenging PDE in Mathematics, which is closely related to OMT since the motion of an incompressible fluid is naturally described, from the kinematical viewpoint, by using *isochoric* (i.e., volume-preserving) transformations. The study of time-dependent isochoric transformations leads in turn to identify the *transport equation*, while the derivation of the Euler equations from the principle of least action leads to consider the minimization of the *action functional* for an incompressible fluid. These last two objects provide the entry point to understand the "Riemannian" (or "infinitesimally Hilbertian") structure of the Wasserstein space, which will be further addressed in Chapter 15. In Section 14.1 we introduce isochoric transformations, while in Section 14.2 we derive the Euler equations from the principle of least action. In Section 14.3 we interpret the principle of least action for an incompressible fluid as the *geodesics equation in the space of isochoric transformations*. We then introduce the idea of geodesics as limits of "iterative projections of mid-points," and then present (as a rather immediate consequence of Theorem 4.2) the Brenier projection theorem, relating the quadratic Monge problem to the L^2-projection on the "manifold" of isochoric transformations.

14.1 Isochoric Transformations of a Domain

We consider an open connected set $\Omega \subset \mathbb{R}^n$, which plays the role of a container completely occupied by an incompressible fluid. An **isochoric transformation** of Ω is a smooth and bijective map $\Phi : \mathrm{Cl}(\Omega) \to \mathrm{Cl}(\Omega)$ such that $\Phi(\partial\Omega) = \partial\Omega$ and $\Phi_{\#}(\mathcal{L}^n \llcorner \Omega) = \mathcal{L}^n \llcorner \Omega$, i.e.,

$$|\Phi^{-1}(E)| = |E| \qquad \text{for every Borel set } E \subset \Omega. \tag{14.1}$$

A by now familiar argument shows that (14.1) is equivalent to

$$|\det \nabla\Phi(x)| = 1 \qquad \text{for every } x \in \Omega, \tag{14.2}$$

see Proposition 1.1. In fact, by connectedness of Ω, we either have $\det \nabla\Phi = 1$ of $\det \nabla\Phi = -1$ on Ω; depending on which condition holds, we have an orientation *preserving* or *inverting* isochoric transformation. We denote by $\mathcal{M}(\Omega)$ the set of isochoric transformations of Ω.

The basic kinematic assumption of classical Fluid Mechanics is that the motion of a fluid can be described by a time dependent family of orientation preserving isochoric transformations of Ω. Precisely, an **incompressible motion in** Ω is a smooth function $\Phi \in C^\infty(\Omega \times [0,\infty); \Omega)$ such that, setting $\Phi_t(x) = \Phi(x,t)$, we have

$$\Phi_0 = \mathbf{id}, \qquad \{\Phi_t\}_{t \geq 0} \subset \mathcal{M}(\Omega).$$

The time derivative $\mathbf{v} = \partial_t \Phi$ is the **velocity field** of the incompressible motion Φ. The fact that $\Phi_t \in \mathcal{M}(\Omega)$ is reflected at the level of $\mathbf{v}$ by the validity of the following two conditions,

$$\mathbf{v}(x,t) \cdot \nu_\Omega(x) = 0, \qquad \forall x \in \partial\Omega, t \geq 0, \tag{14.3}$$

$$\text{trace}(\nabla\Phi_t(x)^{-1}\nabla\mathbf{v}(x,t)) = 0, \qquad \forall x \in \Omega, t \geq 0. \tag{14.4}$$

The validity of (14.3) is immediate from $\Phi(\partial\Omega, t) = \partial\Omega$ and $\mathbf{v} = \partial_t \Phi$. To prove (14.4) we notice that (14.2) and $\nabla\Phi_{t+h}(x) = \nabla\Phi_t(x) + h\,\nabla\mathbf{v}(x,t) + O(h^2)$ imply

$$1 = \det \nabla\Phi_{t+h} = \det \nabla\Phi_t(x) \det\left(1 + h\,\nabla\Phi_t(x)^{-1}\nabla\mathbf{v}(x,t) + O(h^2)\right)$$
$$= 1 + h\,\text{trace}(\nabla\Phi_t(x)^{-1}\nabla\mathbf{v}(x,t)) + O(h^2),$$

where in the last identity we have used $\det(\text{Id} + t\,A) = 1 + h\,\text{trace}(A) + O(t^2)$ for $h \to 0$, see, e.g., [Mag12, Lemma 17.4]. Conversely, one can see that given a smooth vector field $\mathbf{v} \in C^\infty(\Omega \times [0,\infty); \mathbb{R}^n)$ satisfying (14.3) and (14.4), then, by setting $\Phi(x,t) = X(t)$ where $X(t)$ is the solution of the ODE $X'(t) = \mathbf{v}(X(t),t)$ with $X(0) = x$, we define an incompressible motion in Ω with velocity field $\mathbf{v}$.

Both the constraints (14.3) and (14.4) take a simpler form if, rather than on $\mathbf{v}$, we focus on the **Eulerian velocity**

$$\mathbf{u}(y,t) = \mathbf{v}(\Phi_t^{-1}(y),t), \qquad (y,t) \in \Omega \times [0,\infty). \tag{14.5}$$

The relation between $\mathbf{v}$ and $\mathbf{u}$ is simple: while $\mathbf{v}(x,t)$ is the velocity of the fluid particle that at time t occupies position x (Lagrangian viewpoint), $\mathbf{u}(y,t)$ is the velocity of the fluid particle that at time t is transiting through position y (Eulerian viewpoint). We claim that (14.3) and (14.4) are equivalent to

$$\mathbf{u}(y,t) \cdot v_\Omega(x) = 0, \qquad \forall y \in \partial\Omega, t \geq 0, \tag{14.6}$$

$$\operatorname{div}\mathbf{u}(y,t) = 0, \qquad \forall y \in \Omega, t \geq 0. \tag{14.7}$$

The equivalence between (14.3) and (14.6) is immediate, while by differentiating in x the identity $\mathbf{v}(x,t) = \mathbf{u}(\Phi_t(x),t)$ one finds

$$\nabla\Phi_t(x)^{-1}\nabla\mathbf{v}(x,t) = (\nabla\mathbf{u})(\Phi_t(x),t)$$

and deduce by the bijectivity of Φ_t that (14.7) and (14.4) are equivalent.

14.2 The Euler Equations and Principle of Least Action

We now derive the equations of motion for an incompressible fluid. We consider the situation when, at the initial time $t = 0$, the velocity field $\mathbf{v}_0(x)$ and density of mass per unit volume $\rho_0(x)$ of the fluid are known, and consider the problem of determining the future motion of the fluid. We are assuming there are no external forces and that no friction/viscosity effects are in place to dissipate the initial kinetic energy of the fluid, which is given by

$$\frac{1}{2}\int_\Omega \rho_0 |\mathbf{v}_0|^2.$$

Under these idealized conditions, the initial kinetic energy should be conserved, and the fluid be stuck in an endless motion driven by the sole effect of the incompressibility constraint. Indeed, preserving the isochoric character of the motion induces a mechanism such that fluid particles have to continuously move away from their position to make space for incoming particles. The force causing this motion is exerted by the internal pressure of the fluid, which is ultimately due to the intramolecular forces responsible for the validity of the incompressibility constraint.

Equation for the mass density: We denote by $\rho(y,t)$ the density of mass per unit volume of the fluid particle that at time t is transiting through position y. If E is an arbitrary region in Ω, then $\{\Phi_t(E)\}_{t \geq 0}$ describes the motion of the part of the fluid occupying the region E at time $t = 0$. By the Principle of Conservation of Mass, the total mass of $\Phi_t(E)$ must stay constant in time, i.e.,

$$0 = \frac{d}{dt}\int_{\Phi_t(E)} \rho(y,t)\,dy.$$

By using $\det\nabla\Phi_t = 1$, we find that

$$0 = \frac{d}{dt}\int_E \rho(\Phi_t(x),t)\,dx = \int_E (\partial_t\rho)(\Phi_t(x),t) + \nabla\rho(\Phi_t(x),t) \cdot \mathbf{v}(x,t)\,dx.$$

By arbitrariness of E and by $\mathbf{u}(\Phi_t(x),t) = \mathbf{v}(x,t)$, the **transport equation**[1]

$$\partial_t \rho + \nabla \rho \cdot \mathbf{u} = 0 \qquad \text{on } \Omega \times [0,\infty) \tag{14.8}$$

holds. If the Eulerian velocity $\mathbf{u}$ is known, the transport equation, coupled with the initial condition $\rho(\cdot,0) = \rho_0$, allows one to determine the mass density of the fluid during the motion. Notice that if ρ_0 is constant, then $\rho \equiv \rho_0$ solves (14.8): in other words, it makes sense to study the motion of an incompressible fluid under the assumption that the mass density is constant throughout the motion. We also notice that, exactly under the incompressibility constraint $\operatorname{div} \mathbf{u} = 0$, the transport equation (14.8) is equivalent to the PDE

$$\partial_t \rho + \operatorname{div}(\rho \mathbf{u}) = 0 \qquad \text{on } \Omega \times [0,\infty), \tag{14.9}$$

known as the **continuity equation** (and playing a pivotal role in Chapter 15).

Equations for the velocity field: We now derive the equations of motion by exploiting the equivalence between the Newton laws of motion and the principle of least action. In the absence of external or friction/viscosity forces, the **action between times** t_1 **and** $t_2 > t_1$ of a motion Φ is defined as the total kinetic energy associated to the motion between time t_1 and time t_2, i.e.,

$$\mathcal{A}(\Phi; t_1, t_2) = \frac{1}{2} \int_{t_1}^{t_2} dt \int_{\Omega} \rho(\Phi_t^{-1}(x),t) \, |\mathbf{v}(x,t)|^2 \, dx$$

$$= \frac{1}{2} \int_{t_1}^{t_2} dt \int_{\Omega} \rho(y,t) \, |\mathbf{u}(y,t)|^2 \, dy. \tag{14.10}$$

To formulate the principle of least action we need to introduce the notion of variation of an incompressible motion Φ. We say that Ψ is a **variation of Φ over** (t_1, t_2) if $\Psi = \Psi(x,t,s) : \Omega \times [0,\infty) \times (-\varepsilon, \varepsilon) \to \Omega$ is a smooth function such that, setting $\Psi_t^s(x) = \Psi^s(x,t) = \Psi(x,t,s)$, Ψ^s is an incompressible motion of Ω for every $|s| < \varepsilon$, $\Psi^0 = \Phi$ and

$$\Psi_{t_1}^s = \Phi_{t_1} \text{ and } \Psi_{t_2}^s = \Phi_{t_2} \text{ in } \Omega, \forall |s| < \varepsilon. \tag{14.11}$$

The **principle of least action** postulates that Φ is an actual motion of the fluid if and only if it is a critical point of $\mathcal{A}$, i.e., if for every variation Ψ of Φ over (t_1, t_2), we have

$$\frac{d}{ds}\Big|_{s=0} \mathcal{A}(\Psi^s; t_1, t_2) = 0. \tag{14.12}$$

Writing (14.12) just in terms of Φ leads to the Euler equations. We shall do this under the assumption that ρ_0 is constant, and thus, as discussed earlier, that $\rho \equiv \rho_0$ for every time $t \geq 0$. In this way,

---

[1] Every physical quantity conserved along the fluid motion will satisfy a transport equation analogous to (14.8).

$$\frac{d}{ds}\mathcal{A}(\Psi^s;t_1,t_2) = \frac{d}{ds}\frac{\rho_0}{2}\int_{t_1}^{t_2}dt\int_\Omega \left|\partial_t\Psi(x,t,s)\right|^2 dx = \rho_0\int_{t_1}^{t_2}dt\int_\Omega \partial_t\Psi\cdot\partial_{st}\Psi,$$

so that, setting $X(x,t) = \partial_s\Psi(x,t,0)$ and noticing that $\partial_t\Psi(x,t,0) = \partial_t\Phi(x,t) = \mathbf{v}(x,t)$, we find that (14.12) is equivalent to

$$0 = \int_{t_1}^{t_2}dt\int_\Omega \mathbf{v}\cdot\partial_t X = -\int_{t_1}^{t_2}dt\int_\Omega \partial_t\mathbf{v}\cdot X. \qquad (14.13)$$

where we have used (14.11) to claim that $X(\cdot,t_1) = X(\cdot,t_2) = 0$. Now, should there be no constraint on the variations Ψ, then X would be an arbitrary vector field, and (14.13) would imply that $\partial_t\mathbf{v} = 0$ – in other words, the fluid would move by inertial motion. This is the point where the isochoric nature of the fluid motion enters into play. The fact that Φ^s is an incompressible motion implies that $1 = \det\nabla\Phi_t^s$ in Ω for every $|s| < \varepsilon$ and $t \geq 0$. In particular, if we differentiate this identity in s and set $s = 0$, thanks to $\Psi_t^0 = \Phi_t$ we find

$$\text{trace}((\nabla\Phi_t)^{-1}\nabla X) = 0,$$

or equivalently, setting $Y(y,t) = X(\Phi_t^{-1}(y),t)$, $\text{div}\,Y = 0$ on $\Omega\times[0,\infty)$. Therefore, changing variables in (14.13) we get

$$0 = \int_{t_1}^{t_2}dt\int_\Omega (\partial_t\mathbf{v})(\Phi_t^{-1}(y),t)\cdot Y(y,t)\,dy, \qquad (14.14)$$

where Y satisfies

$$\begin{cases} \text{div}\,Y = 0, & \text{on }\Omega\times[0,\infty), \\ Y(\cdot,t_1) = Y(\cdot,t_2) = 0, & \text{on }\Omega, \qquad (14.15) \\ Y\cdot\nu_\Omega = 0, & \text{on }\partial\Omega\times[0,\infty). \end{cases}$$

Viceversa, whenever Y is a vector field such that (14.15), then, setting $X(x,t) = Y(\Phi_t(x),t)$ and defining Ψ as the flow in the s variable generated by X, we obtain a variation Ψ of Φ on (t_1,t_2). Therefore (14.14) holds for every Y satisfying (14.15).

Let us now recall that if a locally integrable vector field $F : \Omega \to \mathbb{R}^n$ satisfies

$$\int_\Omega F\cdot Z = 0,$$

for every smooth test vector field Z such that $\text{div}\,Z = 0$ in Ω and $Z\cdot\nu_\Omega = 0$ on $\partial\Omega$, then the distributional curl of F equals zero, and thus there exists a potential $f : \Omega \to \mathbb{R}$ such that $F = \nabla f$ in Ω. Therefore, coming back to (14.14) we conclude that there exists $p = p(y,t) : \Omega\times[0,\infty) \to \mathbb{R}$ such that

$$\frac{\partial\mathbf{v}}{\partial t}(\Phi_t^{-1}(y),t) = -\nabla p(y,t), \qquad \forall y\in\Omega, t\geq 0. \qquad (14.16)$$

Finally, by differentiating in time $\mathbf{v}(x,t) = \mathbf{u}(\Phi_t(x),t)$, we notice that the acceleration of the fluid particle that at time $t = 0$ is at the position x is

$$\partial_t \mathbf{v}(x,t) = \left(\nabla \mathbf{u}[\mathbf{u}] + \partial_t \mathbf{u}\right)(\Phi_t(x), x).$$

The action of the $n \times n$ matrix $\nabla \mathbf{u}$ on the vector $\mathbf{u}$ is commonly denoted by

$$(\mathbf{u} \cdot \nabla)\mathbf{u} = \left(\sum_{i=1}^n \mathbf{u}^i \, \partial_i \right) \sum_{j=1}^n u^j \, e_j = \sum_{i,j=1}^n \partial_i \, u^j \, \mathbf{u}^i \, e_j = (\nabla \mathbf{u})[\mathbf{u}],$$

so that the acceleration of the particle that at time t is in position y is given by

$$(\partial_t \mathbf{u} + (\mathbf{u} \cdot \nabla)\mathbf{u})(y,t) = \partial_t \mathbf{v}(\Phi_t^{-1}(y),t).$$

This last fact combined with (14.16), (14.3), and (14.7) leads us to the (constant mass density) **Euler equations**

$$\begin{cases} \partial_t \mathbf{u} + (\mathbf{u} \cdot \nabla)\mathbf{u} = -\nabla p, \\ \operatorname{div} \mathbf{u} = 0, \end{cases} \quad \text{in } \Omega \times [0, \infty), \qquad (14.17)$$

which are coupled with the boundary condition $\mathbf{u} \cdot \nu_\Omega = 0$ on $\partial\Omega \times [0, \infty)$ and with the initial condition $\mathbf{u}(\cdot, 0) = \mathbf{u}_0$ in Ω. The function p is called the **pressure** associated with the fluid motion defined by $\mathbf{u}$.

Based on our preliminary physical considerations, we would expect a smooth solution of (14.17) to conserve kinetic energy. And indeed, taking into account (14.16) we find

$$\frac{d}{dt} \frac{1}{2} \int_\Omega |\mathbf{v}|^2 = \int_\Omega \mathbf{v} \cdot \partial_t \mathbf{v} = \int_\Omega \mathbf{u}(y,t) \cdot (\partial_t \mathbf{v})(\Phi_t^{-1}(y),t) \, dy$$

$$= - \int_\Omega \mathbf{u}(y,t) \cdot \nabla p(y,t) \, dy$$

$$= \int_\Omega p \operatorname{div} \mathbf{u} - \int_{\partial\Omega} p \, \mathbf{u} \cdot \nu_\Omega = 0,$$

so that kinetic energy is conserved by smooth solutions of the Euler equations.[2]

---

[2] Constructing solutions of the Euler equations that exist for every $t > 0$ and conserve kinetic energy (e.g., because they are smooth) is an extremely challenging problem. In the planar case $n = 2$ this is possible by a classical result of Yudovich, stating is that if the vorticity $\omega = \nabla \times \mathbf{u}$ is bounded at time $t = 0$, then there is a global-in-time solution, which is unique and conserves kinetic energy. In the physical case $n = 3$, and in higher dimensions, however, the situation is incredibly more complex and many basic questions are open. Our expectation that the motion of an ideal fluid (no friction, no external forces, just the pressure induced by the isochoric constraint) would consist of a never-ending smooth conservative motion clashes with the cold reality that we are actually able to construct weak (non-smooth) solutions of (14.17) which *dissipate* kinetic energy (even at a prescribed rate!) and violate uniqueness.

14.3 The Euler Equations as Geodesics Equations

Based on the above derivation from the principle of least action, Euler equations (14.17) can be naturally interpreted, as originally proposed by Arnold, as the geodesics equation in the space $\mathcal{M}(\Omega)$ of isochoric transformations of Ω. Indeed, constant speed minimizing geodesics in a Riemannian manifold (M,g) (defined over an interval (t_1,t_2)) are minimizers (and thus satisfy the Euler–Lagrange equation) of the action functional

$$\frac{1}{2} \int_{t_1}^{t_2} |\gamma'(t)|^2_{g_{\gamma(t)}} \, dt. \tag{14.18}$$

Interpreting an admissible fluid motion Φ as a curve $\{\Phi_t\}_{t_1 \leq t \leq t_2}$ of isochoric transformations connecting Φ_{t_1} to Φ_{t_2}, and using the $L^2(\Omega; \mathbb{R}^n)$-norm to measure the size of its velocity $\mathbf{v} = \partial_t \Phi$, we can see that the action of Φ over (t_1,t_2), which is defined as

$$\mathcal{A}(\Phi; t_1, t_2) = \frac{1}{2} \int_{t_1}^{t_2} dt \int_\Omega \rho(\Phi_t^{-1}(x),t) \, |\mathbf{v}(x,t)|^2 \, dx \tag{14.19}$$

$$= \frac{1}{2} \int_{t_1}^{t_2} dt \int_\Omega \rho(y,t) \, |\mathbf{u}(y,t)|^2 \, dy,$$

can be understood (up to multiplicative constants) in analogy to the action functional (14.18) for a curve in a manifold. This suggests to interpret solutions to the Euler equations as geodesics in the space of isochoric transformations.

Inspired by this analogy, we now recall that given a manifold M embedded in $\mathbb{R}^n$, there is a classical approach to the construction of geodesics in M (equipped with the Riemannian metric induced by the embedding into $\mathbb{R}^n$), an approach that we may try to reproduce on the Euler equations. The idea goes as follows: given points p_1 and p_2 in M, we can compute their middle–point $(p_1 + p_2)/2$ in $\mathbb{R}^n$, and then project it back on M (which is identified as a subset of $\mathbb{R}^n$), thus defining

$$p_{12} = \text{projection on } M \text{ of } \frac{p_1 + p_2}{2}.$$

(For this construction to make sense, we need of course p_1 and p_2 to be sufficiently close to each other, so that $(p_1 + p_2)/2$ lies in a neighborhood of M where the projection on M is well-defined.) Iterating this basic step, we first project the mid–points between p_1 and p_{12} and between p_{12} and p_2 on M, and then continue indefinitely: at the k-th step of the construction, $O(2^k)$ points on M have been identified, and we expect the probability measure associated to these points to converge toward a geodesic curve with end-points p_1 and p_2.

We would now like to explore the idea of adapting this finite-dimensional construction to $\mathcal{M}(\Omega)$. It seems to natural to consider $\mathcal{M}(\Omega)$ "embedded" in

$L^2(\Omega; \mathbb{R}^n)$, since we are already using the $L^2(\Omega; \mathbb{R}^n)$-norm to measure tangent vectors $\mathbf{v} = \partial_t \Phi$ in defining the action of a curve in $\mathcal{M}(\Omega)$. The projection problem is delicate, because $\mathcal{M}(\Omega)$ is evidently not-closed in $L^2(\Omega; \mathbb{R}^n)$: however, the closure of $\mathcal{M}(\Omega)$ in $L^2(\Omega; \mathbb{R}^n)$ can be characterized[3] as the set of all Borel measurable isochoric transformations of Ω, i.e., as

$$\mathcal{M}^*(\Omega) = \left\{ T : \Omega \to \Omega : T \text{ is Borel measurable and } T_\#(\mathcal{L}^n \llcorner \Omega) = \mathcal{L}^n \llcorner \Omega \right\}.$$

A celebrated theorem of Brenier shows that the projection operator of $L^2(\Omega; \mathbb{R}^n)$ onto $\mathcal{M}^*(\Omega)$ is well-defined, and can be described by composition with Brenier maps.

Theorem 14.1 (Brenier projection theorem)   *Let Ω be a bounded open set in $\mathbb{R}^n$, let $T \in L^2(\Omega; \mathbb{R}^n)$ be such that*

$$T_\#(\mathcal{L}^n \llcorner \Omega) \ll \mathcal{L}^n \llcorner \Omega, \tag{14.20}$$

and set

$$\mu = \frac{T_\#(\mathcal{L}^n \llcorner \Omega)}{|\Omega|}, \qquad \nu = \frac{\mathcal{L}^n \llcorner \Omega}{|\Omega|}.$$

If ∇f is the Brenier map from μ to ν and we set $T_0 = (\nabla f) \circ T$, then

$$T_0 \in \mathcal{M}^*(\Omega), \tag{14.21}$$

$$\|T - T_0\|_{L^2(\Omega)} \le \|T - S\|_{L^2(\Omega)} \qquad \forall S \in \mathcal{M}^*(\Omega). \tag{14.22}$$

In other words, T_0 is a minimizer of $\|(\cdot) - T\|_{L^2(\Omega)}$ on $\mathcal{M}^*(\Omega)$, and is actually uniquely determined by (14.21) and (14.22).

Remark 14.2   As much as projection over an embedded manifold in $\mathbb{R}^n$ is uniquely defined only in a neighborhood of the manifold itself, we do not expect *every* Borel map $T : \Omega \to \Omega$ to have a unique projection over $\mathcal{M}^*(\Omega)$. Condition (14.20) can be interpreted in this sense, as a (mild) proximity condition of T to $\mathcal{M}^*(\Omega)$. A sufficient condition for the validity of (14.20) is that T is Lipschitz continuous with $JT > 0$ $\mathcal{L}^n$-a.e. in Ω. Indeed, by the general form of the area formula

$$\int_\Omega dy \int_{T^{-1}(y)} g(x) \, d\mathcal{H}^0(x) = \int_\Omega g \, JT,$$

---

[3] See [BG03, Corollary 1.1].

which holds for every Borel function $g : \mathbb{R}^n \to [0, \infty]$, we see that

$$
\begin{aligned}
T_\#(\mathcal{L}^n \llcorner \Omega)(E) &= \mathcal{L}^n(\Omega \cap T^{-1}(E)) = \mathcal{L}^n(\Omega \cap T^{-1}(E) \cap \{JT > 0\}) \\
&= \int_\Omega \frac{1_{\{JT>0\}\cap T^{-1}(E)}(x)}{JT(x)} JT(x)\, dx \\
&= \int_\Omega dy \int_{T^{-1}(y)} \frac{1_{\{JT>0\}\cap T^{-1}(E)}(x)}{JT(x)}\, d\mathcal{H}^0(x) \\
&= \int_E dy \int_{T^{-1}(y)} \frac{1_{\{JT>0\}}(x)}{JT(x)}\, d\mathcal{H}^0(x) = 0
\end{aligned}
$$

if $\mathcal{L}^n(E) = 0$. More generally, (14.20) holds whenever we can find a $\mathcal{L}^n$-a.e. Borel partition $\{E_j\}_j$ of Ω such that $\mathrm{Lip}(T; E_j) < \infty$ and $JT > 0$ $\mathcal{L}^n$-a.e. on each E_j. This kind of partition can be constructed, for example, when $T \in BV_{\mathrm{loc}}(\Omega; \mathbb{R}^n)$ and $|\det \nabla T| > 0$ $\mathcal{L}^n$-a.e. in Ω, where $DT = \nabla T\, d[\mathcal{L}^n \llcorner \Omega] + [DT]^s$ is the usual Radon-Nikodym decomposition of DT with respect to $\mathcal{L}^n \llcorner \Omega$.

Remark 14.3 Theorem 14.1 is commonly known[4] as **Brenier's polar factorization** theorem, in reference to the formula

$$
T = (\nabla f^*) \circ T_0 \qquad \mathcal{L}^n\text{-a.e. in } \Omega, \tag{14.23}
$$

which allows us to rearrange a generic map $T \in L^2(\Omega; \mathbb{R}^n)$ with $T_\#(\mathcal{L}^n \llcorner \Omega) \ll \mathcal{L}^n \llcorner \Omega$ into an isochoric Borel transformation of Ω through the composition with the gradient of a convex function. The deduction of (14.23) from the definition of T_0 as $T_0 = (\nabla f) \circ T$ is immediate from Theorem 4.4.

Proof of Theorem 14.1 Since $\mathcal{L}^n(T^{-1}(\Omega)) = \mathcal{L}^n(\Omega)$ and Ω is bounded, we can apply the Brenier theorem (Theorem 4.2) and its corollary (Theorem 4.4) to deduce that if ∇f is the Brenier map from μ to ν, then

$$
\int_{\mathbb{R}^n} |\nabla f(x) - x|^2\, d\mu(x) \leq \int_{\mathbb{R}^n \times \mathbb{R}^n} |x - y|^2\, d\gamma(x, y) \tag{14.24}
$$

whenever $\gamma \in \Gamma(\mu, \nu)$. By recalling the definitions of μ and T_0 we have

$$
\int_{\mathbb{R}^n} |\nabla f(x) - x|^2\, d\mu(x) = \frac{1}{|\Omega|} \int_\Omega |\nabla f(T(y)) - T(y)|^2\, dy = \frac{\|T - T_0\|_{L^2(\Omega)}^2}{|\Omega|}. \tag{14.25}
$$

---

[4] The term "polar factorization" refers to the connection between (14.23) and the polar factorization theorem for real matrices: see [Vil03, Chapter 3.4] for more details.

At the same time, if $S \in \mathcal{M}^*(\Omega)$, then $\gamma_S = |\Omega|^{-1} (T \times S)_\#(\mathcal{L}^n \llcorner \Omega) \in \Gamma(\mu, \nu)$ with

$$\int_{\mathbb{R}^n \times \mathbb{R}^n} |x - y|^2 \, d\gamma_S(x,y) = \frac{1}{|\Omega|} \int_\Omega |T(z) - S(z)|^2 \, dz = \frac{\|T - S\|^2_{L^2(\Omega)}}{|\Omega|}.$$

(14.26)

The combination of (14.24), (14.25) and (14.26) proves (14.22). To show that $T_0 \in \mathcal{M}^*(\Omega)$, as claimed in (14.21), we pick a Borel subset $E \subset \Omega$ and then exploit the transport property of ∇f to get

$$(T_0)_\#(\mathcal{L}^n \llcorner \Omega)(E) = \mathcal{L}^n(\Omega \cap T^{-1}(\nabla f)^{-1}(E))$$
$$= |\Omega| \, \mu((\nabla f)^{-1}(E)) = |\Omega| \, \nu(E) = \mathcal{L}^n(E).$$

Finally, if $T_1 \in \mathcal{M}^*(\Omega)$ is another minimizer of $\|(\cdot) - T\|_{L^2(\Omega)}$ on $\mathcal{M}^*(\Omega)$, then (14.24) and (14.26) show that $\gamma_{S=T_1}$ is an optimal plan in $\mathbf{K}_2(\mu, \nu)$: by the uniqueness statement in Theorem 4.2 we thus have

$$(T \times T_1)_\#(\mathcal{L}^n \llcorner \Omega) = (\mathbf{id} \times \nabla f)_\#(T_\#(\mathcal{L}^n \llcorner \Omega)).$$

Testing this identity with $\varphi(x) \psi(y) : \mathbb{R}^n \times \mathbb{R}^n \to \mathbb{R}$ for $\varphi, \psi \in C^0_c(\mathbb{R}^n)$ we obtain

$$\int_\Omega \varphi(T) \psi(T_1) = \int_\Omega \varphi(T) \psi((\nabla f) \circ T),$$

which by arbitrariness of φ and ψ gives $T_1 = (\nabla f) \circ T = T_0$. $\qquad\square$

15

Action Minimization, Eulerian Velocities, and Otto's Calculus

The Wasserstein space $(\mathcal{P}_2(\mathbb{R}^n), W_2)$ behaves like a path-connected Riemannian manifold, with displacement interpolations associated to optimal transport plans in $K_2(\mu, \nu)$ playing the role of (minimizing) geodesics (see, e.g., Proposition 11.1 and Remark 11.8). We now further develop this analogy, by looking at the facts that, in the Riemannian setting, every constant speed geodesics $\gamma : [0, 1] \to M$ is a minimizer of the action functional

$$A(\gamma) = \frac{1}{2} \int_0^1 |\gamma'(t)|^2_{g_{\gamma(t)}} \, dt,$$

and that the distance between $p, q \in M$ (defined by the length minimization problem (11.4)) can also be characterized by the action minimization problem

$$d_g(p, q)^2 = \inf \{2 A(\gamma) : \gamma(0) = p, \gamma(1) = q\}. \tag{15.1}$$

The main result of this chapter, the Benamou–Brenier formula, translates (15.1) in $(\mathcal{P}_2(\mathbb{R}^n), W_2)$. The first question to answer is: what is γ' when γ is a curve of measures? In Section 15.1, moving from our derivation of the Euler equations, we look at the continuity equation (14.9) to introduce the notions of curve of measures μ in $\mathcal{P}_2(\mathbb{R}^n)$, of Eulerian velocity $\mathbf{u}$ of such a curve (with the caveat that a curve of measures will in general admit multiple Eulerian velocities!), and of action of a pair $(\mu, \mathbf{u})$. In Section 15.2 we show that any sufficiently smooth $\mathbf{u}$ is the Eulerian velocity of a curve of measures μ, while in Section 15.3 we compute the Eulerian velocities of displacement interpolations. In Section 15.4 we prove that every Lipschitz curve of measures admits an Eulerian velocity and in Section 15.5 we finally prove the Benamou–Brenier formula; see Theorem 15.6. This formula suggests the existence of an infinitesimal Hilbertian structure on $(\mathcal{P}_2(\mathbb{R}^n), W_2)$, and a way to informally compute "gradients" with respect to it. The resulting technique, known as Otto's calculus, is presented in Section 15.6.

15.1 Eulerian Velocities and Action for Curves of Measures

The action of an incompressible ideal fluid with Eulerian velocity $\mathbf{u}$ and mass density ρ on a time interval $(t_1, t_2) = (0, 1)$ is defined (see (14.10)) as

$$\frac{1}{2} \int_0^1 dt \int_\Omega \rho(y,t) \, |\mathbf{u}(y,t)|^2 \, dy. \tag{15.2}$$

We want to relate (15.2) to the expression $(1/2) \int_0^1 |\gamma'(t)|^2_{g_{\gamma(t)}} \, dt$ used in the Riemannian setting. The curve of measures involved in (15.2) is definitely

$$\mu_t = \rho(\cdot, t) \, d(\mathcal{L}^n \llcorner \Omega)$$

so that $\|\mathbf{u}(t)\|^2_{L^2(\mu(t))}$ ($\mathbf{u}(t) = \mathbf{u}(\cdot, t)$) plays the role of $|\gamma'(t)|^2_{g_{\gamma(t)}}$ in (15.2). The relation between μ_t and $\mathbf{u}(t)$ is expressed by the continuity equation (14.9), i.e.,

$$\partial_t \rho + \operatorname{div}(\rho \, \mathbf{u}) = 0. \tag{15.3}$$

Equation (15.3) can be formulated in the sense of distributions, and thus recast when, in place of $\rho(\cdot, t) \, d(\mathcal{L}^n \llcorner \Omega)$, we have an arbitrary curve of measures. We now formalize these ideas with a series of definitions: A **curve** μ **in** $\mathcal{P}_2(\mathbb{R}^n)$ is a function $\mu_t : [0, 1] \to \mathcal{P}_2(\mathbb{R}^n)$ such that for every $\varphi \in C^0_c(\mathbb{R}^n)$ the map $t \in [0, 1] \mapsto \int_{\mathbb{R}^n} \varphi(x) \, d\mu_t$ is Borel measurable; whenever this is the case, then

$$\mu[\psi] = \int_0^1 dt \int_{\mathbb{R}^n} \psi(x, t) \, d\mu_t(x), \qquad \psi \in C^0_c(\mathbb{R}^n \times [0, 1]),$$

defines a Radon measure on $\mathbb{R}^n \times [0, 1]$. Given a Borel vector field $\mathbf{u} : \mathbb{R}^n \times [0, 1] \to \mathbb{R}^n$ we say that μ **is transported by** $\mathbf{u}$, or that $\mathbf{u}$ **is an Eulerian velocity for** μ, and write

$$\partial_t \mu_t + \operatorname{div}(\mathbf{u} \, \mu_t) = 0, \tag{15.4}$$

if the following two conditions hold:

(i) Finiteness of the action: $\mathbf{u}(t) = \mathbf{u}(\cdot, t) \in L^2(\mu_t)$ for a.e. $t \in (0, 1)$ and the **action of** $(\mu, \mathbf{u})$, defined by

$$\mathcal{A}(\mu, \mathbf{u}) = \frac{1}{2} \int_0^1 dt \int_{\mathbb{R}^n} |\mathbf{u}(t)|^2 \, d\mu_t, \tag{15.5}$$

is finite;

(ii) Distributional continuity equation: for every $\psi \in C^\infty_c(\mathbb{R}^n \times [0, 1])$,

$$\int_0^1 dt \int_{\mathbb{R}^n} \left(\partial_t \psi + \nabla \psi \cdot \mathbf{u}(t)\right) d\mu_t = \int_{\mathbb{R}^n} \psi(\cdot, 1) \, d\mu_1 - \int_{\mathbb{R}^n} \psi(\cdot, 0) \, d\mu_0. \tag{15.6}$$

We notice that a same curve μ can be transported by multiple fields: for example, if $\mu_t = \rho(\cdot, t) \, d\mathcal{L}^n$ for $\rho > 0$ on $\mathbb{R}^n \times [0, 1]$, $\mathbf{v} \in C^\infty_c(\mathbb{R}^n; \mathbb{R}^n)$ is such that div $\mathbf{v} = 0$ and $\mathbf{u}$ transports μ, then for every $s \in \mathbb{R}$ we have that $\mathbf{u} + s\,[\mathbf{v}/\rho]$ transports μ, since

$$\text{div}\left(\rho\,(\mathbf{u} + s\,[\mathbf{v}/\rho])\right) = \text{div}\,(\rho\,\mathbf{u}) = \partial_t \rho. \tag{15.7}$$

The basic questions that now we want to address are (i) given a vector field $\mathbf{u}$, how to generate curves of measures transported by $\mathbf{u}$? (ii) given a curve of measures, how to find a vector field $\mathbf{u}$ which transports μ? (iii) can W_2 be characterized in terms of minimization of the action defined in (15.5)? Before moving to the analysis of these questions, we make the following remark.

Proposition 15.1 *Given a curve of measures μ in $\mathcal{P}_2(\mathbb{R}^n)$ and a Borel vector field $\mathbf{u} : \mathbb{R}^n \times [0,1] \to \mathbb{R}^n$ such that $\mathbf{u}(t) = \mathbf{u}(\cdot,t) \in L^2(\mu_t)$ for a.e. $t \in (0,1)$ and $\mathcal{A}(\mu,\mathbf{u}) < \infty$, we have that μ is transported by $\mathbf{u}$ if and only if, for every $\varphi \in C_c^\infty(\mathbb{R}^n)$, the map*

$$t \in [0,1] \mapsto \int_{\mathbb{R}^n} \varphi\,d\mu_t \qquad \text{is absolutely continuous on } [0,1] \tag{15.8}$$

and for a.e. $t \in (0,1)$

$$\frac{d}{dt}\left[\int_{\mathbb{R}^n} \varphi\,d\mu_t\right] = \int_{\mathbb{R}^n} \nabla\varphi \cdot \mathbf{u}(t)\,d\mu_t. \tag{15.9}$$

Proof If μ is transported by $\mathbf{u}$, then testing (15.6) with $\psi(x,t) = \varphi(x)\,\zeta(t)$ with $\varphi \in C_c^\infty(\mathbb{R}^n)$ and $\zeta \in C^\infty([0,1])$ we find that

$$\int_0^1 \zeta'(t)\,dt \int_{\mathbb{R}^n} \varphi\,d\mu_t = \left(\zeta(t)\int_{\mathbb{R}^n}\varphi\,d\mu_t\right)\Bigg|_{t=0}^{t=1} - \int_0^1 \zeta(t)\,dt \int_{\mathbb{R}^n}\nabla\varphi\cdot\mathbf{u}(t)\,d\mu_t,$$

from which it follows that $f(t) = \mu_t[\varphi]$ is absolutely continuous on $[0,1]$ with $f'(t) = \int_{\mathbb{R}^n} \nabla\varphi \cdot \mathbf{u}(t)\,d\mu_t$ for a.e. $t \in (0,1)$, that is (15.9). Viceversa, if (15.8) and (15.9) hold, then (15.6) holds for every $\psi(x,t) = \varphi(x)\,\zeta(t)$ corresponding to $\varphi \in C_c^\infty(\mathbb{R}^n)$ and $\zeta \in C^\infty([0,1])$, and by the density of the span of this kind of test functions in $C_c^\infty(\mathbb{R}^n \times [0,1])$ we conclude the proof. $\square$

Not surprisingly, when working with curves of measures, one may need to discuss the differentiability of (15.8) for test functions φ less regular than $C_c^\infty(\mathbb{R}^n)$. The following proposition (which will only be used here in the proof of Theorem 15.6, and can be safely skipped on a first reading) addresses the case when φ is bounded and Lipschitz continuous. The delicate point here is that, since $\nabla\varphi$ may only exists $\mathcal{L}^n$-a.e. on $\mathbb{R}^n$, (15.9) may not even make sense if μ_t is singular with respect to $\mathcal{L}^n$. Nevertheless, (15.8) still holds, and (15.9) can be replaced by an estimate in terms of the asymptotic Lipschitz constant $|\nabla^*\varphi|$ of φ (which, at variance with $\nabla\varphi$, has a precise geometric meaning at *every* point of $\mathbb{R}^n$).

Proposition 15.2 *If μ is a curve of measures in $\mathcal{P}_2(\mathbb{R}^n)$ transported by a vector field $\mathbf{u}$, then there is $I \subset [0,1]$ with $\mathcal{L}^1([0,1] \setminus I) = 0$ with the following property. If $\varphi : \mathbb{R}^n \to \mathbb{R}$ is bounded and Lipschitz continuous, then*

$$\Phi(t) = \int_{\mathbb{R}^n} \varphi \, d\mu_t$$

is absolutely continuous on $[0,1]$, with

$$|\Phi(t) - \Phi(s)| \le \int_t^s d\tau \int_{\mathbb{R}^n} |\nabla^* \varphi| \, |\mathbf{u}(\tau)| \, d\mu_\tau, \quad \forall (t,s) \subset (0,1), \quad (15.10)$$

$$\limsup_{h \to 0} \frac{|\Phi(t+h) - \Phi(t)|}{|h|} \le \int_{\mathbb{R}^n} |\nabla^* \varphi| \, |\mathbf{u}(t)| \, d\mu_t, \quad \forall t \in I. \quad (15.11)$$

*Here, $|\nabla^* \varphi| : \mathbb{R}^n \to [0,\infty)$ is the bounded, upper semicontinuous function called the **asymptotic Lipschitz constant** of φ, and defined by*

$$|\nabla^* \varphi|(x) = \lim_{r \to 0^+} \mathrm{Lip}(\varphi; B_r(x)) = \lim_{r \to 0^+} \sup_{y,z \in B_r(x), y \ne z} \frac{|\varphi(y) - \varphi(z)|}{|y - z|}. \quad (15.12)$$

Proof *Step one*: If η_k is a cut-off function between B_k and B_{k+1} and $v_\varepsilon = v \star \rho_\varepsilon$ denotes the ε-regularization of $v \in L^1_{\mathrm{loc}}(\mathbb{R}^n)$, then we find

$$\left| \Phi(t) - \int_{\mathbb{R}^n} (\eta_k \, \varphi)_\varepsilon \, d\mu_t \right| \le \|\varphi\|_{C^0(\mathbb{R}^n)} \mu_t(\mathbb{R}^n \setminus B_{k+1}) + \|\eta_k \, \varphi - (\eta_k \, \varphi)_\varepsilon\|_{C^0(\mathbb{R}^n)}.$$

By applying (15.9) to $(\eta_k \, \varphi)_\varepsilon \in C_c^\infty(\mathbb{R}^n)$, we thus find that for every $(s,t) \subset (0,1)$

$$\begin{aligned}
|\Phi(s) - \Phi(t)| &= \lim_{k \to \infty} \lim_{\varepsilon \to 0^+} \left| \int_{\mathbb{R}^n} (\eta_k \, \varphi)_\varepsilon \, d\mu_s - \int_{\mathbb{R}^n} (\eta_k \, \varphi)_\varepsilon \, d\mu_t \right| \\
&= \lim_{k \to \infty} \lim_{\varepsilon \to 0^+} \left| \int_t^s d\tau \int_{\mathbb{R}^n} \nabla[(\eta_k \, \varphi)_\varepsilon] \cdot \mathbf{u}(\tau) \, d\mu_\tau \right| \\
&\le \limsup_{k \to \infty} \limsup_{\varepsilon \to 0^+} \int_t^s d\tau \int_{B_{k+1}} |\nabla \varphi_\varepsilon| \, |\mathbf{u}(\tau)| \, d\mu_\tau. (15.13)
\end{aligned}$$

Since, clearly, $|\nabla \varphi_\varepsilon(x)| \le \mathrm{Lip}(\varphi; B_\varepsilon(x))$ we immediately find

$$\limsup_{\varepsilon \to 0^+} |\nabla \varphi_\varepsilon(x)| \le |\nabla^* \varphi|(x), \quad \forall x \in \mathbb{R}^n.$$

By combining this last estimate with Fatou's lemma and (15.13) we conclude that

$$|\Phi(s) - \Phi(t)| \le \limsup_{k \to \infty} \int_t^s d\tau \int_{B_{k+1}} |\nabla^* \varphi| \, |\mathbf{u}(\tau)| \, d\mu_\tau$$

so that (15.10) follows by monotone convergence.

Step two: We prove the existence of $I \subset (0,1)$ with $\mathcal{L}^1((0,1) \setminus I) = 0$ such that if $t \in I$ and $f : \mathbb{R}^n \to \mathbb{R}$ is bounded and upper semicontinuous on $\mathbb{R}^n$, then

$$\limsup_{h \to 0} \frac{1}{h} \int_t^{t+h} d\tau \int_{\mathbb{R}^n} f \, |\mathbf{u}(\tau)| \, d\mu_\tau \le \int_{\mathbb{R}^n} f \, |\mathbf{u}(t)| \, d\mu_t. \tag{15.14}$$

(Then (15.11) will follow by taking $f = |\nabla^* \varphi|$.) To clarify the issue at hand, we first notice that, being f bounded, the map $t \mapsto \int_{\mathbb{R}^n} f \, |\mathbf{u}(t)| \, d\mu_t$ belongs to $L^1(0,1)$; thus, denoting by $I(f)$ the set of the Lebesgue points of this map, we see that (15.14) holds as an identity at every $t \in I(f)$. The goal is thus finding a set of good values of t which is independent of f. To this end, let ζ_k be a smooth cut-off function between B_k and B_{k+1} ($k \in \mathbb{N}$), let $\mathcal{G}$ be a countable dense set in $C_c^0(\mathbb{R}^n)$ (in the local uniform convergence), and let

$$I = \bigcap_{k \in \mathbb{N}} \bigcap_{g \in \mathcal{G}} \left[I(\zeta_k \, g) \cap I(\zeta_k) \cap I(1 - \zeta_k) \right],$$

so that $\mathcal{L}^1((0,1) \setminus I) = 0$. If $t \in I$ and $f \in C_b^0(\mathbb{R}^n)$, then integrating

$$\int_{\mathbb{R}^n} f \, |\mathbf{u}(\tau)| \, d\mu_\tau \le \int_{\mathbb{R}^n} \left(\zeta_k \, g + \|f - g\|_{C^0(B_{k+1})} \, \zeta_k + \|f\|_{C^0(\mathbb{R}^n)} (1 - \zeta_k) \right) |\mathbf{u}(\tau)| \, d\mu_\tau$$

in $d\tau$ over $(t, t+h)$, dividing by h, and then letting $h \to 0$, we find that

$$\limsup_{h \to 0} \frac{1}{h} \int_t^{t+h} d\tau \int_{\mathbb{R}^n} f \, |\mathbf{u}(\tau)| \, d\mu_\tau \tag{15.15}$$

$$\le \int_{\mathbb{R}^n} \left(\zeta_k \, g + \|f - g\|_{C^0(B_{k+1})} \, \zeta_k + \|f\|_{C^0(\mathbb{R}^n)} (1 - \zeta_k) \right) |\mathbf{u}(t)| \, d\mu_t.$$

Taking $\{g_j\}_j \subset \mathcal{G}$ such that $g_j \to f$ locally uniformly on $\mathbb{R}^n$, and letting first $j \to \infty$ and then $k \to \infty$ in (15.15) with $g = g_j$, we deduce the validity of (15.14) whenever $f \in C_b^0(\mathbb{R}^n)$. If now f is merely upper semicontinuous and bounded, then

$$f_s(x) = \sup_{y \in \mathbb{R}^n} f(y) - \frac{|x - y|^2}{2 \, s}, \qquad x \in \mathbb{R}^n, s > 0, \tag{15.16}$$

is such that $f_s \in C_b^0(\mathbb{R}^n)$ (so that (15.14) holds for f_s at every $t \in I$) with $f \le f_s \downarrow f$ on $\mathbb{R}^n$ as $s \to 0^+$ (so that (15.14) for f_s implies (15.14) for f at every $t \in I$). To prove the claimed properties of f_s, we notice that the supremum in (15.16) is always achieved. Denoting by $y_s(x)$ a maximum point, it must be

$$|y_s(x) - x| \le \sqrt{2 \, s \, \text{osc}_{\mathbb{R}^n} f} \tag{15.17}$$

for otherwise

$$f(x) \le f_s(x) = f(y_s(x)) - \frac{|x - y_s(x)|^2}{2 \, s} < f(y_s(x)) - \text{osc}_{\mathbb{R}^n} f \le \inf_{\mathbb{R}^n} f$$

a contradiction. By (15.17), $y_s(x) \to x$ as $s \to 0^+$, so that $f(x) \le f_s(x) \le f(y_s(x))$ and the upper semicontinuity of f imply $f_s \downarrow f$ on $\mathbb{R}^n$ as $s \to 0^+$. Finally, if $x, z \in \mathbb{R}^n$, then we have

$$f_s(x) - f_s(z) \le f_s(x) - f(y_s(x)) + \frac{|z - y_s(x)|^2}{2s} \le \frac{|z - y_s(x)|^2}{2s} - \frac{|x - y_s(x)|^2}{2s},$$

and by the elementary identity $|z - y|^2 - |x - y|^2 = |z - x|^2 + 2(y - x) \cdot (x - z)$ and (15.17) we conclude that

$$f_s(x) - f_s(z) \le \frac{|z - x|^2}{2s} + \frac{|z - x| \sqrt{2 \operatorname{osc}_{\mathbb{R}^n} f}}{\sqrt{s}}.$$

By symmetry in (x, z), we deduce the (local) Lipschitz continuity of f_s. $\qquad \square$

15.2 From Vector Fields to Curves of Measures

We now exploit some classical ODE theory to show how sufficiently regular vector fields can be used to construct curves of measures transported according to the continuity equation. The regularity assumptions used here are not so convenient (e.g., the use of Proposition 15.3 in proving the Benamou–Brenier formula requires a quite delicate approximation argument), nevertheless this result has been included here because of its clear conceptual interest.

Proposition 15.3 *If* $\mathbf{u} : \mathbb{R}^n \times [0, 1] \to \mathbb{R}^n$ *is a Borel vector field such that*

$$\int_0^1 \|\mathbf{u}(t)\|_{C^0(\mathbb{R}^n)} + \operatorname{Lip}(\mathbf{u}(t); \mathbb{R}^n) \, dt = L < \infty, \tag{15.18}$$

then the flow $\Phi : \mathbb{R}^n \times [0, 1] \to \mathbb{R}^n$ *of* $\mathbf{u}$, *defined by*

$$\begin{cases} \partial_t \Phi(x, t) = \mathbf{u}(\Phi(x, t), t), \\ \Phi_0(x) = x, \end{cases} \tag{15.19}$$

is such that $\Phi_t : \mathbb{R}^n \to \mathbb{R}^n$ *is a Lipschitz homeomorphism with*

$$\max\{\operatorname{Lip}\Phi_t, \operatorname{Lip}\Phi_t^{-1}\} \le e^L, \qquad \forall t.$$

Moreover, if $\mu_0 \in \mathcal{P}_2(\mathbb{R}^n)$, *then* $\mu_t = (\Phi_t)_{\#}\mu_0$ *defines a curve of measures* μ *in* $\mathcal{P}_2(\mathbb{R}^n)$ *(with* $\mu_t \in \mathcal{P}_{2, \mathrm{ac}}(\mathbb{R}^n)$ *if* $\mu_0 \in \mathcal{P}_{2, \mathrm{ac}}(\mathbb{R}^n)$*) which is transported by* $\mathbf{u}$ *and is such that*

$$\mathcal{A}(\mu, \mathbf{u}) \ge \frac{1}{2} \mathbf{W}_2(\mu_0, \mu_1)^2. \tag{15.20}$$

Finally, if μ^* *is a curve of measures in* $\mathcal{P}_2(\mathbb{R}^n)$ *transported by* $\mathbf{u}$ *and such that* $\mu_t^* \ll \mathcal{L}^n$ *for a.e. t and* $\mu_0^* = \mu_0 \in \mathcal{P}_{2, \mathrm{ac}}(\mathbb{R}^n)$, *then* $\mu^* = \mu$.

Proof The statements concerning the flow of **u** follow from classical existence and uniqueness theory for ODE, therefore we directly focus on the second part of the statement. Given $\mu_0 \in \mathcal{P}_2(\mathbb{R}^n)$, set $\mu_t = (\Phi_t)_\# \mu_0$. Thanks to (15.18) we easily find

$$\int_{\mathbb{R}^n} |y|^2 \, d\mu_t(y) = \int_{\mathbb{R}^n} |\Phi_t(x)|^2 \, d\mu_0(x) \leq C(L) \int_{\mathbb{R}^n} (1 + |x|^2) \, d\mu_0(x) < \infty$$

so that $\mu_t \in \mathcal{P}_2(\mathbb{R}^n)$ and μ is a curve of measures. If $\varphi \in C_c^\infty(\mathbb{R}^n)$, then

$$\int_{\mathbb{R}^n} \varphi \, d\mu_t = \int_{\mathbb{R}^n} \varphi(\Phi_t) \, d\mu_0$$

implies that $t \mapsto \int_{\mathbb{R}^n} \varphi \, d\mu_t$ is Lipschitz continuous, and thus a.e. differentiable, with (by dominated convergence),

$$\frac{d}{dt} \int_{\mathbb{R}^n} \varphi \, d\mu_t = \frac{d}{dt} \int_{\mathbb{R}^n} \varphi(\Phi_t) \, d\mu_0$$

$$= \int_{\mathbb{R}^n} \nabla\varphi(\Phi_t(x)) \cdot \mathbf{u}(\Phi_t(x), t) \, d\mu_0(x) = \int_{\mathbb{R}^n} \nabla\varphi \cdot \mathbf{u}(t) \, d\mu_t,$$

from which (15.6) follows thanks to Proposition 15.1. We now notice that by Jensen's inequality, by $\mathbf{u}(\Phi_t(x), t) = \partial_t \Phi$, and by $\Phi_0(x) = x$, we have

$$2 \mathcal{A}(\mu, \mathbf{u}) = \int_0^1 dt \int_{\mathbb{R}^n} |\mathbf{u}(t)|^2 \, d\mu_t = \int_0^1 dt \int_{\mathbb{R}^n} |\mathbf{u}(\Phi_t(x), t)|^2 \, d\mu_0(x)$$

$$\geq \int_{\mathbb{R}^n} \left| \int_0^1 \mathbf{u}(\Phi_t(x), t) \, dt \right|^2 d\mu_0(x) = \int_{\mathbb{R}^n} |\Phi_1(x) - x|^2 \, d\mu_0(x)$$

$$\geq \mathbf{W}_2(\mu_0, (\Phi_1)_\# \mu_0)^2 = \mathbf{W}_2(\mu_0, \mu_1)^2,$$

proving (15.20). Finally, let μ^* be a curve of measures in $\mathcal{P}_2(\mathbb{R}^n)$ transported by **u** and such that $\mu_t^* \ll \mathcal{L}^n$. Given $\varphi \in C_c^\infty(\mathbb{R}^n)$ and $\zeta \in C_c^\infty((0,1))$, let

$$\psi(x, t) = \zeta(t) \, \varphi(\Phi_t^{-1}(x)).$$

Clearly, ψ is Lipschitz continuous and compactly supported in $\mathbb{R}^n \times (0, 1)$, therefore $\psi_\varepsilon = \psi \star \rho_\varepsilon$ is admissible in (15.6) and is such that $(\partial_t \psi_\varepsilon, \nabla\psi_\varepsilon)$ is uniformly bounded in ε and is $\mathcal{L}^{n+1}$-a.e. converging on $\mathbb{R}^n \times (0, 1)$ to $(\partial_t \psi, \nabla\psi)$. Since $\mu_t^* \ll \mathcal{L}^n$ for a.e. $t \in (0, 1)$, by dominated convergence, we deduce from the validity of (15.6) on ψ_ε that

$$\int_0^1 dt \int_{\mathbb{R}^n} \left(\partial_t \psi + \nabla\psi \cdot \mathbf{u}(t) \right) d\mu_t^* = 0. \tag{15.21}$$

Now, by differentiating $\psi(\Phi_t(x), t) = \zeta(t) \, \varphi(x)$ we find that

$$(\partial_t \psi + \nabla\psi \cdot \mathbf{u})(\Phi_t(x), t) = \zeta'(t) \, \varphi(x),$$

so that (15.21) boils down to

$$\int_0^1 \zeta'(t)\, dt \int_{\mathbb{R}^n} \varphi(\Phi_t^{-1}(x))\, d\mu_t^* = 0.$$

By the arbitrariness of ζ this shows that, for every $\varphi \in C_c^\infty(\mathbb{R}^n)$, the map $t \mapsto \int_{\mathbb{R}^n} \varphi(\Phi_t^{-1})\, d\mu_t^*$ is constant, i.e.,

$$\int_{\mathbb{R}^n} \varphi(x)\, d\mu_0(x) = \int_{\mathbb{R}^n} \varphi(\Phi_t^{-1}(x))\, d\mu_t^*(x) = \int_{\mathbb{R}^n} \varphi(x)\, d[(\Phi_t^{-1})_\# \mu_t^*](x).$$

This shows $\mu_0 = (\Phi_t^{-1})_\# \mu_t^*$, and since Φ_t is bijective, $\mu_t^* = (\Phi_t)_\# \mu_0$. □

15.3 Displacement Interpolation and the Continuity Equation

In Proposition 11.1, given $\mu_0, \mu_1 \in \mathcal{P}_2(\mathbb{R}^n)$ and an optimal transport plan γ in $\mathbf{K}_2(\mu, \nu)$, we have defined the displacement interpolation from μ_0 to μ_1 as

$$\mu_t = ((1-t)\,\mathbf{p} + t\,\mathbf{q})_\# \gamma, \qquad t \in [0,1]. \tag{15.22}$$

This identity defines a curve of measures μ from μ_0 to μ_1 in the sense of Section 15.1 thanks to the fact that $t \in [0,1] \mapsto \int_{\mathbb{R}^n} \varphi\, d\mu_t = \int_{\mathbb{R}^n} \varphi((1-t)\,x + t\,y)\, d\gamma(x,y)$ is continuous whenever $\varphi \in C_c^0(\mathbb{R}^n)$; actually, μ is a Lipschitz continuous curve of measures since $\mathbf{W}_2(\mu_t, \mu_s) = |t - s|\, \mathbf{W}_2(\mu_0, \mu_1)$, recall (11.2). In the next theorem we relate (15.22) to the continuity equation and the action functional.

Theorem 15.4 *If $\mu_0, \mu_1 \in \mathcal{P}_2(\mathbb{R}^n)$, γ is an optimal transport plan in $\mathbf{K}_2(\mu_0, \mu_1)$, and $h_t = (1-t)\,\mathbf{p} + t\,\mathbf{q}$ for $t \in [0,1]$, then the displacement interpolation $\mu_t = (h_t)_\# \gamma$ defines a Lipschitz curve of measures μ from μ_0 to μ_1. Moreover,*

$$(h_t)_\#[(\mathbf{q} - \mathbf{p})\, d\gamma] = \mathbf{u}(t)\, d\mu_t \tag{15.23}$$

for $\mathbf{u}(t) \in L^2(\mu_t)$ such that $\mathbf{u}$ transports μ and

$$\frac{1}{2}\, \mathbf{W}_2(\mu_0, \mu_1)^2 \geq \mathcal{A}(\mu, \mathbf{u}). \tag{15.24}$$

If, in addition, $\mu_0 \in \mathcal{P}_{2,\mathrm{ac}}(\mathbb{R}^n)$, ∇f is the Brenier map from μ_0 to μ_1, $f_t(x) = (1-t)|x|^2/2 + t\, f(x)$, f_t^* is the Fenchel-Legendre transform of f_t, then the composition $\nabla f(\nabla f_t^*)$ is defined μ_t-a.e. on $\mathbb{R}^n$, the vector field $\mathbf{u}(t)$ satisfies

$$\mathbf{u}(t) = \nabla f(\nabla f_t^*) - \nabla f_t^*, \tag{15.25}$$

and (15.24) holds as an identity.

Proof *Step one*: Given $\varphi \in C_c^0(\mathbb{R}^n; \mathbb{R}^n)$ we see that

$$\int_{\mathbb{R}^n} \varphi \cdot d((h_t)_\# [(\mathbf{q} - \mathbf{p}) \, d\gamma]) = \int_{\mathbb{R}^n} \varphi(h_t) \cdot (\mathbf{q} - \mathbf{p}) \, d\gamma$$

$$\leq \left(\int_{\mathbb{R}^n} |\varphi(h_t)|^2 d\gamma \right)^{1/2} \left(\int_{\mathbb{R}^n} |\mathbf{q} - \mathbf{p}|^2 d\gamma \right)^{1/2} = \mathbf{W}_2(\mu_0, \mu_1) \, \|\varphi\|_{L^2(\mu_t)}.$$

Hence, integration with respect to $(h_t)_\# [(\mathbf{q} - \mathbf{p}) \, d\gamma]$ defines a bounded linear functional on $L^2(\mu_t; \mathbb{R}^n)$, which, by the Riesz theorem, can be represented as $\mathbf{u}(t) \, d\mu_t$ for some $\mathbf{u}(t) \in L^2(\mu_t; \mathbb{R}^n)$ such that $\|\mathbf{u}(t)\|_{L^2(\mu_t)} \leq \mathbf{W}_2(\mu_0, \mu_1)$ for every $t \in [0, 1]$. By this last inequality, (15.24) trivially holds. To see that $\mathbf{u}$ transports μ, let us notice that, by dominated convergence, if $\varphi \in C_c^\infty(\mathbb{R}^n)$, then

$$\frac{d}{dt} \left[\int_{\mathbb{R}^n} \varphi \, d\mu_t \right] = \frac{d}{dt} \left[\int_{\mathbb{R}^n \times \mathbb{R}^n} \varphi((1-t)x + t\, y) \, d\gamma(x, y) \right]$$

$$= \int_{\mathbb{R}^n \times \mathbb{R}^n} \nabla \varphi((1-t)x + t\, y) \cdot (y - x) \, d\gamma$$

$$= \int_{\mathbb{R}^n} \nabla \varphi \cdot d((h_t)_\# [(\mathbf{q} - \mathbf{p}) d\gamma]) = \int_{\mathbb{R}^n} \nabla \varphi \cdot \mathbf{u}(t) \, d\mu_t.$$

Step two: Now let $\mu_0 \in \mathcal{P}_{2,\mathrm{ac}}(\mathbb{R}^n)$. With f and f_t as in the statement, by the uniqueness conclusion in Theorem 4.2 we have $\gamma = (\mathrm{Id} \times \nabla f)_\# \mu_0$, so that

$$\mu_t = ((1-t)\mathbf{p} + t\, \mathbf{q})_\# \gamma = (\nabla f_t)_\# \gamma.$$

where $f_t(x) = (1-t)|x|^2/2 + t \, f(x)$. Since ∇f_t is the Brenier map from μ_0 to μ_t and ∇f_t^* is the Brenier map from $\mu_t \ll \mathcal{L}^n$ (Proposition 10.2) to μ_0, we see that ∇f_t^* is a $\mathcal{L}^n$-a.e. inverse of ∇f_t in the sense of Theorem 4.4, and we can thus compute, for an arbitrary $\varphi \in C_c^0(\mathbb{R}^n)$,

$$\int_{\mathbb{R}^n} \varphi \, d((h_t)_\# [(\mathbf{q} - \mathbf{p}) \, d\gamma]) = \int_{\mathbb{R}^n} \varphi((1-t)\, x + t\, y) \, (y - x) \, d((\mathrm{Id} \times \nabla f)_\# \mu_0)$$

$$= \int_{\mathbb{R}^n} \varphi((1-t)\, x + t \nabla f(x)) \, (\nabla f(x) - x) \, d\mu_0(x)$$

$$= \int_{\mathbb{R}^n} \varphi((1-t)\, \nabla f_t^*(y) + t \nabla f(\nabla f_t^*(y))) \, \mathbf{u}(t)(y) \, d\mu_t(y) = \int_{\mathbb{R}^n} \varphi \, \mathbf{u}(t) d\mu_t.$$

This proves that (15.25) characterizes the vector fields $\mathbf{u}(t)$ defined in (15.23) when $\mu_0 \ll \mathcal{L}^n$. To prove that (15.24) holds as an identity we just have to notice that, once (15.25) is known, it implies

$$\int_{\mathbb{R}^n} |\mathbf{u}(t)|^2 \, d\mu_t = \int_{\mathbb{R}^n} |(\nabla f) \circ (\nabla f_t^*) - \nabla f_t^*|^2 \, d\mu_t$$

$$= \int_{\mathbb{R}^n} |\nabla f(x) - x|^2 \, d\mu_0(x) = \mathbf{W}_2(\mu_0, \mu_1)^2,$$

for every $t \in [0, 1]$, and thus (15.24) follows immediately. $\square$

15.4 Lipschitz Curves Admit Eulerian Velocities

Next, we prove that for every Lipschitz curve of measures μ in $(\mathcal{P}_2(\mathbb{R}^n), \mathbf{W}_2)$ there exists a vector field $\mathbf{u}$ which transports μ.

Theorem 15.5 *If μ is a Lipschitz curve in $(\mathcal{P}_2(\mathbb{R}^n), \mathbf{W}_2)$, i.e., if*

$$\mathbf{W}_2(\mu_t, \mu_s) \le L\,|t - s| \qquad \forall t, s \in [0, 1],$$

then there exists $\mathbf{u}$ which transports μ.

Proof Given $\varphi \in C_c^\infty(\mathbb{R}^n)$ we define a bounded, upper semicontinuous function $R_\varphi : \mathbb{R}^n \times \mathbb{R}^n \to [0, \infty)$ by setting

$$R_\varphi(x, y) = \begin{cases} |\nabla\varphi(x)| & \text{if } x = y, \\ \dfrac{|\varphi(x) - \varphi(y)|}{|x - y|} & \text{if } x \ne y. \end{cases}$$

If $\gamma_{t, t+h}$ is an optimal plan for $\mathbf{W}_2(\mu_t, \mu_{t+h})$, then

$$\left| \int_{\mathbb{R}^n} \varphi \, d\mu_{t+h} - \int_{\mathbb{R}^n} \varphi \, d\mu_t \right| = \left| \int_{\mathbb{R}^n \times \mathbb{R}^n} (\varphi(x) - \varphi(y)) \, d\gamma_{t, t+h}(x, y) \right|$$

$$\le \int_{\mathbb{R}^n \times \mathbb{R}^n} |x - y|\, R_\varphi(x, y) \, d\gamma_{t, t+h}(x, y)$$

$$\le \mathbf{W}_2(\mu_t, \mu_{t+h}) \left(\int_{\mathbb{R}^n \times \mathbb{R}^n} R_\varphi^2 \, d\gamma_{t, t+h} \right)^{1/2}. \quad (15.26)$$

Since $\gamma_{t, t+h}$ narrowly converges to $(\mathbf{id}, \mathbf{id})_\# \mu_t$, by the properties of R_φ we find

$$\limsup_{h \to 0^+} \int_{\mathbb{R}^n \times \mathbb{R}^n} R_\varphi^2 \, d\gamma_{t, t+h} \le \int_{\mathbb{R}^n \times \mathbb{R}^n} R_\varphi^2 \, d[(\mathbf{id}, \mathbf{id})_\# \mu_t] = \|\nabla\varphi\|_{L^2(\mu_t)}^2.$$

Dividing by $|h|$, recalling $\mathbf{W}_2(\mu_t, \mu_{t+h}) \le L\,|h|$, and letting $h \to 0$ in (15.26),

$$\limsup_{h \to 0} \frac{1}{|h|} \left| \int_{\mathbb{R}^n} \varphi \, d\mu_{t+h} - \int_{\mathbb{R}^n} \varphi \, d\mu_t \right| \le L \, \|\nabla\varphi\|_{L^2(\mu_t)}. \quad (15.27)$$

If now $\psi \in C_c^\infty(\mathbb{R}^n \times (0, 1))$, then

$$\int_0^1 dt \int_{\mathbb{R}^n} \partial_t \psi(x, t) \, d\mu_t(x) = \lim_{h \to 0} \int_0^1 dt \int_{\mathbb{R}^n} \frac{\psi(x, t) - \psi(t - h)}{h} \, d\mu_t(x)$$

$$= -\lim_{h \to 0} \int_0^1 dt \, \frac{1}{h} \left\{ \int_{\mathbb{R}^n} \psi(x, t) \, d\mu_{t+h}(x) - \int_{\mathbb{R}^n} \psi(x, t) \, d\mu_t(x) \right\},$$

so that by (15.27)

$$\left| \int_0^1 dt \int_{\mathbb{R}^n} \partial_t \psi(x, t) \, d\mu_t(x) \right| \le L \int_0^1 \|\nabla\psi(\cdot, t)\|_{L^2(\mu_t)} \, dt \le L \, \|\nabla\psi\|_{L^2(\mu)}. \quad (15.28)$$

Let X denote the closure of $\{\nabla\psi : \psi \in C_c^\infty(\mathbb{R}^n \times (0,1))\}$ in $L^2(\mu; \mathbb{R}^n)$. By (15.28) we can define a linear functional $\ell : C_c^\infty(\mathbb{R}^n \times (0,1)) \to \mathcal{E}$ which depends only $\nabla\psi$ by setting

$$\ell(\nabla\psi) = \int_0^1 dt \int_{\mathbb{R}^n} \partial_t \psi(x,t) \, d\mu_t(x), \qquad \psi \in C_c^\infty(\mathbb{R}^n \times (0,1)),$$

and then exploit the Riesz theorem to extend ℓ to a bounded linear functional on X. In particular, the minimization problem

$$\inf\left\{\frac{1}{2}\int_{\mathbb{R}^n \times [0,1]} |\mathbf{u}|^2 \, d\mu - \ell(\mathbf{u}) : \mathbf{u} \in X\right\}$$

admits a minimizer $\mathbf{u}$ which satisfies the Euler–Lagrange equation

$$\int_{\mathbb{R}^n \times [0,1]} \mathbf{u} \cdot \mathbf{v} \, d\mu = \ell(\mathbf{v}) \qquad \forall \mathbf{v} \in X.$$

Testing this equation with $\mathbf{v} = \nabla\psi$ for $\psi \in C_c^\infty(\mathbb{R}^n \times (0,1))$ we find that $\mathbf{u}$ satisfies (15.6). Finally, given a sequence $\{\psi_j\}_j$ in $C_c^\infty(\mathbb{R}^n \times (0,1))$ such that $\nabla\psi_j \to \mathbf{u}$ in $L^2(\mu)$, we see that

$$\int_{\mathbb{R}^n \times [0,1]} |\mathbf{u}|^2 \, d\mu = \lim_{j\to\infty} \int_{\mathbb{R}^n \times [0,1]} (\mathbf{u} \cdot \nabla\psi_j) \, d\mu$$

$$= -\lim_{j\to\infty} \int_{\mathbb{R}^n \times [0,1]} \partial_t \psi_j \, d\mu \le L \lim_{j\to\infty} \|\nabla\psi_j\|_{L^2(\mu)} = L \|\mathbf{u}\|_{L^2(\mu)},$$

where in the last inequality we have used (15.28). This gives

$$\int_0^1 dt \int_{\mathbb{R}^n} |\mathbf{u}(t)|^2 \, d\mu_t = \int_{\mathbb{R}^n \times [0,1]} |\mathbf{u}|^2 \, d\mu \le L^2.$$

thus proving that $\mathbf{u}(t) \in L^2(\mu_t)$ for a.e. $t \in [0,1]$. $\square$

15.5 The Benamou–Brenier Formula

We finally put together the various results of this section to prove a variational characterization of the Wasserstein distance.

Theorem 15.6 (Benamou–Brenier formula)   *If $\mu_0, \mu_1 \in \mathcal{P}_2(\mathbb{R}^n)$, then*

$$\frac{1}{2} \mathbf{W}_2(\mu_0, \mu_1)^2 = \inf_{(\mu,\mathbf{u})\in\mathcal{Y}} \mathcal{A}(\mu, \mathbf{u}),$$

where $\mathcal{Y}$ is the class of all Lipschitz curves μ in $\mathcal{P}_2(\mathbb{R}^n)$ and all vector fields $\mathbf{u}$ transporting μ such that $\mu_{t=0} = \mu_0$, $\mu_{t=1} = \mu_1$.

Proof   In Theorem 15.4, given $\mu_0, \mu_1 \in \mathcal{P}_2(\mathbb{R}^n)$ we have constructed $(\mu, \mathbf{u}) \in \mathcal{Y}$ such that

$$\frac{1}{2}\mathbf{W}_2(\mu_0,\mu_1)^2 \geq \mathcal{A}(\pmb{\mu},\mathbf{u}) \geq \inf_{\mathcal{Y}}\mathcal{A}.$$

Therefore it remains to show that, if $\mu_0,\mu_1 \in \mathcal{P}_2(\mathbb{R}^n)$, then

$$\frac{1}{2}\mathbf{W}_2(\mu_0,\mu_1)^2 \leq \mathcal{A}(\pmb{\mu},\mathbf{u}), \qquad \forall(\pmb{\mu},\mathbf{u}) \in \mathcal{Y}. \qquad (15.29)$$

To this end we set $c(x,y) = |x-y|^2/2$, and recall that, by the "improved" Kantorovich duality formula for the quadratic cost (i.e., the one using bounded Lipschitz continuous test functions, see Theorem 3.20), we have

$$\frac{1}{2}\mathbf{W}_2(\mu_0,\mu_1)^2 = \mathbf{K}_c(\mu_0,\mu_1) = \sup_f \int_{\mathbb{R}^n}(-f)d\mu_0 + \int_{\mathbb{R}^n}(-f^c)\,d\mu_1, \qquad (15.30)$$

where $\quad f^c(x) = \sup_{y \in \mathbb{R}^n}\left\{-f(y) - \dfrac{|x-y|^2}{2}\right\},$

where $\sup_f$ ranges among $f : \mathbb{R}^n \to \mathbb{R}$ bounded and Lipschitz continuous. We then define the **Hopf-Lax semigroup of** f, $\mathrm{HL}f : \mathbb{R}^n \times [0,1] \to \mathbb{R}$, by setting

$$\mathrm{HL}f(x,t) = \mathrm{HL}_t f(x) = \begin{cases} \inf\limits_{y \in \mathbb{R}^n}\left\{f(y) + \dfrac{|x-y|^2}{2t}\right\}, & (x,t) \in \mathbb{R}^n \times (0,1], \\ f(x), & x \in \mathbb{R}^n, t = 0, \end{cases}$$

and notice that (15.29) will follow by proving the following three claims holds for every bounded and Lipschitz continuous function $f : \mathbb{R}^n \to \mathbb{R}$:

(i) $\mathrm{HL}f$ is a bounded and Lipschitz continuous function on $\mathbb{R}^n \times [0,1]$;
(ii) for every $(x,t) \in \mathbb{R}^n \times (0,1]$ we have

$$\frac{d^+}{dt}\mathrm{HL}_t f(x) + \frac{1}{2}|\nabla^*\mathrm{HL}_t f|(x)^2 \leq 0,$$

where $|\nabla^*\mathrm{HL}_t f|$ is the asymptotic Lipschitz constant of $\mathrm{HL}_t f$ (see (15.12)), and where

$$\frac{d^+}{dt}\mathrm{HL}_t f(x) = \limsup_{h \to 0^+}\frac{\mathrm{HL}_{t+h}f(x) - \mathrm{HL}_t f(x)}{h},$$

is the right upper derivative of $\mathrm{HL}f(x,\cdot)$.
(iii) if $(\pmb{\mu},\mathbf{u}) \in \mathcal{Y}$, then $\lambda(t) = \int_{\mathbb{R}^n}\mathrm{HL}_t f\,d\mu_t$ is absolutely continuous on $[0,1]$ with

$$\lambda'(t) \leq \int_{\mathbb{R}^n}\frac{d^+}{dt}\mathrm{HL}_t f\,d\mu_t + \int_{\mathbb{R}^n}|\nabla^*\mathrm{HL}_t f|\,|\mathbf{u}(t)|\,d\mu_t \qquad (15.31)$$

for $\mathcal{L}^1$-a.e. $t \in [0,1]$.

Indeed, by using, in the order, (i), (iii), $ab \le (a^2 + b^2)/2$, and (ii), we find that, for every bounded Lipschitz function $f : \mathbb{R}^n \to \mathbb{R}$,

$$
\int_{\mathbb{R}^n} (-f) d\mu_0 + \int_{\mathbb{R}^n} (-f^c) d\mu_1 = \int_{\mathbb{R}^n} \mathrm{HL}_1 f \, d\mu_1 - \int_{\mathbb{R}^n} \mathrm{HL}_0 f \, d\mu_0 = \int_0^1 \lambda'(t) \, dt
$$

$$
\le \int_0^1 dt \left\{ \int_{\mathbb{R}^n} \frac{d^+}{dt} \mathrm{HL}_t f \, d\mu_t + \int_{\mathbb{R}^n} |\nabla^* \mathrm{HL}_t f| \, |\mathbf{u}(t)| \, d\mu_t \right\}
$$

$$
\le \int_0^1 dt \int_{\mathbb{R}^n} \left\{ \frac{d^+}{dt} \mathrm{HL}_t f + \frac{|\nabla^* \mathrm{HL}_t f|^2}{2} \right\} d\mu_t + \mathcal{A}(\mu,\mathbf{u}) \le \mathcal{A}(\mu,\mathbf{u}),
$$

which, by (15.30), implies (15.29). We thus turn to the proof of (i), (ii), and (iii).

To prove (ii): It is easily seen that for every $t \in (0,1]$ and $x \in \mathbb{R}^n$ there exists $z_t(x) \in \mathbb{R}^n$ such that

$$
\mathrm{HL}_t f(x) = f(z_t(x)) + \frac{|z_t(x) - x|^2}{2t}. \tag{15.32}
$$

Therefore, for every $x \in \mathbb{R}^n$ and $t, s \in (0,1)$, we have

$$
\mathrm{HL}_t f(x) - \mathrm{HL}_s f(x) \ge f(z_t(x)) + \frac{|z_t(x) - x|^2}{2t} - f(z_t(x)) - \frac{|z_t(x) - x|^2}{2s}
$$

$$
= \frac{s - t}{t s} \frac{|z_t(x) - x|^2}{2}, \tag{15.33}
$$

as well as

$$
\mathrm{HL}_t f(x) - \mathrm{HL}_s f(x) \le f(z_s(x)) + \frac{|z_s(x) - x|^2}{2t} - f(z_s(x)) - \frac{|z_s(x) - x|^2}{2s}
$$

$$
= \frac{s - t}{t s} \frac{|z_s(x) - x|^2}{2}. \tag{15.34}
$$

By (15.33) we immediately deduce

$$
\frac{d^+}{dt} \mathrm{HL}_t f(x) \le -\frac{|z_t(x) - x|^2}{2t^2}. \tag{15.35}
$$

Similarly, if $y, w \in \mathbb{R}^n, t \in (0,1]$, then

$$
\mathrm{HL}_t f(y) - \mathrm{HL}_t f(w) \le \frac{|z_t(w) - y|^2}{2t} - \frac{|z_t(w) - x|^2}{2t}
$$

$$
= \frac{|y - w|^2}{2t} + \frac{1}{t} (y - w) \cdot (w - z_t(w)) \tag{15.36}
$$

so that for every $x \in \mathbb{R}^n$ and $t \in (0,1]$,

$$
|\nabla^* \mathrm{HL}_t f|(x) = \lim_{r \to 0^+} \mathrm{Lip}(\mathrm{HL}_t f; B_r(x)) \le \limsup_{r \to 0^+} \max_{w \in B_r(x)} \frac{|w - z_t(w)|}{t}.
$$

Now, if $w_j \to x$, then, up to extracting subsequences, $z_t(w_j)$ converges to some $z_t(x)$ (i.e., to a minimum point for $\mathrm{HL}_t f(x)$), and we thus find

$$|\nabla^* \mathrm{HL}_t f|(x) \le \frac{|x - z_t(x)|}{t}. \tag{15.37}$$

Combining (15.35) and (15.37) we immediately find (ii).

To prove (i): We start proving that

$$\lim_{t \to 0^+} \mathrm{HL}_t f(x) = f(x), \qquad \forall x \in \mathbb{R}^n. \tag{15.38}$$

(Compare with step two in the proof of Proposition 15.2.) Indeed, by (15.32), we have

$$\frac{|z_t(x) - x|^2}{2t} = \mathrm{HL}_t f(x) - f(z_t(x)) \le f(x) - f(z_t(x)) \le \mathrm{Lip}(f)\,|z_t(x) - x|,$$

so that

$$|z_t(x) - x| \le 2t\,\mathrm{Lip}(f), \qquad \forall (x,t) \in \mathbb{R}^n \times (0,1], \tag{15.39}$$

and hence (15.38) follows by $f(x) \ge \mathrm{HL}_t f(x) \ge f(z_t(x))$. Let us now consider $\eta_x(t) = \mathrm{HL}_t f(x)$, which is increasing on $[0,1]$. By (15.34) with $0 < t < s < 1$ and by (15.39), we find

$$\frac{|\eta_x(s) - \eta_x(t)|}{|s - t|} = \frac{\eta_x(t) - \eta_x(s)}{s - t} \le \frac{|z_s(x) - x|^2}{2ts} \le 2\,\mathrm{Lip}(f)\,\frac{s}{t},$$

so that η_x is locally Lipschitz in $(0,1]$ (with $\mathrm{Lip}(\eta_x; [t,1]) \le 2\mathrm{Lip}(f)/t$ for every $t \in (0,1]$). In fact, if s is a differentiability point of η_x, then, by (15.34),

$$\eta_x'(s)\,(t - s) + \mathrm{o}(|t - s|) = \eta_x(t) - \eta_x(s)$$
$$\le \frac{s - t}{ts}\,\frac{|z_s(x) - x|^2}{2} = -(t - s)\,\frac{|z_s(x) - x|^2}{2s^2} + \mathrm{o}(|t - s|)$$

so that

$$\eta_x'(s) = -\frac{|z_s(x) - x|^2}{2s^2} \qquad \text{for a.e. } s \in (0,1). \tag{15.40}$$

By combining (15.40) with (15.39) we have thus proved

$$|\mathrm{HL}_t f(x) - \mathrm{HL}_s f(x)| \le 2\,\mathrm{Lip}(f)\,|t - s|, \qquad \forall x \in \mathbb{R}^n, t, s \in [0,1]. \tag{15.41}$$

To prove the Lipschitz bound in the spatial variable we notice that (15.36) implies, for every $x, y \in \mathbb{R}^n$ and $t \in (0,1]$,

$$|\mathrm{HL}_t f(y) - \mathrm{HL}_t f(x)| \le \frac{|y - x|^2}{2t} + \frac{|y - x|}{t}\,\max\{|x - z_t(x)|, |y - z_t(y)|\}$$
$$\le \frac{|y - x|^2}{2t} + 2\,\mathrm{Lip}(f)\,|y - x|$$

so that $HL_t f$ is locally Lipschitz on $\mathbb{R}^n$. At a differentiability point x of $HL_t f$, again by (15.36) we have

$$\nabla HL_t f(x) \cdot (y - x) + o(|y - x|) = HL_t f(y) - HL_t f(x)$$
$$\leq \frac{|y - x|^2}{2t} + \frac{1}{t}(y - x) \cdot (x - z_t(x)) = \frac{x - z_t(x)}{t} \cdot (y - x) + o(|y - x|),$$

thus giving that, if $t \in (0, 1]$, then

$$\nabla HL_t f(x) = \frac{x - z_t(x)}{t}, \qquad \text{for } \mathcal{L}^n\text{-a.e. } x \in \mathbb{R}^n. \tag{15.42}$$

In particular, by (15.39),

$$|HL_t f(x) - HL_t f(y)| \leq 2 \operatorname{Lip}(f) |x - y|, \qquad \forall x, y \in \mathbb{R}^n, t[0, 1], \tag{15.43}$$

and the proof of (i) is complete by combining (15.41) and (15.43).

Proof of (iii): If $0 < s < t < 1$, by (15.41) and by (15.10) in Proposition 15.2 (which can be applied thanks to claim (i)), we have

$$|\lambda(t) - \lambda(s)| \leq \int_{\mathbb{R}^n} |HL_t f - HL_s f| \, d\mu_s + \left| \int_{\mathbb{R}^n} HL_t f \, d\mu_t - \int_{\mathbb{R}^n} HL_t f \, d\mu_s \right|$$
$$\leq 2 \operatorname{Lip}(f) |t - s| + \int_t^s d\tau \int_{\mathbb{R}^n} |\nabla^* HL_t f| \, |\mathbf{u}(\tau)| \, d\mu_\tau.$$

Since (15.43) implies $|\nabla^* HL_t f| \leq 2 \operatorname{Lip}(f)$, and since $\mathcal{A}(\mu, \mathbf{u}) < \infty$, we conclude that λ is absolutely continuous on $[0, 1]$. To estimate λ' it is convenient to consider $\Lambda(t, s) = \int_{\mathbb{R}^n} HL_s f \, d\mu_t$, so that $\lambda(t) = \Lambda(t, t)$, and then notice that for every $\zeta \in C_c^1(0, 1)$, $\zeta \geq 0$, and for h small enough depending on $\operatorname{dist}(\{0, 1\}, \operatorname{spt} \zeta)$, we have

$$\int_0^1 \lambda(t) \frac{\zeta(t + h) - \zeta(t)}{h} \, dt = \int_0^1 \zeta(t) \frac{\lambda(t) - \lambda(t - h)}{h} \, dt$$
$$= \int_0^1 \zeta(t) \frac{\Lambda(t, t) - \Lambda(t - h, t)}{h} \, dt + \int_0^1 \zeta(t) \frac{\Lambda(t - h, t) - \Lambda(t - h, t - h)}{h} \, dt$$
$$= \int_0^1 \zeta(t) \frac{\Lambda(t, t) - \Lambda(t - h, t)}{h} \, dt + \int_0^1 \zeta(t + h) \frac{\Lambda(t, t + h) - \Lambda(t, t)}{h} \, dt.$$

Now, by Fatou's lemma, property (i) of $HL f$, and the definition of d^+/dt, we see that

$$\limsup_{h \to 0} \left| \frac{\Lambda(t, t) - \Lambda(t - h, t)}{h} \right| \leq \int_{\mathbb{R}^n} \frac{d^+}{dt} HL_t f \, d\mu_t,$$

while if I denotes the good set associated by Proposition 15.2 to $(\mu, \mathbf{u})$, then, for every $t \in I$ we have, thanks to (15.11),

$$\limsup_{h \to 0^+} \left| \frac{\Lambda(t, t+h) - \Lambda(t,t)}{h} \right| \leq \int_{\mathbb{R}^n} |\nabla^* \mathrm{HL}_t f| \, |\mathbf{u}(t)| \, d\mu_t .$$

Hence, by $\zeta \geq 0$, Fatou's lemma, and $\mathcal{L}^1((0,1) \setminus I) = 0$, we find that

$$\left| \int_0^1 \lambda' \, \zeta \right| = \lim_{h \to 0} \left| \int_0^1 \lambda(t) \, \frac{\zeta(t+h) - \zeta(t)}{h} \, dt \right|$$

$$\leq \int_0^1 \zeta(t) \, dt \int_{\mathbb{R}^n} \frac{d^+}{dt} \mathrm{HL}_t f + |\nabla^* \mathrm{HL}_t f| \, |\mathbf{u}(t)| \, d\mu_t .$$

If t is a Lebesgue point of λ' and of $t \mapsto \int_{\mathbb{R}^n} (d^+/dt) \mathrm{HL}_t f + |\nabla^* \mathrm{HL}_t f| \, |\mathbf{u}(t)| \, d\mu_t$, then by taking $\zeta = \zeta_j \to 1_{(t-r, t-r)}/(2r)$ as $j \to \infty$, and then sending $r \to 0^+$, we obtain (15.31), as required. The proof of Theorem 15.6 is thus complete. $\qquad\qquad\square$

15.6 Otto's Calculus

We finally discuss, very informally, how the Benamou–Brenier formula points at a way to describe the Riemannian/infinitesimal Hilbertian structure of the Wasserstein space. These informal considerations are the basis of Otto's original insights on why pursuing the implementation of the minimizing movements scheme for the displacement convex internal energies introduced in Chapter 10, and are the starting point of more formal treatments of the subject.

Consider a curve of measures μ such that

$$\mu_t = \rho(\cdot, t) \, d\mathcal{L}^n$$

where $\rho = \rho(x,t)$ is smooth and positive on $\mathbb{R}^n$. By the Benamou–Brenier formula,

$$\mathbf{W}_2(\mu_0, \mu_1)^2 = \inf \int_0^1 dt \int_{\mathbb{R}^n} |\mathbf{u}(x,t)|^2 \, \rho(x) \, dx ,$$

where the infimum ranges among all $\mathbf{u}$ transporting μ. As noticed in (15.7), if $\mathrm{div} \, \mathbf{v} = 0$ on $\mathbb{R}^n$, then $\mathbf{u} + \varepsilon \, [\mathbf{v}/\rho]$ is also transporting μ, therefore an optimal choice of $\mathbf{u}$ for μ must satisfy

$$\int_{\mathbb{R}^n} |\mathbf{u}(x,t)|^2 \, \rho(x) \, dx \leq \int_{\mathbb{R}^n} \left| \mathbf{u} + \varepsilon \left(\frac{\mathbf{v}}{\rho} \right) \right|^2 \rho(x) \, dx ,$$

that is

$$\int_{\mathbb{R}^n} \mathbf{u} \cdot \mathbf{v} = 0 \qquad \text{whenever } \mathrm{div} \, \mathbf{v} = 0 .$$

In other words, it must be $\mathbf{u} = \nabla f$ for a scalar potential f, so that

$$\mathbf{W}_2(\mu_0, \mu_1)^2 = \inf\left\{ \int_0^1 dt \int_{\mathbb{R}^n} |\nabla f|^2 \rho : \partial_t \rho + \operatorname{div}(\rho \nabla f) = 0 \right\}. \quad (15.44)$$

The point of view in (15.44) is that for a given $\rho = \rho(x,t)$ defining a curve $\mathcal{P}_2(\mathbb{R}^n)$ we must have $\int_{\mathbb{R}^n} \partial_t \rho = 0$ (to preserve $\int_{\mathbb{R}^n} \rho = 1$ for every t), so that the elliptic PDEs (parameterized over $t \in [0,1]$)

$$-\operatorname{div}(\rho \nabla f) = \partial_t \rho \qquad \text{on } \mathbb{R}^n,$$

are solvable and determine $f = f(x,t)$.

We are thus invited to consider every function $\tau : \mathbb{R}^n \to \mathbb{R}$ with $\int_{\mathbb{R}^n} \tau = 0$ as a possible tangent direction to $\mu = \rho\, d\mathcal{L}^n$ in $\mathcal{P}_2(\mathbb{R}^n)$, and to introduce a metric on $\mathcal{P}_2(\mathbb{R}^n)$ by setting

$$g_\rho(\tau_1, \tau_2) = \int_{\mathbb{R}^n} (\nabla f_1 \cdot \nabla f_2)\, \rho,$$

where the functions f_k are related to τ_k by the elliptic PDEs

$$-\operatorname{div}(\rho \nabla f_k) = \tau_k \qquad \text{on } \mathbb{R}^n.$$

With this structure at hand, we can try to compute gradients. For example, let

$$S(\rho) = \int_{\mathbb{R}^n} \rho \log \rho,$$

then the differential of U at ρ along the direction τ is given by

$$dS_\rho[\tau] = \int_{\mathbb{R}^n} (\log(\rho) + 1)\, \tau = \int_{\mathbb{R}^n} \tau \log \rho.$$

Therefore, $\sigma = \nabla^{\mathbf{W}_2} S(\rho)$ must be a function with $\int_{\mathbb{R}^n} \sigma = 0$ such that

$$\int_{\mathbb{R}^n} \tau \log \rho = \langle \tau, \sigma \rangle_\rho = g_\rho(\tau, \sigma) = \int_{\mathbb{R}^n} (\nabla f \cdot \nabla g)\, \rho$$

provided f and g are potentials such that $-\operatorname{div}(\rho \nabla f) = \tau$ and $-\operatorname{div}(\rho \nabla g) = \sigma$ on $\mathbb{R}^n$; we thus find,

$$\int_{\mathbb{R}^n} \tau \log \rho = \int_{\mathbb{R}^n} (\nabla f \cdot \nabla g)\, \rho = \int_{\mathbb{R}^n} \nabla g \cdot (\rho \nabla f) = -\int_{\mathbb{R}^n} g \operatorname{div}(\rho \nabla f) = \int_{\mathbb{R}^n} g\, \tau.$$

By arbitrariness of τ with $\int_{\mathbb{R}^n} \tau = 0$, $g = (\log \rho) + \text{constant}$, $\nabla g = (\nabla \rho)/\rho$, and thus

$$\nabla^{\mathbf{W}_2} S(\rho) = \sigma = -\operatorname{div}(\rho \nabla g) = -\Delta \rho.$$

This is the celebrated computation indicating in which sense *the Laplace operator is the gradient of the (physical) entropy functional*, a statement of clear physical appeal.

Of course, the above computation is not specific to the entropy functional and can be informally carried over for generic functionals. Again informally, we can establish in great generality a relation between the L^2-gradient and $\mathbf{W}_2$-gradient of an abstract functional $\mathcal{F} : L^2(\mathbb{R}^n) \cap \mathcal{P}_2(\mathbb{R}^n) \to \mathbb{R}$; see (15.45). Indeed, denoting by $\nabla^{L^2}\mathcal{F}(\rho) \in L^2(\mathbb{R}^n)$ the gradient of $\mathcal{F}$ at ρ with respect to the L^2-scalar product, so that, by definition,

$$d\mathcal{F}_\rho[\tau] = \int_{\mathbb{R}^n} \tau \, \nabla^{L^2}\mathcal{F}(\rho) \qquad \forall \tau \in L^2(\mathbb{R}^n) \text{ s.t. } \int_{\mathbb{R}^n} \tau = 0,$$

holds, then we have

$$\nabla^{\mathbf{W}_2}\mathcal{F}(\rho) = -\text{div}\,(\rho \, \nabla[\nabla^{L^2}\mathcal{F}(\rho)]). \tag{15.45}$$

Indeed, denoting $\sigma = \nabla^{\mathbf{W}_2}\mathcal{F}(\rho)$, if f and g satisfy the equations $-\text{div}\,(\rho\nabla f) = \tau$ and $-\text{div}\,(\rho\nabla g) = \sigma$, then, by arguing as before, we get

$$d\mathcal{F}_\rho[\tau] = g_\rho(\tau,\sigma) = \int_{\mathbb{R}^n} (\nabla f \cdot \nabla g)\, \rho = \int_{\mathbb{R}^n} \nabla g \cdot (\rho\nabla f)$$

$$= -\int_{\mathbb{R}^n} g\, \text{div}\,(\rho\nabla f) = \int_{\mathbb{R}^n} g\,\tau;$$

by arbitrariness of τ with $\int_{\mathbb{R}^n} \tau = 0$, we find $g = \nabla^{L^2}\mathcal{F}(\rho) + \text{constant}$, and hence $\sigma = -\text{div}\,(\rho\nabla g) = \sigma$ boils down to (15.45). To give an example, let us consider the case when $\mathcal{F}$ is a generic potential energy $\mathcal{U}$, that is, $\mathcal{F}(\rho) = \mathcal{U}(\rho) = \int_{\mathbb{R}^n} U(\rho(x))\, dx$. In that case,

$$d\mathcal{U}_\rho[\tau] = \lim_{t\to 0} \int_{\mathbb{R}^n} \frac{U(\rho + t\tau) - U(\rho)}{t} = \int_{\mathbb{R}^n} U'(\rho)\,\tau,$$

so that $\nabla^{L^2}\mathcal{U}(\rho) = U'(\rho)$ and we obtain the formula

$$\nabla^{\mathbf{W}_2}\mathcal{U}(\rho) = -\text{div}\,(\rho\,\nabla[U'(\rho)]),$$

for the $\mathbf{W}_2$-gradient of the internal energy with density U.

PART IV

Solution of the Monge Problem with
Linear Cost: The Sudakov Theorem

16

Optimal Transport Maps on the Real Line

In this chapter, after introducing the notion of cumulative distribution function of a measure, we address the construction of optimal transport maps in dimension one, see Theorem 16.1. This result is one the two basic building blocks used in Sudakov's argument for the solution of the Monge problem, the other one being the disintegration theorem presented in Chapter 17.

16.1 Cumulative Distribution Functions

The **cumulative distribution function** $M_\mu : \mathbb{R} \to [0,1]$ of a probability measure μ on $\mathbb{R}$ is the right-continuous and increasing function defined by

$$M_\mu(x) = \mu((-\infty, x]), \qquad x \in \mathbb{R}.$$

Since the σ-algebra generated by $\{(-\infty, x] : x \in \mathbb{R}\}$ is $\mathcal{B}(\mathbb{R})$, the knowledge of M_μ completely characterizes μ. The fundamental property of M_μ in relation to transport problems is that, if μ **has no atoms** (i.e., if $\mu(\{x\}) = 0$ for every $x \in \mathbb{R}$), then M_μ pushes-forward μ into the Lebesgue measure over $[0,1]$, i.e.,

$$(M_\mu)_\# \mu = \mathcal{L}^1 \llcorner [0,1] \qquad \text{(for } \mu \text{ with no atoms).} \qquad (16.1)$$

Postponing for a moment the proof of (16.1), we notice that if the target measure ν (into which we want to transport μ) is such that M_ν is invertible, with inverse M_ν^{-1}, then, based on (16.1), we may expect to have

$$(M_\nu^{-1})_\# \mathcal{L}^1 \llcorner [0,1] = \nu. \qquad (16.2)$$

Whenever (16.1) and (16.2) hold, we can push-forward μ into ν by first using M_μ to get from μ to $\mathcal{L}^1 \llcorner [0,1]$, and then by using M_ν^{-1} to get from $\mathcal{L}^1 \llcorner [0,1]$ to ν. The composition $T = (M_\nu^{-1}) \circ M_\mu$ is then increasing (the inverse M_ν^{-1} of the increasing function M_ν is increasing too), transports μ into ν, and is thus a solid candidate as an optimal transport map (for a large class of natural transport costs). This strategy works even better than expected, in the sense that, even when M_ν fails to be invertible (e.g., because the support of ν is disconnected), and even when ν has atoms (and thus (16.1) does not hold for ν), we can nevertheless define a *generalized inverse* of M_ν, denoted by W_ν, which is increasing and has the property of pushing-forward $\mathcal{L}^1 \llcorner [0,1]$ into ν. In summary, by exploiting basic properties of cumulative distribution functions, we will be able to construct monotone transport maps from non-atomic measures $\mu \in \mathcal{P}(\mathbb{R})$ into arbitrary target measures $\nu \in \mathcal{P}(\mathbb{R})$. In Theorem 16.1 we shall prove that these maps are the *unique* minimizers of the Monge problem with transport cost $c(x,y) = h(|x - y|)$ if h is strictly convex, increasing and such that $h(0) = 0$; and that they also solve the Monge problem with linear cost (although, of course, non-uniquely, as shown by the book-shifting example of Section 1.3). We now prove (16.1), introduce the notion of generalized inverse of a cumulative distribution function, and prove the general validity of (16.2) if inversion is intended in this generalized sense.

Proof that if μ has no atoms, then (16.1) *holds*: Since μ has no atoms, M_μ is continuous on $\mathbb{R}$: therefore, by monotonicity of M_μ, for every $t \in (0,1)$ there exists $x_t \in \mathbb{R}$ such that

$$\{x : M_\mu(x) \le t\} = (-\infty, x_t], \qquad M_\mu(x_t) = t.$$

In particular,

$$((M_\mu)_{\#}\mu)((0,t]) = \mu((-\infty, x_t]) = M_\mu(x_t) = t,$$

that is to say, $(M_\mu)_{\#}\mu$ is equal to $\mathcal{L}^1 \llcorner [0,1]$ on every interval of the form $(0,t]$ with $t \in (0,1)$. Since this class of intervals generates the Borel subsets of $[0,1]$, we have proved (16.1).

Generalized inverse of a cumulative distribution function: The cumulative distribution function M_μ of $\mu \in \mathcal{P}(\mathbb{R})$ may fail to be invertible because of possible "gaps" in the support of μ: for example, if $\mu = \mathcal{L}^1 \llcorner [(0,1/2) \cup (1,3/2)]$, then $M_\mu(x) = 1/2$ for every $x \in (1/2,1)$. It is however possible to define a "left-continuous inverse" W_μ of M_μ, $W_\mu : (0,1) \to \mathbb{R}$, by setting

$$W_\mu(t) = \inf\{x \in \mathbb{R} : M_\mu(x) \ge t\}, \qquad t \in (0,1), \tag{16.3}$$

see Figure 16.1. It is easily seen that W_μ is increasing, left-continuous, and, as implicitly stated above, such that $W_\mu(t) \in \mathbb{R}$ for every $t \in (0,1)$. Geometrically,

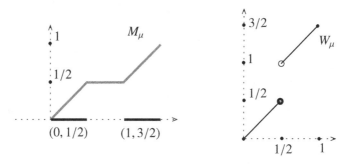

Figure 16.1 M_μ and W_μ for $\mu = \mathcal{L}^1 \llcorner [(0, 1/2) \cup (1, 3/2)]$. Notice that $W_\mu(1/2) = 1/2$ while $W_\mu(t) \to 1$ as $t \to (1/2)^+$.

the extended graph[1] of W_μ is obtained by reflecting the extended graph of M_μ with respect to the bisectrix of $\mathbb{R}^2$, so that plateaus of M_μ correspond to jumps in W_μ, and jumps of M_μ correspond to plateaus of W_μ. While the equivalence "$W_\mu(t) = x$ if and only if $M_\mu(x) = t$" does not hold for arbitrary values of x and t, we still have that for every $x \in \{0 < M_\mu < 1\}$ and $t \in (0, 1)$,

$$W_\mu(t) \le x \qquad \text{if and only if} \qquad M_\mu(x) \ge t. \tag{16.4}$$

We can then use (16.4) to prove the crucial property

$$\mu = (W_\mu)_\# (\mathcal{L}^1 \llcorner [0, 1]), \qquad \forall \mu \in \mathcal{P}(\mathbb{R}), \tag{16.5}$$

(which of course boils down to (16.2) for invertible cumulative distribution functions). To prove (16.5), let $\mu_0 = (W_\mu)_\# (\mathcal{L}^1 \llcorner [0, 1])$: by (16.4), for every $x \in \{0 < M_\mu < 1\}$ we have

$$\{t \in (0, 1) : W_\mu(t) \le x\} = \{t \in (0, 1) : M_\mu(x) \ge t\} = (0, M_\mu(x)],$$

which gives $\mu_0((-\infty, x]) = \mathcal{L}^1(\{W_\mu \le x\}) = M_\mu(x) = \mu((-\infty, x])$. Since $\{(-\infty, x] : x \in \mathbb{R}\}$ generates $\mathcal{B}(\mathbb{R})$, we have indeed proved (16.5).

16.2 Optimal Transport on the Real Line

We now state and prove the main result of this chapter. We shall use the following statement, to be proved later on as Proposition 17.8 in Chapter 17:

if $\gamma \in \mathcal{P}(\mathbb{R}^n \times \mathbb{R}^n)$, $T : E \to \mathbb{R}^n$ is a Borel map,
μ is concentrated on E and γ on $G = \{(x, T(x)) : x \in E\}$, $\qquad$ (16.6)
then γ is induced by T, in the sense that $\gamma = (\mathbf{id} \times T)_\# \mu$.

---

[1] The extended graph of an increasing function M is the planar closed set obtained by adding a closed interval to the graph of M in correspondence to each jump point of M; the endpoints of the segment are the left and right limits of M at the corresponding jump point.

It is also convenient to introduce the notation

$$M_\mu^\circ(x) = \mu((-\infty, x)), \qquad x \in \mathbb{R},$$

for the "left-continuous version" of M_μ. Notice that $M_\mu^\circ \le M_\mu$ on $\mathbb{R}$, and that $M_\mu^\circ = M_\mu$ at all but countably many points in $\mathbb{R}$.

Theorem 16.1 (Optimal transport on the real line) *If $\mu, \nu \in \mathcal{P}(\mathbb{R})$, then the following holds:*

(i) *if $\gamma \in \Gamma(\mu, \nu)$ and spt γ is increasing in the sense that*

$$\text{if } (x, y), (x^*, y^*) \in \text{spt } \gamma \text{ and } x < x^*, \text{ then } y \le y^*, \tag{16.7}$$

then γ is uniquely determined on $\mathcal{B}(\mathbb{R} \times \mathbb{R})$ by the following property: for every $x < x^*$ and $y < y^*$,

$$\gamma([x, x^*] \times [y, y^*]) = \min\{M_\mu(x^*), M_\nu(y^*))\} - \min\{M_\mu^\circ(x), M_\nu^\circ(y)\}; \tag{16.8}$$

(ii) *if $\gamma \in \Gamma(\mu, \nu)$ is c-cyclically monotone for a transport cost $c(x, y) = c[h](x, y) = h(|y - x|)$ corresponding to a strictly convex and increasing $h : [0, \infty) \to [0, \infty)$ with $h(0) = 0$, then spt γ is increasing and thus uniquely determined by (16.8);*

(iii) *If μ as no atoms, $I_\mu = \{0 < M_\mu < 1\}$, and $T : \to \mathbb{R}$ is defined by setting*

$$T(x) = W_\nu(M_\mu(x)), \qquad x \in \{0 < M_\mu < 1\}. \tag{16.9}$$

then μ is concentrated on I_μ, I_μ is an open interval, T is increasing on I_μ, $T_\#\mu = \nu$, and T takes values into the smallest closed interval containing spt ν;

(iv) *if μ has no atoms, and T is as in (16.9), then $S = T$ μ-a.e. on $\mathbb{R}$ whenever $S : I_\mu \to \mathbb{R}$ is an increasing map such that $S_\#\mu = \nu$; moreover, $\gamma = (\text{id} \times T)_\#\mu$ whenever $\gamma \in \Gamma(\mu, \nu)$ with spt γ increasing;*

(v) *if μ has no atoms and T is as in (16.9), then, as soon as $\mu, \nu \in \mathcal{P}_1(\mathbb{R})$, we have that $\gamma_T = (\text{id} \times T)_\#\mu$ is an optimal transport plan in $\mathbf{K}_1(\mu, \nu)$ and T is an optimal transport map in $\mathbf{M}_1(\mu, \nu)$; similarly, $\gamma_T = (\text{id} \times T)_\#\mu$ is the unique optimal transport plan in $\mathbf{K}_c(\mu, \nu)$, and T is the unique optimal transport map in $\mathbf{M}_c(\mu, \nu)$, whenever $c = c[h] \in L^1(\mu \times \nu)$ for h as in statement (ii);*

(vi) *if $\mu, \nu \in \mathcal{P}_1(\mathbb{R})$ have no atoms and T is as in (16.9), then $W_\mu \circ M_\nu : I_\nu \to \mathbb{R}$ is the inverse of T and an optimal transport map in $\mathbf{M}_1(\nu, \mu)$.*

Remark 16.2 By Theorem 16.1-(iv), when μ has no atoms there are no *generic* monotone optimal transport plans from μ to ν, but there is *only* $T = W_\nu \circ M_\mu$, which can then be called **the monotone transport plan from μ to ν**, and which is automatically optimal in $\mathbf{M}_c(\mu, \nu)$ for the transport costs

c described in Theorem 16.1-(v). Notice also that, in the special case when $\mu = \rho\, d\mathcal{L}^1$ and $\nu = \sigma\, d\mathcal{L}^1$, the monotone transport plan T satisfies the infinitesimal transport condition

$$\rho(x) = \sigma(T(x))\, T'(x) \qquad \mathcal{L}^1\text{-a.e. on } \{\rho > 0\}.$$

Remark 16.3 Recalling that when μ has atoms there may be no transport maps from μ to ν (e.g., $\mu = \delta_0$ and $\nu = (\delta_1 + \delta_{-1})/2$), Theorem 16.1 provides a complete picture concerning optimal transport maps in one dimension.

Remark 16.4 Statements (i) and (ii) combined show that, in the presence of strict convexity, uniqueness of minimizers holds, at the level of transport plans, even when μ has atoms.

Remark 16.5 (Non-uniqueness in the Monge problem with linear cost) We already know from the "book-shifting" example of Section 1.3 that we should not expect non-uniqueness of optimal transport maps in the Monge problem with linear cost. To get a better sense of how dramatic such "level of non-uniqueness" really is, let us first of all show that *if $\mu, \nu \in \mathcal{P}_1(\mathbb{R}^n)$, and* spt μ *and* spt ν *are contained into disjoint intervals, then all the transport maps from μ to ν* (assuming that there is at least one, e.g., because μ has not atoms) *have the same transportation cost*. To see this, assuming without loss of generality that spt $\mu \subset (-\infty, 0)$ and spt $\nu \subset (0, \infty)$, we argue that if T is a transport map from μ to ν, then $T(x) > 0 > x$ for μ-a.e. $x \in \mathbb{R}$, and thus

$$\int_{\mathbb{R}} |T(x) - x|\, d\mu(x) = \int_{\mathbb{R}} T(x)\, d\mu(x) - \int_{\mathbb{R}} x\, d\mu(x) = \int_{\mathbb{R}} y\, d\nu(y) - \int_{\mathbb{R}} x\, d\mu(x),$$

thus proving our assertion. This simple remark becomes striking when we realize that, under the above assumptions on μ and ν, there are actually lots of transport maps. For example, assuming (in addition to the above assumptions) that $\mu, \nu \ll \mathcal{L}^1$, we can canonically construct a countable family of transport maps from μ to ν as follows: Given $N \geq 2$, there are two *disjoint* families of open intervals $\{I_N^i\}_{i=1}^N \subset (-\infty, 0)$ and $\{J_N^i\}_{i=1}^N \subset (0, \infty)$ such that $\mu(I_N^i) = \nu(J_N^i) = 1/N$ for each i, and whose unions are $\mathcal{L}^1$-equivalent, respectively, to spt μ and to spt ν. Given a bijection $\sigma : \{1, \ldots, N\} \to \{1, \ldots, N\}$, we can define a map T^σ $\mathcal{L}^1$-a.e. on spt μ by asking that T^σ, on I_N^i, is the monotone increasing transport map from $N\,\mu\llcorner I_N^i$ to $N\,\nu\llcorner J_N^{\sigma(i)}$ constructed in Theorem 16.1. In this way T^σ is a transport map from μ to ν. Moreover, if $\sigma \neq \sigma'$, then $T^\sigma \neq T^{\sigma'}$, and, for each $N \geq 2$, there is exactly one σ such that T^σ is the monotone transport map from μ to ν constructed in Theorem 16.1. By the above statements, all the maps in $\{T^\sigma\}_{\sigma, N}$ are optimal in $\mathbf{M}_1(\mu, \nu)$. From this viewpoint, it is evident why, in solving the Monge problem in higher

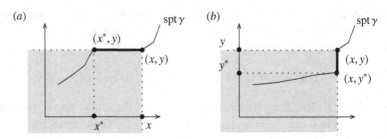

Figure 16.2 Proof of (16.12): (a) the case when there is $x^* < x$ such that $(x^*, y) \in$ spt γ, and thus the second inclusion in (16.10) may be strict; (b) the case when there is $y^* < y$ such that $(x, y^*) \in$ spt γ, and thus the second inclusion in (16.11) may be strict.

dimension, we will be interested not just in constructing *any* optimal transport map, but we will actually care of constructing optimal transport maps with a nice structure (see the discussion at the beginning of Chapter 18 for more details).

Proof of Theorem 16.1   *Step one*: We prove statement (i). Given $\gamma \in \Gamma(\mu, \nu)$ with spt γ increasing we want to prove that (16.8) holds. To this end, given $x, y \in \mathbb{R} \cup \{+\infty\}$, let us set

$$H^\circ(x, y) = (-\infty, x) \times (-\infty, y), \qquad H^c(x, y) = (-\infty, x] \times (-\infty, y].$$

We then notice that if $(x, y) \in$ spt γ and there is $x^* < x$ such that $(x^*, y) \in$ spt γ, then, since spt γ is increasing, we have

$$\text{spt } \gamma \cap H^\circ(x, y) = \text{spt } \gamma \cap H^\circ(+\infty, y) \subsetneq \text{spt } \gamma \cap H^\circ(x, +\infty); \quad (16.10)$$

while, if there is $y^* < y$ such that $(x, y^*) \in$ spt γ, then

$$\text{spt } \gamma \cap H^\circ(x, y) = \text{spt } \gamma \cap H^\circ(x, +\infty) \subsetneq \text{spt } \gamma \cap H^\circ(+\infty, y), \quad (16.11)$$

see Figure 16.2; and, if neither of these possibilities occur, then we simply have

$$\text{spt } \gamma \cap H^\circ(x, y) = \text{spt } \gamma \cap H^\circ(+\infty, y) = \text{spt } \gamma \cap H^\circ(x, +\infty).$$

Therefore, for every $\gamma \in \Gamma(\mu, \nu)$ with spt γ increasing it must be

$$\gamma(H^\circ(x, y)) = \min\{M_\mu^\circ(x), M_\nu^\circ(y)\}, \tag{16.12}$$

and by an entirely analogous argument, we see that

$$\gamma(H^c(x, y)) = \min\{M_\mu(x), M_\nu(y)\}, \tag{16.13}$$

so that (16.8) descends from (16.12), (16.13), and $[x, x^*] \times [y, y^*] = H^c(x^*, y^*) \setminus H^\circ(x, y)$ for every $x < x^*$ and $y < y^*$.

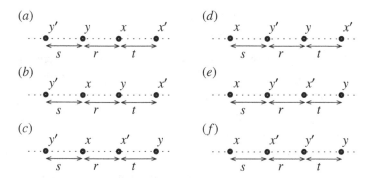

Figure 16.3 Proof that $x < x'$ and (16.14) imply $y \le y'$.

Step two: We prove statement (ii). Let us consider $h : [0, \infty) \to [0, \infty)$ strictly convex, increasing and with $h(0) = 0$, we set $c(x, y) = c[h](x, y) = h(|y - x|)$, and consider $\gamma \in \Gamma(\mu, \nu)$ such that spt γ is c-cyclically monotone. This latter assumption implies in particular that, for every $(x, y), (x', y') \in$ spt γ,

$$h(|x - y|) + h(|x' - y'|) \le h(|x - y'|) + h(|x' - y|). \tag{16.14}$$

We want to deduce from (16.14) that if $(x, y), (x', y') \in$ spt γ with $x < x'$, then $y \le y'$ (i.e., that spt γ is increasing). To this end, let us notice that, if $x < x'$ and $y' < y$, then x, x', y and y' can be arranged in six different configurations[2] (see Figure 16.3), in correspondence of which (16.14) takes the following forms (with $r, s, t \ge 0$ plus additional strict sign conditions depending on the case under examination):

$$h(r) + h(r + s + t) \le h(r + s) + h(r + t) \quad \text{in cases (a) and (f),} \tag{16.15}$$
(with $t, s > 0$);

$$h(r) + h(r + s + t) \le h(s) + h(t) \qquad \text{in cases (b) and (e),} \tag{16.16}$$
(with $r + t, r + s > 0$);

$$h(r + s) + h(r + t) \le h(s) + h(t) \qquad \text{in cases (c) and (d),} \tag{16.17}$$
(with $r > 0$).

However: (16.15) cannot hold, since h' strictly increasing and $s > 0$ give $h(r + s + t) - h(r + s) > h(r + t) - h(r)$; (16.16) cannot hold, since h increasing, h' strictly increasing and $r + s > 0$ give

---

[2] A more geometric (and, possibly, a conceptually clearer) way to see this argument is that, by the properties of h, the difference between the right-hand side and left-hand side of (16.14) can be *decreased* by moving to the right $\min\{x, y'\}$ (which is the left-most point in $\{x, x', y, y'\}$), or by moving to the left $\max\{x', y\}$ (which is the right-most point of the four). One can do this up to a position where either $x = x'$ or $y = y'$, and thus (16.14) becomes an identity.

$$h(r + s + t) = (h(r + s + t) - h(r + s)) + (h(r + s) - h(s)) + h(s)$$
$$\geq (h(r + s + t) - h(r + s)) + h(s) > (h(t) - h(0)) + h(s) = h(t) + h(s);$$

and, finally, (16.17) cannot hold, since $r > 0$ and h is strictly increasing. We have thus proved that $x < x'$ and (16.14), imply $y \leq y'$, as claimed.

Step three: We now assume that μ has no atoms, and set $I_\mu = \{0 < M_\mu < 1\}$ and $T = W_\nu \circ M_\mu$. To prove that μ is concentrated on I_μ it is enough to notice that, by (16.1),
$$\mu(\{x : M_\mu(x) \in \{0, 1\}\}) = \mathcal{L}^1(\{0, 1\}) = 0.$$

The set I_μ is in interval in general, and an open interval since μ has no atoms. Evidently T is defined and increasing on I_μ, while the fact that $T_\#\mu = \nu$ is immediate by combining (16.1) with (16.5) (applied to ν). If J is the smallest closed interval containing spt ν and $T(x) < \inf J$ for some $x \in \{0 < M_\mu < 1\}$, then the interval $\{0 < M_\mu < M_\mu(x)\}$, which has positive μ-measure, is mapped by T on a ν-null set; we rule out similarly that $T(x) > \sup J$ for some $x \in \{0 < M_\mu < 1\}$. This proves statement (iii).

We begin the proof of statement (iv) by considering an increasing map $S : I_\mu \to \mathbb{R}$ such that $S_\#\mu = \nu$. For every $x \in I_\mu = (a, b)$ we have
$$(a, x] \subset \{y \in (a, b) : S(y) \leq S(x)\},$$

so that $S_\#\mu = \nu$ gives
$$M_\mu(x) \leq \mu(\{y \in (a, b) : S(y) \leq S(x)\}) = \nu((-\infty, S(x)]) = M_\nu(S(x)).$$

Since $W_\nu(M_\nu(y)) \leq y$ for every $y \in \mathbb{R}$ (recall (16.4)), by definition of T, by the previous inequality, and since W_ν is increasing we find that
$$T(x) = W_\nu(M_\mu(x)) \leq W_\nu(M_\nu(S(x))) \leq S(x), \qquad \forall x \in I_\mu. \quad (16.18)$$

Now pick $x \in (a, b)$ such that, for some $\varepsilon_x > 0$, $T(x) \leq S(x) - \varepsilon_x$. By applying M_ν to both sides of this identity we find $M_\mu(x) \leq M_\nu(S(x) - \varepsilon)$ for every $\varepsilon \in (0, \varepsilon_x)$, while $S^{-1}((-\infty, S(x) - \varepsilon)) \subset (-\infty, x)$ gives $M_\nu(S(x) - \varepsilon) \leq M_\mu(x)$, and hence $M_\mu(x) = M_\nu(S(x) - \varepsilon)$ for every $\varepsilon \in (0, \varepsilon_x)$. This shows that if $x \in (a, b)$ is such that $T(x) < S(x)$, then there is an open interval I_x such that M_ν is constantly equal to $M_\mu(x)$ on I_x. Evidently, there can be at most countably many disjoint intervals on which M_ν is constant, therefore $\{T < S\}$ is at most countable, and since μ has no atoms, we conclude that $T = S$ μ-a.e. on $\mathbb{R}$.

We complete the proof of statement (iv) by considering $\gamma \in \Gamma(\mu, \nu)$ such that spt γ is increasing, and proving that $\gamma = (\mathbf{id} \times T)_\#\mu$. Indeed, if we set
$$Z := \Big\{x \in \mathbb{R} : \text{there exist } y_1 \neq y_2 \text{ such that } (x, y_1), (x, y_2) \in \text{spt } \gamma\Big\},$$

then to each $x \in Z$ we can associate an open interval I_x, with the property that if $x_1, x_2 \in Z$, $x_1 \neq x_2$, then $I_{x_1} \cap I_{x_2} = \emptyset$. Therefore Z is at most countable, and thus μ-negligible. We conclude that μ is concentrated on a set $E \subset \mathbb{R}$ such that for every $x \in E$ there is a unique $S(x) \in \mathbb{R}$ such that $(x, S(x)) \in \mathrm{spt}\, \gamma$, and such that if $x < x'$ ($x, x' \in E$), then $S(x) \leq S(x')$. Since γ is concentrated on $G = \{(x, S(x)) : x \in E\}$, we are in a position to exploit (16.6) to conclude that $\gamma = (\mathbf{id} \times S)_{\#}\mu$. As a consequence, $\nu = \mathbf{q}_{\#}\gamma = S_{\#}\mu$, and thus by the first part of (iv), $S = T$ μ-a.e. on $\mathbb{R}$, and thus $\gamma = (\mathbf{id} \times T)_{\#}\mu$.

To prove statement (v) let us first consider h as in statement (ii). If $c[h] \in L^1(\mu \times \nu)$, then there is an optimal transport plan γ in $\mathbf{K}_c(\mu, \nu)$, which is $c[h]$-cyclically monotone by Theorem 3.9, and thus such that $\mathrm{spt}\, \gamma$ is increasing thanks to statement (ii). Hence statement (iv) implies that $\gamma = \gamma_T = (\mathbf{id} \times T)_{\#}\mu$, and in particular that $\mathbf{K}_c(\mu, \nu) = \mathbf{M}_c(\mu, \nu)$ with T optimal in $\mathbf{M}_c(\mu, \nu)$. Now consider $h_\varepsilon(r) = \sqrt{\varepsilon^2 + r^2}$ and $c_\varepsilon(x, y) = h_\varepsilon(|x - y|)$. If $\mu, \nu \in \mathcal{P}_1(\mathbb{R})$, then $c_\varepsilon \in L^1(\mu \times \nu)$, and hence γ_T is optimal in $\mathbf{K}_{c_\varepsilon}(\mu, \nu)$. In particular, γ_T is c_ε-cyclically monotone, a property which is easily seen to imply that γ_T is c_0-cyclically monotone, where $c_0(x, y) = |x - y|$ is the linear transport cost. Hence γ_T is optimal in $\mathbf{K}_1(\mu, \nu)$, and T is optimal in $\mathbf{M}_1(\mu, \nu)$.

Finally, to prove statement (vi) it is enough to check that when both μ and ν have no atoms, then $W_\mu \circ M_\nu$ is the inverse function of $W_\nu \circ M_\mu$. We omit the simple details of this proof. $\qquad\square$

17

Disintegration

The disintegration theorem (Theorem 17.1) is the second main ingredient in Sudakov's strategy for the solution of the Monge problem, and can be introduced as a generalization of Fubini's theorem. Given integers $n = m + k$, Fubini's theorem provides a decomposition of $\mathcal{L}^n$ as a *superposition* (i.e., as a sum/integral) of an m-dimensional family of k-dimensional measures concentrated on a partition (disjoint covering) of $\mathbb{R}^n$ by k-dimensional planes obtained as counter-images $\{\mathbf{p}^{-1}(y)\}_{y \in \mathbb{R}^m}$ of the projection map $\mathbf{p} : \mathbb{R}^n \to \mathbb{R}^m$. Generalizing on this example, we consider the problem of decomposing a finite Borel measure μ on $\mathbb{R}^n$ as a superposition of probability measures μ_σ on $\mathbb{R}^n$, concentrated on the counter-images $P^{-1}(\sigma)$ of a Borel function $P : \mathbb{R}^n \to S$. The superposition operation is performed[1] through the Radon measure $P_\# \mu$ on S, and the resulting **disintegration formula** takes the form

$$\int_{\mathbb{R}^n} \varphi \, d\mu = \int_S d(P_\# \mu)(\sigma) \int_{P^{-1}(\sigma)} \varphi \, d\mu_\sigma \qquad \forall \varphi \in C_c^0(\mathbb{R}^n). \qquad (17.1)$$

In many applications one can simply take $S = \mathbb{R}^m$, but in our case we will need S to be the space of (possibly unbounded) oriented segments in $\mathbb{R}^n$ (to be introduced in Section 18.1). For this reason, we will allow S to be a locally compact and separable metric space.

The chapter is organized as follows. In Section 17.1 we state the disintegration theorem (Theorem 17.1), relate it to Fubini's theorem and to the coarea formula, and make some other important remarks concerning the localization and reparametrization of disintegrations. In Section 17.2 we prove the disintegration theorem, and, finally, in Section 17.3 we discuss some interesting implications of the disintegration theorem concerning the relation between the notions of transport map and transport plan.

---

[1] The choice of $P_\# \mu$ here is essentially the only possible one, see Theorem 17.1-(ii).

17.1 Statement of the Disintegration Theorem and Examples

To formally state the disintegration theorem we need to introduce the notion of **Borel measurable map with values in the space of Borel measures**: given a Borel set Σ in a locally compact and separable metric space S, and a collection of Borel measures $\{\mu_\sigma\}_{\sigma \in \Sigma}$, we say that **the map $\sigma \in \Sigma \mapsto \mu_\sigma$ is Borel measurable** if for every $E \in \mathcal{B}(\mathbb{R}^n)$, the map $\sigma \in \Sigma \mapsto \mu_\sigma(E)$ is Borel measurable; or, equivalently, if for every bounded Borel function $\varphi : \mathbb{R}^n \times S \to \mathbb{R}$, the map

$$\sigma \in \Sigma \mapsto \int_{\mathbb{R}^n} \varphi(x, \sigma) \, d\mu_\sigma(x),$$

is Borel measurable. Notice that whenever η is a Radon measure concentrated on a Borel set $\Sigma \subset S$, and $\sigma \in \Sigma \mapsto \mu_\sigma$ is a Borel measurable map, then the formula

$$\int_{\mathbb{R}^n} \varphi \, d(\mu_\sigma \otimes \eta) = \int_\Sigma d\eta(\sigma) \int_{\mathbb{R}^n} \varphi(x) \, d\mu_\sigma(x) \qquad \varphi \in C_c^0(\mathbb{R}^n),$$

defines a Radon measure $\mu_\sigma \otimes \eta$ on $\mathbb{R}^n$.

Theorem 17.1 (Disintegration) *Let S be a locally compact and separable metric space, let μ be a finite Borel measure concentrated on a Borel set E, and let $P : E \to S$ be a Borel map. Then the following properties hold:*

(i) *there exists a Borel set $\Sigma \subset S$ of full $(P_\# \mu)$-measure such that for every $\sigma \in \Sigma$ there exists $\mu_\sigma \in \mathcal{P}(\mathbb{R}^n)$ such that μ_σ is concentrated on $P^{-1}(\sigma)$, $\sigma \in \Sigma \mapsto \mu_\sigma$ is Borel measurable, and*

$$\int_{\mathbb{R}^n} \varphi \, d\mu = \int_\Sigma d(P_\# \mu)(\sigma) \int_{\mathbb{R}^n} \varphi \, d\mu_\sigma, \tag{17.2}$$

for every bounded Borel function $\varphi : \mathbb{R}^n \to \mathbb{R}$, i.e., $\mu = \mu_\sigma \otimes [P_\# \mu]$;

(ii) *if F is a Borel subset of E, η is a finite Borel measure concentrated on a Borel set $\Sigma' \subset S$, and for every $\sigma \in \Sigma'$ there exists a finite Borel measure μ_σ^* such that $\sigma \in \Sigma' \mapsto \mu_\sigma^*$ is a Borel measurable map, and*

$$\mu_\sigma^*(\mathbb{R}^n) > 0, \tag{17.3}$$

$$\mu_\sigma^* \text{ is concentrated on } F \cap P^{-1}(\sigma), \tag{17.4}$$

$$\int_F \varphi \, d\mu = \int_{\Sigma'} d\eta(\sigma) \int_{\mathbb{R}^n} \varphi \, d\mu_\sigma^*, \tag{17.5}$$

for every bounded Borel function $\varphi : \mathbb{R}^n \to \mathbb{R}$, then

$$\eta \ll P_\# \mu, \qquad \mu_\sigma \llcorner F = h(\sigma) \, \mu_\sigma^* \qquad \text{for } P_\# \mu\text{-a.e. } \sigma \in S, \tag{17.6}$$

where h is the Radon–Nikodym derivative of η with respect to $P_\# \mu$.

Remark 17.2 (Uniqueness of disintegrations) By statement (ii) applied with $\eta = P_{\#}\mu$ and $F = E$ it makes sense to call $\{\mu_\sigma\}_\sigma$ (where σ is tacitly meant to be ranging over an unspecified Borel set Σ of full $(P_{\#}\mu)$-measure in S) **the disintegration of μ with respect to P**. When needed, we shall refer to a Borel set Σ as in statement (i) as to an **index set** for the disintegration $\{\mu_\sigma\}_\sigma$.

Remark 17.3 (Localization of disintegrations) Let F be a Borel subset of E and let $P|_F$ be the restriction of P to F, so that $(P|_F)^{-1}(\sigma) = \{x \in F : P(x) = \sigma\} = F \cap P^{-1}(\sigma)$. Let $\{\mu_\sigma\}_\sigma$ and $\{\mu_\sigma^F\}_\sigma$ denote, respectively, the disintegrations of μ and of $\mu \llcorner F$ with respect to P and to $P|_F$. Then $\eta = (P|_F)_{\#}(\mu \llcorner F) = P_{\#}(\mu \llcorner F) \ll P_{\#}\mu$ and $\mu_\sigma^* = \mu_\sigma^F$ satisfies (17.3), (17.4), and (17.5). By the second conclusion in (17.6) and since $\mu_\sigma^F(\mathbb{R}^n) = 1$, denoting by h_F the Radon–Nykodim derivative of $P_{\#}(\mu \llcorner F)$ with respect to $P_{\#}\mu$, we see that $h_F(\sigma) = \mu_\sigma(F)$, and conclude that

$$\mu_\sigma \llcorner F = \mu_\sigma(F)\, \mu_\sigma^F \qquad \text{for } P_{\#}\mu\text{-a.e. } \sigma \in S. \tag{17.7}$$

(In particular, $\mu_\sigma \llcorner F \ll \mu_\sigma^F$.) Also, when $\mu_\sigma(F) > 0$ we can divide by $\mu_\sigma(F)$ in (17.7) to conclude that *the disintegration of the localization is (modulo the necessary renormalization) the localization of the disintegration.*

The measures μ_σ can sometimes be characterized in terms of μ and P. This is discussed in the following two remarks.

Remark 17.4 (Disintegration and Fubini's theorem) We claim that if $S = \mathbb{R}^m$, $\mathbf{q}: \mathbb{R}^n = \mathbb{R}^\ell \times \mathbb{R}^m \to \mathbb{R}^m$ with $\mathbf{q}(x, y) = y$, and $\mu = \mathcal{L}^n \llcorner E$ for a Borel set $E \subset \mathbb{R}^n$, then, denoting by y rather than by σ the generic point of $S = \mathbb{R}^m$ and by $\{\mu_y\}_y$ the disintegration of μ with respect to $P = \mathbf{q}|_E$, we have

$$P_{\#}\mu = \mathcal{H}^\ell(E \cap \mathbf{q}^{-1}(y))\, dy, \qquad \mu_y = \frac{\mathcal{H}^\ell \llcorner [E \cap \mathbf{q}^{-1}(y)]}{\mathcal{H}^\ell(E \cap \mathbf{q}^{-1}(y))}, \tag{17.8}$$

for $(P_{\#}\mu)$-a.e. $y \in \mathbb{R}^m$ (i.e., $\mathcal{L}^m$-a.e. $y \in \mathbf{q}(E)$). Indeed, let $E_y = \{x \in \mathbb{R}^\ell : (x, y) \in E\}$, so that $\mathcal{L}^\ell(E_y) = \mathcal{H}^\ell(E \cap \mathbf{q}^{-1}(y))$. By Fubini's theorem, for every Borel set $Y \subset \mathbb{R}^m$ we have

$$(P_{\#}\mu)(Y) = \mathcal{L}^n(E \cap \mathbf{q}^{-1}(Y)) = \int_Y \mathcal{L}^\ell(E_y)\, dy = \int_Y \mathcal{H}^\ell(E \cap \mathbf{q}^{-1}(y))\, dy,$$

and the first identity in (17.8) follows. Again by Fubini's theorem, if we denote by μ_y^* the right-hand side of the second identity in (17.8), then

$$\int_{\mathbb{R}^n} \varphi\, d\mu = \int_E \varphi = \int_{\mathbb{R}^m} dy \int_{E_y} \varphi(x, y)\, dx = \int_{\mathbb{R}^m} d(P_{\#}\mu) \int_{\mathbb{R}^n} \varphi\, d\mu_y^*,$$

for every $\varphi \in C_c^0(\mathbb{R}^n)$, where μ_y^* is concentrated on $E \cap \mathbf{q}^{-1}(y)$ for every $y \in \mathbb{R}^n$, and $\mu_y^*(\mathbb{R}^n) = 1$ for $y \in Y_* = \{y \in \mathbb{R}^m : \mathcal{L}^\ell(E_y) > 0\}$. Since Y_* has full $P_\#\mu$-measure, we conclude by Theorem 17.1-(ii) (with $\eta = P_\#\mu$, and thus $h = 1$) that $\mu_y^* = \mu_y$ for $(P_\#\mu)$-a.e. $y \in \mathbb{R}^n$, as claimed in the second identity in (17.8).

Remark 17.5 (Disintegration and the coarea formula)   Generalizing the situation considered in Remark 17.4, we now consider the case of Theorem 17.1 when $S = \mathbb{R}^m$, $P : E \subset \mathbb{R}^n \equiv \mathbb{R}^\ell \times \mathbb{R}^m \to \mathbb{R}^m$ is a Lipschitz function and $\mu = \rho \, d\mathcal{L}^n$ is concentrated on E. In this setting, the **coarea factor of** P is defined as the Borel function $\mathbf{C}P : E \to [0, \infty]$ given by

$$\mathbf{C}P(x) = \begin{cases} \det(\nabla P(x) \, \nabla P(x)^*)^{1/2}, & \text{if } P \text{ is differentiable at } x, \\ +\infty, & \text{else.} \end{cases}$$

(Recall that $A^* \in \mathbb{R}^n \otimes \mathbb{R}^m$ is the transpose of $A \in \mathbb{R}^m \otimes \mathbb{R}^n$.) It turns out that $P^{-1}(y)$ is a countably $\mathcal{H}^\ell$-rectifiable set for $\mathcal{L}^m$-a.e. $y \in \mathbb{R}^m$, and that, for every Borel measurable function $\varphi : \mathbb{R}^n \to [0, \infty]$, one has the **coarea formula**

$$\int_{\mathbb{R}^m} dy \int_{P^{-1}(y)} \varphi \, d\mathcal{H}^\ell = \int_E \varphi \, \mathbf{C}P; \tag{17.9}$$

see Appendix A.14–A.15. When ∇P is non-degenerate,[2] i.e., if $\mathbf{C}P(x,y) > 0$ for $\mathcal{L}^n$-a.e. $(x, y) \in E$, we can apply (17.9) to $\varphi = (\rho/\mathbf{C}P)\,\psi$ for an arbitrary Borel function $\psi : \mathbb{R}^n \to [0, \infty]$, to find that

$$\int_{\mathbb{R}^n} \psi \, d\mu = \int_E \varphi \, \mathbf{C}P = \int_{\mathbb{R}^m} dy \int_{P^{-1}(y)} \varphi \, d\mathcal{H}^\ell. \tag{17.10}$$

Taking $\psi = 1_{P^{-1}(Y)}$ for an arbitrary Borel set $Y \subset \mathbb{R}^m$ we find

$$(P_\#\mu)(Y) = \int_Y dy \int_{P^{-1}(y)} \frac{\rho}{\mathbf{C}P} \, d\mathcal{H}^\ell, \quad \text{i.e.,} \quad P_\#\mu = \Big[\int_{P^{-1}(y)} \frac{\rho}{\mathbf{C}P} \, d\mathcal{H}^\ell \Big] \, dy.$$

Again by (17.10), for every Borel function $\psi : \mathbb{R}^n \to [0, \infty]$ we have

$$\int_{\mathbb{R}^n} \psi \, d\mu = \int_{\mathbb{R}^m} d(P_\#\mu)(y) = \int_{P^{-1}(y)} \psi \, \frac{\rho/\mathbf{C}P}{\int_{P^{-1}(y)} (\rho/\mathbf{C}P) \, d\mathcal{H}^\ell} \, d\mathcal{H}^\ell,$$

which we compare to (17.5): by arguing as in Remark 17.4, we also verify (17.3) and (17.4), and conclude by Theorem 17.1-(ii) that

$$\mu_y = \frac{(\rho/\mathbf{C}P) \, d\mathcal{H}^\ell \llcorner [P^{-1}(y)]}{\int_{P^{-1}(y)} (\rho/\mathbf{C}P) \, d\mathcal{H}^\ell}, \tag{17.11}$$

for $(P_\#\mu)$-a.e. $y \in \mathbb{R}^m$. Identity (17.11) shows, in particular, that μ_y is absolutely continuous with respect to $\mathcal{H}^\ell \llcorner [P^{-1}(y)]$, a fact that will be crucially used in the solution of the Monge problem with linear cost (Theorem 18.1).

---

[2] If $A \in \mathbb{R}^m \otimes \mathbb{R}^n$ with $n \geq m$, the condition $\det(A\,A^*) > 0$ is equivalent to $\text{rank}(A) = m$.

Remark 17.6 (Reparameterizing a disintegration)   With E, μ, P, and S as in the statement of Theorem 17.1, let us now consider another separable, locally compact metric space $\mathcal{T}$ and let

$$\eta : P(E) \to \mathcal{T} \quad \text{be an injective Borel map.} \tag{17.12}$$

Then, the Borel map $Q : E \to \mathcal{T}$ defined by setting $Q = \eta \circ P$ is such that

$$\mu_\sigma^P = \mu_{\eta(\sigma)}^Q \qquad \text{for } (P_\# \mu)\text{-a.e. } \sigma \in S. \tag{17.13}$$

where $\{\mu_\sigma^P\}_\sigma$ and $\{\mu_\tau^Q\}_\tau$ the disintegrations of μ with respect to P and Q respectively. Indeed, since $Q = \eta \circ P$ gives $Q_\# \mu = \eta_\#(P_\# \mu)$, if $\varphi \in C_c^0(\mathbb{R}^n)$, then

$$\int_{\mathbb{R}^n} \varphi \, d\mu = \int_{\mathcal{T}} d(Q_\# \mu)(\tau) \int_{\mathbb{R}^n} \varphi \, d\mu_\tau^Q = \int_{\mathcal{T}} d(\eta_\#(P_\# \mu))(\tau) \int_{\mathbb{R}^n} \varphi \, d\mu_\tau^Q$$

$$= \int_S d(P_\# \mu)(\sigma) \int_{\mathbb{R}^n} \varphi \, d\mu_{\eta(\sigma)}^Q,$$

so that (17.5) holds with $F = E$, $\eta = P_\# \mu$ (thus $h = 1$) and $\mu_\sigma^* = \mu_{\eta(\sigma)}^Q$. Now let $\Sigma \subset S$ and $\Theta \subset \mathcal{T}$ denote index sets for $\{\mu_\sigma^P\}_\sigma$ and $\{\mu_\tau^Q\}_\tau$. Since $P_\# \mu$ is concentrated on both $\eta^{-1}(\Theta)$ and Σ, so it is on $\eta^{-1}(\Theta) \cap \Sigma$. We thus deduce (17.3) since, if $\sigma \in \eta^{-1}(\Theta) \cap \Sigma$, then

$$\mu_\sigma^*(\mathbb{R}^n) = \mu_{\eta(\sigma)}^Q(\mathbb{R}^n) = 1 \qquad (\text{since } \eta(\sigma) \in \Theta).$$

Moreover, $\mu_\sigma^* = \mu_{\eta(\sigma)}^Q$ is concentrated on $Q^{-1}(\eta(\sigma)) = P^{-1}(\eta^{-1}(\eta(\sigma))) = P^{-1}(\sigma)$, where in the last identity we have used that η is injective. In particular (17.4) holds too. We can thus deduce (17.13) by Theorem 17.1-(ii).

17.2 Proof of the Disintegration Theorem

Proof of Theorem 17.1   *Step one*: Denote by $\mathbf{p}$ and $\mathbf{p}^S$ the projections of $\mathbb{R}^n \times S$ onto $\mathbb{R}^n$ and S. We prove that if γ is a finite Borel measure on $\mathbb{R}^n \times S$, concentrated on a Borel set G, then there is a Borel set $\Sigma \subset S$ of full $(\mathbf{p}_\#^S \gamma)$-measure such that for every $\sigma \in \Sigma$ there is a probability measure γ_σ on $\mathbb{R}^n$, concentrated on $\{x \in \mathbb{R}^n : (x, \sigma) \in G\}$, and such that, for every $\zeta \in C_c^0(\mathbb{R}^n \times S)$,

$$\sigma \in \Sigma \mapsto \int_{\mathbb{R}^n \times S} \zeta(x, \sigma) \, d\gamma_\sigma(x) \qquad \text{is Borel measurable,}$$

and

$$\int_{\mathbb{R}^n \times S} \zeta \, d\gamma = \int_\Sigma d(\mathbf{p}_\#^S \gamma)(\sigma) \int_{\mathbb{R}^n} \zeta(x, \sigma) \, d\gamma_\sigma. \tag{17.14}$$

We start by noticing that for each $\varphi \in C_c^0(\mathbb{R}^n)$, the correspondence

$$\psi \in C_c^0(S) \mapsto \ell_\varphi[\psi] = \int_{\mathbb{R}^n \times S} \varphi(x)\,\psi(\sigma)\,d\gamma(x,\sigma)$$

defines a bounded linear functional

$$\ell_\varphi : L^1(\mathbf{p}_\#^S \gamma) \to \mathbb{R} \qquad \text{with} \quad \|\ell\|_{(L^1)^*} \le \|\varphi\|_{C^0(\mathbb{R}^n)}.$$

By applying the Riesz theorem to each functional ℓ_φ, we construct a linear map $m : C_c^0(\mathbb{R}^n) \to L^\infty(\mathbf{p}_\#^S \gamma)$ so that

$$\|m[\varphi]\|_{L^\infty(\mathbf{p}_\#^S \gamma)} \le \|\varphi\|_{C^0(\mathbb{R}^n)} \qquad \forall \varphi \in C_c^0(\mathbb{R}^n),$$

and

$$\int_{\mathbb{R}^n \times S} \varphi(x)\,\psi(\sigma)\,d\gamma(x,\sigma) = \ell_\varphi[\psi] = \int_{\mathbb{R}^n} m[\varphi](\sigma)\,\psi(\sigma)\,d(\mathbf{p}_\#^S \gamma)(\sigma) \tag{17.15}$$

for every $(\varphi,\psi) \in C_c^0(\mathbb{R}^n) \times C_c^0(S)$. Given $\varphi \in C_c^0(\mathbb{R}^n)$, let Σ_φ be a set of full $(\mathbf{p}_\#^S \gamma)$-measure in S such that

$$|m[\varphi](\sigma)| \le \|m[\varphi]\|_{L^\infty(\mathbf{p}_\#^S \gamma)} \le \|\varphi\|_{C^0(\mathbb{R}^n)} \qquad \forall \sigma \in \Sigma_\varphi.$$

If $\mathcal{F}$ is a countable dense subset of $C_c^0(\mathbb{R}^n)$ and $\Sigma = \bigcap\{\Sigma_\varphi : \varphi \in \mathcal{F}\}$, then Σ is a set of full $(\mathbf{p}_\#^S \gamma)$-measure in S such that $|m[\varphi](\sigma)| \le \|\varphi\|_{C^0(\mathbb{R}^n)}$ for every $\sigma \in \Sigma$ and $\varphi \in \mathcal{F}$. Therefore the linear functional $\gamma_\sigma : C_c^0(\mathbb{R}^n) \to \mathbb{R}$ defined by setting $\gamma_\sigma[\varphi] = m[\varphi](\sigma)$ is such that

$$\sup\left\{\gamma_\sigma[\varphi] : \varphi \in \mathcal{F}, \|\varphi\|_{C^0(\mathbb{R}^n)} \le 1\right\} \le 1, \qquad \forall \sigma \in \Sigma.$$

In particular, again by the Riesz theorem, if $\sigma \in \Sigma$, then γ_σ is a Radon measure on $\mathbb{R}^n$ with $\gamma_\sigma(\mathbb{R}^n) \le 1$, and, thanks to (17.15), with

$$\int_{\mathbb{R}^n \times S} \varphi(x)\,\psi(\sigma)\,d\gamma(x,\sigma) = \int_\Sigma \psi(\sigma)\,d(\mathbf{p}_\#^S \gamma)(\sigma) \int_{\mathbb{R}^n} \varphi(x)\,d\gamma_\sigma(x), \tag{17.16}$$

for every $(\varphi,\psi) \in C_c^0(\mathbb{R}^n) \times C_c^0(S)$. By approximating $1_{\mathbb{R}^n}(x)$ and $1_S(\sigma)$ with C_c^0-functions $\varphi(x)$ and $\psi(\sigma)$, we deduce from (17.16) that

$$\gamma(\mathbb{R}^n \times S) = \int_\Sigma \gamma_\sigma(\mathbb{R}^n)\,d(\mathbf{p}_\#^S \gamma)(\sigma) \le \mathbf{p}_\#^S \gamma(S) = \gamma(\mathbb{R}^n \times S),$$

so that $\gamma_\sigma(\mathbb{R}^n) = 1$ for $(\mathbf{p}_\#^S \gamma)$-a.e. $\sigma \in \Sigma$. Denoting by $C_c^0(\mathbb{R}^n) \otimes C_c^0(S)$ the subset of those $\zeta \in C_c^0(\mathbb{R}^n \times S)$ such that $\zeta(x,\sigma) = \varphi(x)\psi(\sigma)$ for $(\varphi,\psi) \in C_c^0(\mathbb{R}^n) \times C_c^0(S)$, and noticing that the span of $C_c^0(\mathbb{R}^n) \otimes C_c^0(S)$ is dense in $C_c^0(\mathbb{R}^n \times S)$, we see that (17.16) implies

$$\int_{\mathbb{R}^n \times S} \zeta\,d\gamma = \int_\Sigma d(\mathbf{p}_\#^S \gamma)(\sigma) \int_{\mathbb{R}^n} \zeta(x,\sigma)\,d\gamma_\sigma(x), \tag{17.17}$$

for every $\zeta \in C_c^0(\mathbb{R}^n \times S)$. By approximating $1_{(\mathbb{R}^n \times S) \setminus G}(x, \sigma)$ with C_c^0-functions $\zeta(x, \sigma)$ in (17.17), we find

$$0 = \int_\Sigma \gamma_\sigma(\{x \in \mathbb{R}^n : (x, \sigma) \notin G\}) \, d(\mathbf{p}_\#^S \mu)(\sigma),$$

so that for $(\mathbf{p}_\#^S \gamma)$-a.e. $\sigma \in \Sigma$, γ_σ is concentrated on $\{x \in \mathbb{R}^n : (x, \sigma) \in G\}$.

Step two: We now let μ be a finite Borel measure on $\mathbb{R}^n$, concentrated on a Borel set E, and given a Borel map $P : E \to S$, we consider the Borel map $g : E \to \mathbb{R}^n \times S$ defined by setting $g(x) = (x, P(x))$, $x \in E$. We apply the claim proved in step one to the Radon measure $\gamma = g_\# \mu$, which is concentrated on the graph $G = \{(x, \sigma) \in E \times S : \sigma = P(x)\}$ of P over E, and is such that

$$\mathbf{p}_\# \gamma = \mu, \qquad \mathbf{p}_\#^S \gamma = P_\# \mu. \tag{17.18}$$

Hence, for $(\mathbf{p}_\#^S \gamma) = (P_\# \mu)$-a.e. $\sigma \in S$, there is $\gamma_\sigma \in \mathcal{P}(\mathbb{R}^n)$, concentrated on $\{x \in E : (x, \sigma) \in G\} = P^{-1}(\sigma)$, and such that (17.14) holds. By applying (17.14) to $\zeta(x, \sigma) = \varphi(x)$, $\varphi \in C_c^0(\mathbb{R}^n)$ (an approximation argument using the finiteness of γ is needed here, since ζ has not compact support) and by (17.18),

$$\int_{\mathbb{R}^n} \varphi \, d\mu = \int_{\mathbb{R}^n \times S} \zeta \, d\gamma = \int_S d(\mathbf{p}_\#^S \gamma)(\sigma) \int_{\mathbb{R}^n} \zeta(x, \sigma) \, d\gamma_\sigma$$

$$= \int_S d(P_\# \mu)(\sigma) \int_{\mathbb{R}^n} \varphi(x) \, d\gamma_\sigma(x),$$

Therefore $\mu_\sigma = \gamma_\sigma$ has the required properties.

Step three: Now let F, η, Σ', and μ_σ^* be as in (ii). If Ω is a $P_\# \mu$-negligible Borel set in S, then $P^{-1}(\Omega)$ is a μ-negligible Borel set in $\mathbb{R}^n$. By applying (17.5) to $\varphi = 1_{P^{-1}(\Omega)}$, and recalling that $\mu_\sigma^*(P^{-1}(\sigma)) = \mu_\sigma^*(\mathbb{R}^n)$, we find that

$$0 = \int_{\Sigma'} \mu_\sigma^*(P^{-1}(\Omega)) \, d\eta(\sigma) = \int_{\Sigma'} \mu_\sigma^*(P^{-1}(\sigma) \cap P^{-1}(\Omega)) \, d\eta(\sigma)$$

$$= \int_{\Sigma' \cap \Omega} \mu_\sigma^*(\mathbb{R}^n) \, d\eta(\sigma).$$

In particular, since $\mu_\sigma^*(\mathbb{R}^n) > 0$ for every $\sigma \in \Sigma'$, and η is concentrated on Σ', we find $\eta(\Omega) = 0$, thus proving $\eta \ll P_\# \mu$. Let h be the Radon–Nikodym derivative of η with respect to $P_\# \mu$ (so that $h = 0$ on $S \setminus \Sigma'$). Given $\varphi \in C_c^0(\mathbb{R}^n)$, we can apply first (17.2) to $1_F \varphi$, and then (17.5) to φ, to find

$$\int_\Sigma d(P_\# \mu)(\sigma) \int_F \varphi \, d\mu_\sigma = \int_F \varphi \, d\mu = \int_S d\eta(\sigma) \int_{\mathbb{R}^n} \varphi \, d\mu_\sigma^*$$

$$= \int_S h(\sigma) \, d(P_\# \mu)(\sigma) \int_{\mathbb{R}^n} \varphi \, d\mu_\sigma^*.$$

Hence, the second conclusion in (17.6) follows by arbitrariness of φ. $\qquad \square$

17.3 Stochasticity of Transport Plans

In this section we present a first consequence of the disintegration theorem: *a transport plan is induced by a transport map if and only if it is concentrated on the graph of that map.* To this result, formulated in Proposition 17.8, we premise the following remark, which clarifies the kind of conceptual shift introduced in Chapter 3 with the notion of transport plan.

Remark 17.7 (Transport maps vs transport plans: deterministic vs stochastic) Given $\mu \in \mathcal{P}(\mathbb{R}^n)$ and $\gamma \in \mathcal{P}(\mathbb{R}^n \times \mathbb{R}^n)$ with $\mathbf{p}_\# \gamma = \mu$, $\mathbf{p}(x,y) = x$, let us consider the disintegration $\{\gamma_x\}_x$ of γ with respect to $\mathbf{p}$, and prove that, if γ is induced by a transport map (i.e., if $\gamma = (\mathbf{id} \times T)_\# \mu$ for a Borel map $T : E \to \mathbb{R}^n$ such that μ is concentrated on the Borel set E), then

$$\gamma = \gamma_x \otimes \mu, \qquad \text{with } \gamma_x = \delta_{(x,T(x))} \text{ for } \mu\text{-a.e. } x \in \mathbb{R}^n. \qquad (17.19)$$

In particular, when γ is induced by T, the map $x \mapsto \gamma_x = \delta_{(x,T(x))}$ is equivalent to the "deterministic" map $x \mapsto T(x)$, and thus loses its "stochastic" meaning (compare to the case when, say, γ_x is a discrete probability measure concentrated at multiple points). To prove (17.19) we notice that if $\varphi : \mathbb{R}^n \times \mathbb{R}^n \to \mathbb{R}$ is a bounded Borel function, then, by Theorem 17.1,

$$\int_{\mathbb{R}^n} d\mu(x) \int_{\mathbb{R}^n \times \mathbb{R}^n} \varphi(x,y) \, d\gamma_x(y) = \int_{\mathbb{R}^n \times \mathbb{R}^n} \varphi(x,y) \, d\gamma(x,y)$$

$$= \int_{\mathbb{R}^n \times \mathbb{R}^n} \varphi(x,y) \, d\big((\mathbf{id} \times T)_\# \mu\big)(x,y) = \int_{\mathbb{R}^n} \varphi(x,T(x)) \, d\mu(x) \quad (17.20)$$

so that (17.19) follows by Remark 17.2.

Proposition 17.8 *Let $E \subset \mathbb{R}^n$ be a Borel set, $T : E \to \mathbb{R}^n$ be a Borel map, and $\gamma \in \mathcal{P}(\mathbb{R}^n \times \mathbb{R}^n)$. Then, γ is concentrated on the graph of T over E,*

$$G = \big\{(x,y) \in \mathbb{R}^n \times \mathbb{R}^n : y = T(x), x \in E\big\}, \qquad (17.21)$$

if and only if $\mu = \mathbf{p}_\# \gamma$ is concentrated on E and $\gamma = (\mathbf{id} \times T)_\# \mu$.

Remark 17.9 Notice that this proposition was used in step three of the proof of Theorem 16.1. Of course there was no circular reasoning in doing so, since Theorem 16.1 is not involved in the proof of Proposition 17.8.

Proof of Proposition 17.8 Let $\{\gamma_x\}_x$ be the disintegration of γ with respect to $\mathbf{p}$. To prove the "if" part of the statement just take φ equal to the characteristic function of $1_{(\mathbb{R}^n \times \mathbb{R}^n) \setminus G}$ in (17.20), to deduce that γ is concentrated on G.

To prove the "only if" part of the statement, using the general fact that, for μ-a.e. $x \in \mathbb{R}^n$, γ_x is concentrated on $\mathbf{p}^{-1}(x) = \{x\} \times \mathbb{R}^n$, we see that if γ is concentrated on G, then, $\mu = \mathbf{p}_\# \gamma$ is concentrated on $E = \mathbf{p}(G)$, and, by Theorem 17.1,

$$0 = \gamma((\mathbb{R}^n \times \mathbb{R}^n) \setminus G) = \int_{\mathbb{R}^n} \gamma_x \big(\mathbf{p}^{-1}(x) \cap ((\mathbb{R}^n \times \mathbb{R}^n) \setminus G) \big) \, d\mu(x)$$

$$= \int_E \gamma_x (\{x\} \times (\mathbb{R}^n \setminus \{T(x)\})) \, d\mu(x).$$

In particular, for μ-a.e. $x \in E$, γ_x is concentrated on $(x, T(x))$, and since $\gamma_x \in \mathcal{P}(\mathbb{R}^n)$, it must be $\gamma_x = \delta_{(x,T(x))}$. Therefore, $\gamma = \delta_{(x,T(x))} \otimes \mu = (\mathbf{id} \times T)_\# \mu$. $\square$

17.4 $\mathbf{K}_c = \mathbf{M}_c$ for Nonatomic Origin Measures

We present here a more advanced application of Theorem 17.1, to the problem of establishing when the general inequality $\mathbf{M}_c \geq \mathbf{K}_c$ holds as an identity.[3] We begin with the following remark, where we provide an informal justification of a basic approximation result in the theory of disintegrations.

Remark 17.10 (Approximation of stochastic by deterministic)   The following approximation theorem holds: *if $\gamma \in \mathcal{P}(\mathbb{R}^n \times \mathbb{R}^n)$ and $\mu = \mathbf{p}_\# \gamma$ has not atoms ($\mathbf{p}(x, y) = x$), then there is a sequence $\{T_j\}_j$ of Borel measurable maps $T_j :$ $\mathbb{R}^n \to \mathbb{R}^n$ such that*

$$\gamma_{T_j} = (\mathbf{id} \times T_j)_\# \mu = \delta_{(x, T_j(x))} \otimes \mu \overset{*}{\rightharpoonup} \gamma \qquad \textit{as Radon measures on } \mathbb{R}^n \times \mathbb{R}^n,$$

and such that $(T_j)_\# \mu$ has no atoms. A sketch of the proof is as follows. We proceed by defining successive approximations of γ in the weak-star convergence: **(i)** by truncation to compact sets (followed by renormalization), we reduce to the case when spt γ is compact; **(ii)** we consider the disintegration $\{\gamma_x\}_x$ of γ with respect to $\mathbf{p}$, approximate the Borel map $x \in \mathbb{R}^n \mapsto \gamma_x \in \mathcal{P}_1(\mathbb{R}^n)$ with its ε-regularization by convolution (which is Lipschitz continuous with values in the metric space $(\mathcal{P}_1(\mathbb{R}^n), \mathbf{K}_1)$ since spt $\gamma_x \subset \mathbf{q}(\text{spt } \gamma)$ for μ-a.e. $x \in \mathbb{R}^n$), and thus reduce to the case when $\gamma = \gamma_x \otimes \mu$ for some $x \in \mathbb{R}^n \mapsto \gamma_x$ with $\mathbf{K}_1(\gamma_x, \gamma_y) \leq C |x - y|$; **(iii)** considering a covering of $\mathbb{R}^n$ by cubes Q_ε of side length ε and disjoint interiors, we use the Lipschitz property of $x \in \mathbb{R}^n \mapsto \gamma_x$ to reduce to the case when, given $Q \in Q_\varepsilon$, we have $\gamma_x = \gamma^Q$ for every $x \in Q$ (where $\gamma^Q = \gamma_{x(Q)}$ if $x(Q)$ is the center of Q); **(iv)** for each $Q \in Q_\varepsilon$, we

---

[3]   As this discussion is not needed in the solution of the Monge problem, which is the main focus of this part of the book, this section can be safely omitted on a first reading.

consider the standard approximation of γ^Q by discrete probability measures (see, e.g., step one in the proof of Theorem 3.16)

$$\gamma^Q \approx \sum_i^{N(Q)} \lambda_i^Q \delta_{y_i^Q}, \qquad \text{with } 0 \le \lambda_i^Q \le 1, \sum_i^{N(Q)} \lambda_i^Q = 1, N(Q) \in \mathbb{N};$$

(v) finally, we use the fact that μ has no atoms to find a partition $\{E_i^Q\}_i$ of Q such that $\mu(E_i^Q) = \lambda_i^Q \mu(Q)$, and then require T to map in a Borel and injective way E_i^Q into $B_\varepsilon(y_i^Q)$. In this way, given $\varphi, \psi \in C_c^0(\mathbb{R}^n)$, and setting $(\varphi \otimes \psi)(x,y) = \varphi(x)\psi(y)$, we find

$$\int_{\mathbb{R}^n \times \mathbb{R}^n} \varphi \otimes \psi \, d\gamma_T = \sum_{Q \in \mathcal{Q}_\varepsilon} \sum_{i=1}^{N(Q)} \int_{E_i^Q} \varphi(x)\psi(T(x)) \, d\mu(x)$$

$$\approx \sum_{Q \in \mathcal{Q}_\varepsilon} \sum_{i=1}^{N(Q)} \psi(y_i^Q) \int_{E_i^Q} \varphi \, d\mu$$

$$\approx \sum_{Q \in \mathcal{Q}_\varepsilon} \sum_{i=1}^{N(Q)} \lambda_i^Q \psi(y_i^Q) \int_Q \varphi \, d\mu = \sum_{Q \in \mathcal{Q}_\varepsilon} \int_Q \varphi \, d\mu \int_{\mathbb{R}^n} \psi \, d\Big[\sum_{i=1}^{N(Q)} \lambda_i^Q \delta_{y_i^Q}\Big]$$

$$\approx \sum_{Q \in \mathcal{Q}_\varepsilon} \int_Q \varphi \, d\mu \int_{\mathbb{R}^n} \psi \, d\gamma^Q \approx \int_{\mathbb{R}^n} \varphi(x) \, d\mu(x) \int_{\mathbb{R}^n} \psi \, d\gamma_x = \int_{\mathbb{R}^n \times \mathbb{R}^n} \varphi \otimes \psi \, d\gamma,$$

and thus conclude, by density of the span of $C_c^0(\mathbb{R}^n) \otimes C_c^0(\mathbb{R}^n)$ into $C_c^0(\mathbb{R}^n \times \mathbb{R}^n)$, that γ can be approximated with the desired precision by map-induced measures γ_T. Notice that, by construction, every $y \in \mathbb{R}^n$ has at most countably many counter-images under T (at most one in each E_i^Q), so that $T_\# \mu$ has no atoms. We also notice that, given $\delta > 0$, we can tune in the parameter $\varepsilon > 0$ used in the construction so to obtain that T takes values in $I_\delta(\mathbf{q}(\text{spt } \gamma))$, and thus that $\text{spt } \gamma_T \subset \text{spt } \mu \times \text{spt } (T_\# \mu) \subset I_\delta(\text{spt } \gamma) = \{y \in \mathbb{R}^n : \text{dist}(y, \text{spt } \gamma) \le \delta\}$.

Coming to the criterion for equality in (3.9), we have the following general result by Ambrosio:[4] *if* $\mu, \nu \in \mathcal{P}(\mathbb{R}^n)$, μ *has not atoms,* $c : \mathbb{R}^n \to [0,\infty)$ *is continuous, and* $\mathbf{K}_c(\mu,\nu) < \infty$, *then* $\mathbf{K}_c(\mu,\nu) = \mathbf{M}_c(\mu,\nu)$. Here we limit ourselves to present a proof in the case when μ and ν have compact supports and c is locally Lipschitz continuous (in particular, the linear and quadratic costs are admissible here).

---

[4] See [Amb03, Theorem 2.1].

Theorem 17.11 (Equality of Monge's and Kantorovich's transport costs)   *If*
$\mu, \nu \in \mathcal{P}(\mathbb{R}^n)$ *have compact supports,* μ *has no atoms, and* $c : \mathbb{R}^n \rightarrow [0, \infty)$ *is*
locally Lipschitz continuous, then $\mathbf{K}_c(\mu, \nu) = \mathbf{M}_c(\mu, \nu)$.

Proof   Thanks to (3.9) we only need to prove that $\mathbf{K}_c(\mu, \nu) \geq \mathbf{M}_c(\mu, \nu)$. Setting $K = \text{spt}\,\mu \times \text{spt}\,\nu$, and given $i \in \mathbb{N}$, by applying Remark 17.10 to μ, ν,
and γ an optimal plan in $\mathbf{K}_c(\mu, \nu)$, we can find Borel maps $T_1^i : \mathbb{R}^n \rightarrow \mathbb{R}^n$
such that $(T_1^i)_\#\mu$ has no atoms, T_1^i takes values in a 2^{-i}-neighborhood of
$\mathbf{q}(\text{spt}\,\gamma) \subset \mathbf{q}(K)$, and the plan $(\mathbf{id} \times T_1^i)_\#\mu$, whose second marginal satisfies

$$\mathbf{q}_\#((\mathbf{id} \times T_1^i)_\#\mu) = (T_1^i)_\#\mu,$$

is sufficiently close to γ in weak-star convergence to entail that

$$\mathbf{K}_1(\nu, (T_1^i)_\#\mu) < \frac{1}{2^i}, \qquad \left| \mathbf{K}_c(\mu, \nu) - \int_{\mathbb{R}^n \times \mathbb{R}^n} c\,d[(\mathbf{id} \times T_1^i)_\#\mu] \right| < \frac{1}{2^i}. \tag{17.22}$$

Here we have used that $\mathbf{K}_c(\mu, \nu) = \int_{\mathbb{R}^n \times \mathbb{R}^n} c\,d\gamma$ and c is bounded on

$$\bigcup_i \text{spt}\,[(\mathbf{id} \times T_1^i)_\#\mu] \subset I_1(K) = \{\text{dist}(\cdot, K) \leq 1\},$$

as well as the fact that, if $\lambda_i \xrightarrow{n} \lambda$ in $\mathcal{P}(\mathbb{R}^n \times \mathbb{R}^n)$ with uniformly bounded
supports, then $\mathbf{K}_1(\mathbf{q}_\#\lambda_i, \mathbf{q}_\#\lambda) \rightarrow 0$ as $i \rightarrow \infty$ (compare with Theorem 11.10).
Would $(T_1^i)_\#\mu = \nu$, we could easily conclude the proof letting $i \rightarrow \infty$. We
therefore start an iterative application of Remark 17.10 aimed at decreasing the
size of $\mathbf{K}_1(\nu, (T_1^i)_\#\mu)$.

Since $(T_1^i)_\#\mu$ has no atoms, we can apply Remark 17.10 to $(T_1^i)_\#\mu$, ν, and γ_1^i
an optimal plan[5] in $\mathbf{K}_1((T_1^i)_\#\mu, \nu)$. We can thus find a Borel map $T_2^i : \mathbb{R}^n \rightarrow \mathbb{R}^n$
such that, if setting $S_1^i = T_1^i$ and $S_2^i = T_2^i \circ T_1^i$, then $(S_2^i)_\#\mu = (T_2^i)_\#[(T_1^i)_\#\mu]$
has no atoms, S_2^i takes values $2^{-(i+1)}$-neighborhood of $\mathbf{q}(\text{spt}\,\gamma_1^i)$, and the map-
induced plan

$$(\mathbf{id} \times T_2^i)_\#((T_1^i)_\#\mu) = (S_1^i \times S_2^i)_\#\mu,$$

which has second marginal

$$\mathbf{q}_\#((S_1^i \times S_2^i)_\#\mu) = (S_2^i)_\#\mu,$$

---

[5]   Notice carefully that γ_1^i is taken optimal in $\mathbf{K}_1$, and not in $\mathbf{K}_c$. While, obviously, the starting
plan γ had to be taken optimal in $\mathbf{K}_c$, in this iteration we work with $\mathbf{K}_1$ since we aim at reducing
the size of $\mathbf{K}_1(\nu, (T_1^i)_\#\mu)$. We can do this while keeping track of c-transport costs since c is
locally Lipschitz continuous.

satisfies

$$\mathbf{K}_1\left(\nu, (S_2^i)_\# \mu\right) < \frac{1}{2^{i+1}}, \quad \left|\mathbf{K}_1((T_1^i)_\# \mu, \nu) - \int_{\mathbb{R}^n \times \mathbb{R}^n} |x - y| \, d[(S_1^i \times S_2^i)_\# \mu]\right| < \frac{1}{2^{i+1}}.$$

(Here we have used $\mathbf{q}_\# \gamma_1^i = \nu$ and $\int_{\mathbb{R}^n \times \mathbb{R}^n} |x-y| \, d\gamma_1^i = \mathbf{K}_1((T_1^i)_\# \mu, \nu)$.) We then apply Remark 17.10 to $(S_2^i)_\# \mu$, ν, and an optimal plan γ_2^i in $\mathbf{K}_1((S_2^i)_\# \mu, \nu)$, and iterate the procedure. In this way, we eventually define a sequence of Borel maps $\{T_j^i\}_{j=1}^\infty$ and of transport plans $\{\gamma_j^i\}_{j=1}^\infty$ such that, setting $S_j^i = T_j^i \circ \cdots \circ T_1^i$, we have (i) each γ_j^i is an optimal plan in $\mathbf{K}_1(\nu, (S_j^i)_\# \mu)$, and each S_j^i takes values in a $2^{-(i+j)}$-neighborhood of $\mathbf{q}(\mathrm{spt}\,\gamma_j^i)$ and satisfies $(S_j^i)_\# \mu$ has no atoms; (ii) the following estimates hold:

$$\mathbf{K}_1\left(\nu, (S_{j+1}^i)_\# \mu\right) < \frac{1}{2^{i+j}}, \quad \left|\mathbf{K}_1((S_j^i)_\# \mu, \nu) - \int_{\mathbb{R}^n \times \mathbb{R}^n} |x - y| \, d[(S_j^i \times S_{j+1}^i)_\# \mu]\right| < \frac{1}{2^{i+j}}.$$

By property (ii) we find that $\{S_j^i\}_j$ is a Cauchy sequence in $L^1(\mu)$: indeed,

$$\int_{\mathbb{R}^n} |S_{j+1}^i - S_j^i| \, d\mu = \int_{\mathbb{R}^n \times \mathbb{R}^n} |x - y| \, d[(S_j^i \times S_{j+1}^i)_\# \mu] \le \mathbf{K}_1((S_j^i)_\# \mu, \nu) + \frac{1}{2^{i+j}}.$$

If S^i denotes the $L^1(\mu)$-limit of S_j^i, we have $S_\#^i \mu = \nu$, since $(S_j^i)_\# \mu \overset{*}{\rightharpoonup} S_\# ^i \mu$ as $j \to \infty$ (with uniformly bounded supports thanks to (i)) and $\mathbf{K}_1(\nu, (S_{j+1}^i)_\# \mu) < 2^{-(i+j)}$. By $S_\#^i \mu = \nu$, by the local Lipschitz property of c, and by the uniform boundedness implied by (i), setting $\delta_j^i = \|S^i - S_j^i\|_{L^1(\mu)}$, for some $L > 0$ we find

$$\mathbf{M}_c(\mu, \nu) \le \int_{\mathbb{R}^n} c(x, S^i(x)) \, d\mu(x) \le \int_{\mathbb{R}^n} c(x, S_{j+1}^i(x)) \, d\mu(x) + L\delta_{j+1}^i$$

$$\le \int_{\mathbb{R}^n} c(x, T_1^i(x)) \, d\mu(x) + L \sum_{\ell=1}^j \int_{\mathbb{R}^n} |S_{\ell+1}^i - S_\ell^i| \, d\mu + L\delta_{j+1}^i$$

$$\le \int_{\mathbb{R}^n \times \mathbb{R}^n} c \, d[(\mathbf{id} \times T_1^i)_\# \mu] + C(L)\, 2^{-i} + L\delta_{j+1}^i \le \mathbf{K}_c(\mu, \nu) + C(L)\, 2^{-i} + L\delta_{j+1}^i,$$

where in the last inequality we have used (17.22) (which exploited the optimality of γ in $\mathbf{K}_c(\mu, \nu)$). We let $j \to \infty$ and then $i \to \infty$ to conclude. $\square$

18

Solution to the Monge Problem with Linear Cost

This chapter is devoted to the proof of the following theorem concerning the existence of solutions to the Monge problem with linear cost, namely,

$$\mathbf{M}_1(\mu, \nu) = \inf \left\{ \int_{\mathbb{R}^n} |T(x) - x| \, d\mu(x) : T_\# \mu = \nu \right\}, \qquad \mu, \nu \in \mathcal{P}_1(\mathbb{R}^n).$$
(18.1)

Theorem 18.1 (Sudakov theorem)   *If $\mu, \nu \in \mathcal{P}_1(\mathbb{R}^n)$ and $\mu \ll \mathcal{L}^n$, then there is a transport map T from μ to ν such that $(\mathbf{id} \times T)_\# \mu$ is an optimal transport plan in $\mathbf{K}_1(\mu, \nu)$, and T is an optimal transport map in $\mathbf{M}_1(\mu, \nu)$. Moreover, if we also have $\nu \ll \mathcal{L}^n$, then we can find an optimal transport map T in $\mathbf{M}_1(\mu, \nu)$ and an optimal transport map S in $\mathbf{M}_1(\nu, \mu)$ so that $S \circ T = \mathbf{id}$ μ-a.e. on $\mathbb{R}^n$ and $T \circ S = \mathbf{id}$ ν-a.e. on $\mathbb{R}^n$.*

Since the proof of Theorem 18.1 is quite long, it is convenient to first sketch its various parts. We start by noticing that, as already seen in Remark 16.5, there could be many different (and arbitrarily "wild") optimal transport maps in $\mathbf{M}_1(\mu, \nu)$. Therefore, in solving the Monge problem, one does not only want to prove the existence of an optimal transport map T but actually wants to describe a general construction that leads to define optimal transport maps with nice structural properties. When $n = 1$, the role of "optimal transport map with a nice structure" is of course taken by the *monotone* transport map constructed in Theorem 16.1. Coming to dimension $n \geq 2$, **Sudakov's strategy for the solution of the Monge problem** is, chronologically, the first of a few approaches[1] that had been devised for constructing "structured" solution to the Monge problem, and the one adopted in this book. Sudakov's strategy moves from the linear cost case of the Kantorovich duality (Theorem 3.17): if $\mu, \nu \in \mathcal{P}_1(\mathbb{R}^n)$, then

$$\mathbf{K}_1(\mu, \nu) = \inf \left\{ \int_{\mathbb{R}^n \times \mathbb{R}^n} |x - y| \, d\gamma(x, y) : \gamma \in \Gamma(\mu, \nu) \right\}$$
(18.2)

[1]   See the Bibliographical Notes.

is finite and has minimizers; moreover, there is a **Kantorovich potential** from μ to ν, that is, a (non-uniquely determined) 1-Lipschitz $f : \mathbb{R}^n \to \mathbb{R}$ such that

$$\gamma \text{ is an optimal plan for } \mathbf{K}_1(\mu, \nu) \tag{18.3}$$
$$\Leftrightarrow f(y) = f(x) + |x - y| \text{ for every } (x, y) \in \operatorname{spt} \gamma.$$

In particular, the following theorem holds:

Theorem 18.2 *If $\mu, \nu \in \mathcal{P}_1(\mathbb{R}^n)$, f is a Kantorovich potential from μ to ν, T transports μ into ν, and the plan $\gamma_T = (\mathbf{id} \times T)_{\#}\mu$ induced by T satisfies*

$$f(y) = f(x) + |x - y| \qquad \forall (x, y) \in \operatorname{spt} \gamma_T, \tag{18.4}$$

then T is an optimal transport map in $\mathbf{M}_1(\mu, \nu)$.

Proof   Indeed, by (18.3) and (18.4), γ_T is optimal in $\mathbf{K}_1(\mu, \nu)$. Hence, by (3.9),

$$\mathbf{M}_1(\mu, \nu) \geq \mathbf{K}_c(\mu, \nu) = \int_{\mathbb{R}^n \times \mathbb{R}^n} |x - y| \, d\gamma_T = \int_{\mathbb{R}^n} |x - T| \, d\mu \geq \mathbf{M}_1(\mu, \nu),$$

and thus T is optimal in $\mathbf{M}_1(\mu, \nu)$. $\square$

To understand how to use (18.4) in constructing optimal transport maps, let us notice that if (18.4) holds for a continuous[2] transport map T, then we have

$$f(T(x)) = f(x) + |T(x) - x|, \qquad \forall x \in \operatorname{spt} \mu. \tag{18.5}$$

Anticipating some notation[3] from Section 18.1, (18.5) says that f saturates its own 1-Lipschitz bound along each open segment $]x, T(x)[$ with $x \in \operatorname{spt} \mu$. This suggest to look at the **transport set** $\mathcal{T}(f)$ **of** f, defined as the union of those open segments along which f saturates its 1-Lipschitz bound, namely

$$\mathcal{T}(f) = \bigcup \Big\{]x, y[\, : \, x, y \in \mathbb{R}^n \text{ s.t. } x \neq y, \, f(y) = f(x) + |y - x| \Big\}. \tag{18.6}$$

---

[2] Continuity of transport maps is irrelevant in the actual construction. It is mentioned here only for the purpose of deriving from (18.4) the validity of (18.5) not just at μ-a.e. $x \in \mathbb{R}^n$ but actually at every $x \in \operatorname{spt} \mu$. In turn, this choice streamlines a bit our introductory discussion by getting rid of many "for μ-a.e. x"s that should have otherwise been included.

[3] Given $x, y \in \mathbb{R}^n$, $x \neq y$, we set $]x, y[= \{(1 - t) x + t y : t \in (0, 1)\}$ for the open (non-oriented) segment with end-points x and y, and $]\!]x, y[\![$ for the open *oriented* segment from x to y, intended as the map $t \in (0, 1) \mapsto (1 - t) x + t y$. S° is the set of the (possibly unbounded) oriented segments in $\mathbb{R}^n$, and, given $\sigma \in S^\circ$, $\Phi(\sigma) \subset \mathbb{R}^n$ is the non-oriented version, or realization, of σ (e.g., $\Phi(]\!]x, y[\![) =]x, y[$, etc.).

If $]x,y[$ is one of such segments, then f is differentiable at every $z \in]x,y[$, with ∇f constant along $]x,y[$. Thus, any two *different* segments $]x,y[$ and $]x',y'[$ saturating the 1-Lipschitz bound of f are either disjoint or, if intersecting, they do so by overlapping along a same open subsegment, and sharing the same orientation (the one given by ∇f). The validity of this "non-crossing condition" allows us to define, for each $x \in \mathcal{T}(f)$, *the* **maximal transport ray through** x, denoted by $P_f(x)$, as the (possibly unbounded) *maximal and oriented* segment $\sigma \in \mathcal{S}^\circ$ whose realization $\Phi(\sigma)$ contains x.

In this notation and terminology, (18.5) says that, if $x \in \mathcal{T}(f) \cap \operatorname{spt} \mu$ and the transport map T induces an optimal plan γ_T in $\mathbf{K}_1(\mu,\nu)$, then $]x,T(x)[\subset \Phi(P_f(x))$ with $T(x)$ "to the right" of x in the orientation of $]x,T(x)[$ induced by ∇f. We thus look for our candidate "nicely structured optimal transport map" in the class of those T that are monotone increasing along maximal transport rays $P_f(x)$ with $x \in \mathcal{T}(f) \cap \operatorname{spt} \mu$. We may thus break down Sudakov's strategy in four distinct steps:

Step one: disintegrate μ and ν with respect to $P_f : \mathcal{T}(f) \to \mathcal{S}^\circ$;

Step two: for every[4] $\sigma \in P_f(\mathcal{T}(f))$, consider the monotone transport map T_σ from μ_σ to ν_σ;

Step three: define $T : \mathcal{T}(f) \to \mathbb{R}^n$ by gluing the various T_σ, that is, define $T = T_\sigma$ on $\Phi(\sigma)$ for each $\sigma \in P_f(\mathcal{T}(f))$;

Step four: use the disintegration theorem to check that T transports μ into ν (thanks to $(T_\sigma)_\# \mu_\sigma = \nu_\sigma$ for each σ), and conclude that T is optimal in $\mathbf{M}_1(\mu,\nu)$ since the monotonicity of each T_σ ensures the validity of (18.5) for T.

The attentive reader should have already spotted two problems with this plan. First, if μ and ν are not concentrated on $\mathcal{T}(f)$, where P_f is defined, then the disintegrations $\{\mu_\sigma\}_\sigma$ and $\{\nu_\sigma\}_\sigma$ will not take into account the entirety of μ and ν, and the gluing procedure of step three will not result in a transport map T from μ to ν. Second, in the implementation of step two, we need to verify that μ_σ has no atoms – least the risk of being in a situation where there is *no* transport map T_σ from μ_σ to ν_σ. Both problems are solved by the key assumption that $\mu \ll \mathcal{L}^n$.

The first problem will be better discussed with some additional terminology in place, so we postpone its discussion to the end of Section 18.1. Concerning the second problem, $\mu \ll \mathcal{L}^n$ implies, as one may somehow expect from the examples discussed in Remarks 17.4 and 17.5, that

$$\mu_\sigma \ll \mathcal{H}^1 \llcorner P_f^{-1}(\sigma). \tag{18.7}$$

---

[4] Again, for the sake of clarity, in this preliminary description we are ignoring all the required a.e.s.

Interestingly, this is a point where Sudakov's original work contains a gap.[5] Precisely, Sudakov claims that this "preservation of absolute continuity through disintegration" is a general fact that holds whenever we disintegrate $\mu \ll \mathcal{L}^n$ through a foliation of segments. However, as explained in the Bibliographical Notes, very fine counterexamples show that this is not correct in the generality claimed by Sudakov. At the same time, as firstly pointed out by Ambrosio, (18.7) holds in the specific setting of Sudakov's strategy (i.e., for foliations by segments induced by a Kantorovich potential) since, in that case, one can prove that the *direction map* ∇f of the segments has a "countable Lipschitz property" – a property that in turn enables the use of Remark 17.5, and thus the deduction of (18.7).

The chapter is organized as follows. In Section 18.1 we formally introduce transport rays and sets, and review our sketch of Sudakov's strategy. In Section 18.2 we formalize the construction of Sudakov's maps, and prove (Theorem 18.7) that to *every* transport plan $\gamma \in \Gamma(\mu, \nu)$ whose support satisfies suitable "non-crossing" and "countable Lipschitz" properties, we can associate a transport map T from μ to ν so that $\gamma_T = (\mathbf{id} \times T)_\# \mu$ has *lower* transport cost than γ. In Section 18.3, see Theorem 18.9, we exploit (18.3) and the regularity of Kantorovich's potentials to check that the assumptions of Theorem 18.7, including the countable Lipschitz property, hold when γ is optimal in $\mathbf{K}_1(\mu, \nu)$. In Section 18.4 we put all the pieces together and present a complete proof of Theorem 18.1. Finally, Section 18.5 contains the proofs of some technical, measure-theoretic statements that had been initially omitted in Section 18.1.

18.1 Transport Rays and Transport Sets

In this section we formally introduce the fundamental concepts of transport ray, transport set and maximal transport ray map associated to a set $G \subset \mathbb{R}^n \times \mathbb{R}^n$, the main cases of interest being $G = \operatorname{spt} \gamma$ for $\gamma \in \Gamma(\mu, \nu)$, and $G(f) = \{(x, y) : f(y) = f(x) + |x - y|\}$ for f 1-Lipschitz. Then, with the aid of these concepts, we revisit our sketch of Sudakov's argument in a more formal way.

Segments and oriented segments: We will distinguish between *oriented* and *non-oriented* segments in $\mathbb{R}^n$. Given $x, y \in \mathbb{R}^n$ with $x \neq y$, we denote by

$$[\![x, y]\!] \tag{18.8}$$

---

[5] Because of this gap in Sudakov's original work, the attribution to Sudakov of Theorem 18.1 is not completely satisfactory, even though this is really more a case of "having a wrong explanation for a correct assertion" than anything else. See the Bibliographical Notes for more information on the first complete proofs of what here is called, for the sake of simplicity, the Sudakov theorem.

the **closed oriented segment from** x **to** y, that is to say, the map $t \in [0,1] \mapsto (1-t)x + ty \in \mathbb{R}^n$; and by

$$[x,y] \qquad\qquad (18.9)$$

the **closed segment with endpoints** x **and** y, i.e., the subset of $\mathbb{R}^n$ defined as $[x,y] = \{tx + (1-t)y : t \in [0,1]\}$. In particular, $[x,y] = [y,x]$, but $[\![x,y]\!] \neq [\![y,x]\!]$. Open, left-closed/right-open and left-open/right-closed variants are denoted by

$$]\!]x,y[\![, \qquad [\![x,y[\![, \qquad]\!]x,y]\!]$$

in the oriented case, and by

$$]x,y[, \qquad [x,y[, \qquad]x,y]$$

in the non-oriented case. We will also need to consider unbounded segments, i.e., segments for which either or both endpoints cannot be specified, and that, therefore, cannot be represented with the notation we have just introduced. Without getting into the weeds of clumsy, formal definitions, we denote by

$$S^c, \qquad S^\circ, \qquad S^\ell, \qquad S^r, \qquad\qquad (18.10)$$

the families of the (possibly unbounded) oriented segments in $\mathbb{R}^n$, not coinciding with single points, that are, respectively, closed, open, left-closed/right-open and left-open/right-closed; moreover we use the symbol

$$S^\star$$

in statements that hold indifferently for each of the four classes considered in (18.10) (that is to say, $\star$ ranges over $\{\circ, c, \ell, r\}$, unless otherwise stated). We denote by

$$\Phi$$

the **realization map** that associates to an oriented segment its image in $\mathbb{R}^n$: in other words, we set

$$\Phi([\![x,y]\!]) = [x,y], \qquad \Phi(]\!]x,y[\![) =]x,y[, \qquad\qquad (18.11)$$

etc., and define as expected $\Phi(\sigma)$ for all the other possible kind of $\sigma \in S^\star$. Given two oriented segments $\sigma, \sigma^* \in S^\star$ we say that σ is a **sub-segment** of σ^*, or equivalently that σ^* is a **super-segment** of σ, if $\Phi(\sigma) \subset \Phi(\sigma^*)$ and the orientation of σ agrees with that of σ^*. We also say that $x \in \mathbb{R}^n$ is "contained" in $\sigma \in S^\star$ if $x \in \Phi(\sigma)$. Moreover, we let

$$\mathbb{R}_\sigma \subset \mathbb{R}^n$$

be the realization of the oriented line corresponding to $\sigma \in S^\star$. On $\mathbb{R}_\sigma$ we have notions of monotone function, right and left halfline, and right and left continuity based on the orientation induced by σ. In particular, Theorem 16.1 can be applied to measures $\mu_\sigma, \nu_\sigma \in \mathcal{P}(\mathbb{R}^n)$ supported in $\mathbb{R}_\sigma$.

We endow $S^\star$ with a metric structure. When dealing with bounded closed oriented segments[6] this can be naturally done by setting

$$\sqrt{|x - x^*|^2 + |y - y^*|^2},$$

for the distance between $[\![x, y]\!]$ and $[\![x^*, y^*]\!]$; however, one needs a different definition to deal with half-lines and lines. We thus introduce the following metric on $S^\star$: whenever $\sigma, \sigma^* \in S^\star$, we set

$$\mathrm{dist}(\sigma, \sigma^*) = \sum_{k=1}^\infty \frac{1}{2^k} \frac{|x_k - x_k^*| + |y_k - y_k^*|}{1 + |x_k - x_k^*| + |y_k - y_k^*|},$$

provided $x_k, y_k \in \mathbb{R}^n$ are such that $\mathrm{Cl}(\Phi(\sigma)) \cap \mathrm{Cl}(B_k) = [x_k, y_k]$, and x_k^*, y_k^* are similarly defined for σ^*. We omit the simple proof that $(S^\star, \mathrm{dist})$ *is a locally compact and separable metric space*. The following statement will be useful in addressing several measurability issues related to the construction of transport maps (see Section 18.5 for a proof).

Proposition 18.3   *If Σ is a countable union of closed subsets of $S^\star$, then $\Phi(\Sigma)$, the image of Σ through the realization map Φ, is a Borel set in $\mathbb{R}^n$.*

Transport rays and transport sets: As recalled at the beginning of this chapter, if $\mu, \nu \in \mathcal{P}_1(\mathbb{R}^n)$ and f is a Kantorovich potential from μ to ν, then γ is optimal in $\mathbf{K}_1(\mu, \nu)$ if and only if $\mathrm{spt}\,\gamma \subset G(f)$, where

$$G(f) = \left\{ (x, y) \in \mathbb{R}^n \times \mathbb{R}^n : f(y) = f(x) + |x - y| \right\}. \tag{18.12}$$

We have then moved to describe Sudakov's strategy, and in doing so we have stressed the importance of the transport set $\mathcal{T}(f)$ of f, which was defined in (18.6) as the union of the open segments $]\!]x, y[\![$ corresponding to pairs $(x, y) \in G(f)$ with $x \neq y$. We now apply the same procedure leading from $G(f)$ to $\mathcal{T}(f)$ to an arbitrary subset G of $G(f)$, thus defining a notion of transport set $\mathcal{T}(G)$ for every such G. We also introduce, in a similar vein, several other useful, related notions which will play an important role in the sequel. The cases $G = \mathrm{spt}\,\gamma$ for γ an optimal plan of $\mathbf{K}_1(\mu, \nu)$ and $G = G(f)$ for f

---

[6] When discussing the Monge problem for μ and ν compactly supported on $\mathbb{R}^n$ one can indeed identify S° with $\{(x, y) \in \mathbb{R}^n \times \mathbb{R}^n : x \neq y\}$, and thus simplify the presentation of Sudakov's strategy.

a Kantorovich potential, should both be kept in mind to gain insights on the geometric meaning of the various objects introduced below.

Given $G \subset \mathbb{R}^n \times \mathbb{R}^n$ we say that

$$]\!]x, y[\![\text{ is a \textbf{transport ray} of } G \text{ if } (x, y) \in G \text{ and } x \neq y,}$$

and denote the **set of transport rays** by

$$\Theta(G) = \left\{ \sigma \in S^\circ : \sigma \text{ is a transport ray of } G \right\}.$$

A (possibly unbounded) oriented open segment $\sigma \in S^\circ$ is a **maximal transport ray** of G if: **(i)** for each $z \in \Phi(\sigma)$ there exists $\sigma' \in \Theta(G)$ such that $z \in \Phi(\sigma')$ and σ' is a sub-segment of σ; and **(ii)** if $\sigma^* \in S^\circ$ is a super-segment of σ and satisfies property (i), then $\sigma^* = \sigma$. We set

$$\Theta^{\max}(G) = \left\{ \sigma \in S^\circ : \sigma \text{ is a maximal transport ray of } G \right\}.$$

In a funny terminology blunder, *a maximal transport ray of G may not be a transport ray of G*. A trivial reason is that transport rays are necessarily bounded, while $\Theta^{\max}(G)$ could contain unbounded segments; but bounded examples are also easy to construct: e.g., when $n = 1$, $\sigma =]\!]0, 1[\![$ is a maximal transport ray, but not a transport ray, of $G = \{(1/(t + 1), 1/t) \in \mathbb{R} \times \mathbb{R} : t \geq 1\}$.

The **transport set of** G is the subset of $\mathbb{R}^n$ defined as the union of the realizations of all the transport rays of G,

$$\mathcal{T}(G) = \bigcup_{(x,y) \in G, x \neq y}]\!]x, y[= \bigcup_{\sigma \in \Theta^{\max}(G)} \Phi(\sigma) = \Phi(\Theta^{\max}(G)). \tag{18.13}$$

We can also define the **transport sets with endpoints, with left-endpoints and with right-endpoints of** G by setting

$$\mathcal{T}_\star(G) = \bigcup_{\sigma \in \Theta^{\max}(G)} \Phi(\rho^\star(\sigma)), \qquad \star \in \{c, \ell, r\}. \tag{18.14}$$

Here we are using the natural bijections between the various classes $S^\star$: for example,

$$\rho^c : (S^\circ \cup S^\ell \cup S^r) \to S^c, \tag{18.15}$$

is the natural identification between open, left-closed/right-open or left-open/right-closed segments σ and their "closed version" $\rho^c(\sigma)$. Analogous maps ρ°, ρ^ℓ and ρ^r are similarly defined with values in S°, S^ℓ, and S^r. In this way,

$$\mathcal{T}_c(G) \setminus \mathcal{T}(G), \qquad \mathcal{T}_\ell(G) \setminus \mathcal{T}(G), \qquad \mathcal{T}_r(G) \setminus \mathcal{T}(G),$$

denote, respectively, the **sets of endpoints, of left endpoints and of right endpoints of maximal transport rays of** G.

It is convenient to introduce the notion of **set of points fixed by** $G \subset \mathbb{R}^n \times \mathbb{R}^n$, or **motionless set of** G, defined by setting

$$\mathcal{ML}(G) = \left\{ x \in \mathbb{R}^n : (x,x) \in G \text{ and } (x,y) \in G \text{ implies } y = x \right\}$$
$$= \{ x \in \mathbb{R}^n : \mathbf{p}^{-1}(x) \cap G = \{(x,x)\} \} \subset \mathbf{p}(G), \tag{18.16}$$

(where, as usual, $\mathbf{p}(x,y) = x$).

Remark 18.4 (Mass storage on the motionless set and on the endpoints sets) If $\gamma \in \Gamma(\mu, \nu)$, then the information $(x,y) \in \operatorname{spt} \gamma$, which is equivalent to $\gamma(B_r(x) \times B_r(y)) > 0$ for every $r > 0$, means that, under the transport instructions contained in γ, some portion of the mass stored by μ nearby $x \in \operatorname{spt} \mu$ has to be sent nearby $y \in \operatorname{spt} \nu$. Therefore $x \in \mathcal{ML}(G)$ with $G = \operatorname{spt} \gamma$ means that $x \in \operatorname{spt} \mu \cap \operatorname{spt} \nu$ and $\gamma(B_r(x) \times B_r(y)) = 0$ for every $y \neq x$ and r small enough – in other words, the transport instructions in γ are to keep in place the (non-trivial amount of) mass stored by μ around x. More formally, for every $G \subset \mathbb{R}^n \times \mathbb{R}^n$ we can notice the validity of the inclusions

$$\mathbf{p}(G) \subset \mathcal{T}_\ell(G) \cup \mathcal{ML}(G), \qquad \mathcal{T}_\ell(G) \cap \mathcal{ML}(G) = \emptyset, \tag{18.17}$$
$$\mathbf{q}(G) \subset \mathcal{T}_r(G) \cup \mathcal{ML}(G), \qquad \mathcal{T}_r(G) \cap \mathcal{ML}(G) = \emptyset, \tag{18.18}$$

which imply that whenever $\gamma \in \Gamma(\mu, \nu)$ is concentrated on G, then

$$\mu \text{ is concentrated on } \mathcal{T}_\ell(G) \cup \mathcal{ML}(G), \tag{18.19}$$
$$\nu \text{ is concentrated on } \mathcal{T}_r(G) \cup \mathcal{ML}(G). \tag{18.20}$$

In particular, μ could store a non-trivial amount of mass both on the set of left endpoints $\mathcal{T}_\ell(G) \setminus \mathcal{T}(G)$ and on the motionless set $\mathcal{ML}(G)$.

While every transport ray is clearly contained in a *unique* maximal transport ray, when G is an arbitrary subset of $\mathbb{R}^n \times \mathbb{R}^n$ nothing prevents a point in $\mathcal{T}(G)$ to belong to the realizations of multiple maximal transport rays.[7] The following **non-crossing condition** (which, by Proposition 18.6, holds for the choices of G that are relevant in the analysis of the Monge problem) eliminates this ambiguity:

$$\begin{gathered} \text{if distinct transport rays }]\!] x, y [\![\text{ and }]\!] x', y' [\![\text{ of } G \\ \text{are such that } [x,y] \cap [x',y'] \neq \emptyset \\ \text{then either } \frac{y-x}{|y-x|} = \frac{y'-x'}{|y'-x'|} \text{ (same orientation),} \\ \text{or } x = x' \text{ or } y = y' \text{ (at least one common endpoint).} \end{gathered} \tag{18.21}$$

---

[7] For example, setting $x_0 = (1,0)$, $y_0 = (0,1)$, and $G = \{(-x_0, x_0), (-y_0, y_0)\} \subset \mathbb{R}^2 \times \mathbb{R}^2$ we see that $(0,0)$ belongs to $\mathcal{T}(G)$ and to the realizations of both the maximal transport rays $]\!] - x_0, x_0 [\![$ and $]\!] - y_0, y_0 [\![$.

(The two possibilities are not mutually exclusive of course.) When G satisfies (18.21), then to every point $x \in \mathcal{T}(G)$ there corresponds a *unique maximal transport ray* containing x, that we denote by $P_G(x)$. We can thus define the **maximal transport ray map**

$$P_G : \mathcal{T}(G) \to \mathcal{S}^\circ,$$

and the **direction map** τ_G of P_G, $\tau_G : \mathcal{T}(G) \to \mathbb{S}^{n-1}$, by setting

$$\tau_G(x) = \text{direction of } P_G(x), \qquad x \in \mathcal{T}(G).$$

Of course, even under the non-crossing condition (18.21), a point $x \in \mathcal{T}_\ell(G) \setminus \mathcal{T}(G)$ could be the left endpoint of multiple maximal transport rays, and, *mutatis mutandis*, a point $y \in \mathcal{T}_r(G) \setminus \mathcal{T}(G)$ could be the right endpoint of multiple maximal transport rays. Therefore, it may be impossible to extend P_G (and thus τ_G) in a geometrically meaningful way to the whole of $\mathcal{T}_c(G)$.

The Borel measurability of the sets and maps introduced earlier is discussed in the following proposition, whose proof is postponed to Section 18.5.

Proposition 18.5 *If G is a closed subset of $\mathbb{R}^n \times \mathbb{R}^n$, then $\mathcal{T}(G)$, $\mathcal{T}_c(G)$, $\mathcal{T}_\ell(G)$, $\mathcal{T}_r(G)$ and $\mathcal{ML}(G)$ are Borel sets in $\mathbb{R}^n$. Moreover, if the non-crossing condition (18.21) holds, then the map $P_G : \mathcal{T}(G) \to \mathcal{S}^\circ$ is well-defined and Borel measurable.*

In the next proposition we review the concepts introduced so far in the case $G = G(f)$ for f a 1-Lipschitz function.

Proposition 18.6 (Transport sets of a 1-Lipschitz function) *If $f : \mathbb{R}^n \to \mathbb{R}$ is a 1-Lipschitz function, then $G(f) = \{(x,y) : f(y) = f(x) + |x - y|\}$ is closed and satisfies the non-crossing condition (18.21), the transports sets*

$$\mathcal{T}(f) := \mathcal{T}(G(f)), \quad \mathcal{T}_\star(f) := \mathcal{T}_\star(G(f)), \qquad \star \in \{r, c, \ell\},$$

are Borel measurable, and the maps $P_f := P_{G(f)} : \mathcal{T}(f) \to \mathcal{S}^\circ$ and $\tau_f := \tau_{G(f)} : \mathcal{T}(f) \to \mathbb{S}^{n-1}$ are Borel measurable. Moreover, if F denotes the set of differentiability points of f, then

$$\mathcal{T}(f) \subset F, \qquad \tau_f = \nabla f \text{ on } \mathcal{T}(f), \tag{18.22}$$

and we can obtain a well-defined Borel measurable extension of P_f to $\mathcal{T}_c(f) \cap F$ by setting,

$$P_f(x) = \text{unique } \sigma \in \Theta^{\max}(f) \text{ s.t. } x \in \Phi(\rho^\ell(\sigma)), \tag{18.23}$$

if $x \in (\mathcal{T}_\ell(f) \setminus \mathcal{T}(f)) \cap F$, and by setting,

$$P_f(y) = \text{unique } \sigma \in \Theta^{\max}(f) \text{ s.t. } y \in \Phi(\rho^r(\sigma)), \tag{18.24}$$

if $y \in (\mathcal{T}_r(f) \setminus \mathcal{T}(f)) \cap F$, where

$$\Theta^{\max}(f) = \Theta^{\max}(G(f)) = P_f(\mathcal{T}(f)).$$

Finally, the corresponding extension of τ_f to $\mathcal{T}_c(f) \cap F$ is also Borel measurable, since it satisfies $\tau_f = \nabla f$ on $\mathcal{T}_c(f) \cap F$.

Proof Since $G(f)$ is obviously closed, the Borel measurability of the transport sets follows by Proposition 18.5 applied to $G = G(f)$. To prove the differentiability of f on $\mathcal{T}(f)$, let $z \in \mathcal{T}(f)$, so that $z \in]\!]x, y[\![$ for some $x, y \in G(f)$ with $x \neq y$, and set $e_0 = (y-x)/|y-x|$. Since $\mathrm{Lip}(f) \leq 1$ and $f(y) = f(x) + |x - y|$ we easily see that $f(x + t\,e_0)$ is affine with unit derivative in t. In particular, there exists $\varepsilon > 0$ such that $f(z + t\,e_0) = f(z) + t$ for every $|t| < \varepsilon$. Let us now consider a point $w = z + t\,e_0 + s\,e$ for some unit vector e such that $e_0 \cdot e = 0$, so that $|z - w|^2 = t^2 + s^2$. Provided t and s are such that $|t| + \sqrt{|s|} < \varepsilon$, we find that

$$
\begin{aligned}
f(w) &- f(z) - (w - z) \cdot e_0 \\
&= f(w) - f(z + (t + \sqrt{|s|})\,e_0) + f(z + (t + \sqrt{|s|})\,e_0) - f(z) - t \\
&\geq -\sqrt{s^2 + |s|} + \sqrt{|s|} = \mathrm{O}(|s|^{3/2}) = \mathrm{o}(|z - w|),
\end{aligned}
$$

and, similarly,

$$
\begin{aligned}
f(w) &- f(z) - (w - z) \cdot e_0 \\
&= f(w) - f(z + (t - \sqrt{|s|})\,e_0) + f(z + (t - \sqrt{|s|})\,e_0) - f(z) - t \\
&\leq \sqrt{s^2 + |s|} - \sqrt{|s|} = \mathrm{O}(|s|^{3/2}) = \mathrm{o}(|z - w|),
\end{aligned}
$$

so that f is differentiable at z with $\nabla f(z) = e_0$, as claimed. To prove the non-crossing condition, let $]\!]x, y[\![$ and $]\!]x', y'[\![$ be transport rays of $G(f)$ with $[x, y] \cap [x', y'] \neq \emptyset$. If these rays do not meet at least one of their endpoints, i.e., if $x \neq x'$ and $y \neq y'$, then there exists a point z that belongs to both $]x, y[$ and $]x', y'[$, and, in particular, f is differentiable at z with

$$\frac{y' - x'}{|y' - x'|} = \nabla f(z) = \frac{y - x}{|y - x|},$$

thus proving that $]\!]x, y[\![$ and $]\!]x', y'[\![$ have the same orientation.

Finally, let P_f^c denote[8] the extension of P_f from $\mathcal{T}(f)$ to $\mathcal{T}_c(f) \cap F$ defined by (18.23) and (18.24). Given $\Sigma \subset S^\circ$ closed, we have that

$$(P_f^c)^{-1}(\Sigma) = \mathcal{T}_c(f) \cap F \cap \Phi(\rho^c(\Sigma)). \tag{18.25}$$

Since $\rho^c(\Sigma)$ is closed in S^c, by Proposition 18.3, $\Phi(\rho^c(\Sigma))$ is a Borel set in $\mathbb{R}^n$, and thus P_f^c is Borel measurable. □

---

[8] This notation is used only in this short argument, where it is convenient to distinguish between P_f and P_f^c. Everywhere else, the extension of P_f to $F \cap \mathcal{T}_c(f)$ will be simply denoted by P_f.

Revisiting Sudakov's strategy: We finally revisit the sketch of Sudakov's argument given at the beginning of the chapter. Given an optimal plan γ in $\mathbf{K}_1(\mu, \nu)$, by (18.4) there is a Kantorovich potential f from μ to ν such that γ is concentrated on $G(f) = \{(x, y) : f(y) = f(x) + |x - y|\}$. By Proposition 18.6, $G(f)$ satisfies the non-crossing condition (18.21), and thus the maximal transport ray map P_f is well-defined on $\mathcal{T}_r(f) \cap F$, F the set of differentiability points of f. Now, we would like to consider the disintegrations of $\{\mu_\sigma\}_\sigma$ and $\{\nu_\sigma\}$ of μ and ν with respect to P_f, and then solve $\mathbf{M}_1(\mu_\sigma, \nu_\sigma)$ (as a transport problem on $\mathbb{R}_\sigma$) by Theorem 16.1. However, considering those disintegrations requires checking that μ and ν are concentrated on $\mathcal{T}(f)$, whereas (recall (18.19) and (18.20)) μ and ν may store mass outside of $\mathcal{T}(f)$, precisely either on the motionless set

$$\mathcal{ML}(f) = \mathcal{ML}(G(f)) = \{x \in \mathbb{R}^n : f(y) > f(x) + |x - y|, \quad \forall y \ne x\}, \tag{18.26}$$

or on the set of left endpoints $\mathcal{T}_\ell(f) \setminus \mathcal{T}(f)$ (in the case of μ), or on the set of right endpoints $\mathcal{T}_r(f) \setminus \mathcal{T}(f)$ (in the case of ν). Since, at least intuitively, the first issue should be easily solvable by the setting $T = \mathbf{id}$ on $\mathcal{ML}(f)$, for the rest of this discussion we assume to be in a situation where[9]

$$\mathcal{ML}(f) = \emptyset.$$

Under this assumption, having extended P_f to $\mathcal{T}_r(f) \cap F$, if $\mu \ll \mathcal{L}^n$, then μ is concentrated on $\mathcal{T}_r(f) \cap F$ (as, by Rademacher's theorem, $\mathcal{L}^n(\mathbb{R}^n \setminus F) = 0$). Therefore it makes sense to consider the disintegration $\{\mu_\sigma\}_\sigma$ of μ with respect to P_f. We will then prove that the foliation by segments induced by P_f satisfies a certain "countable Lipschitz property" (see Theorem 18.9), which in turn implies (see Theorem 18.7) that, for $(P_f)_\#\mu$-a.e. σ,

$$\mu_\sigma \ll \mathcal{H}^1 \llcorner P_f^{-1}(\sigma). \tag{18.27}$$

By combining (18.27) with $\mu = \mu_\sigma \otimes [(P_f)_\#\mu]$ we thus find

$$\mu(\mathcal{T}_\ell(f) \setminus \mathcal{T}(f)) = 0,$$

so that μ is concentrated on $\mathcal{T}(f)$. Now, by entirely analogous considerations, we could assert that $\nu(\mathcal{T}_r(f) \setminus \mathcal{T}(f)) = 0$ if $\nu \ll \mathcal{L}^n$, thus becoming able to consider the disintegration $\{\nu_\sigma\}_\sigma$ of ν with respect to P_f. However, on a closer inspection, this is not really what we want to do: indeed, the disintegrations $\mu = \mu_\sigma \otimes [(P_f)_\#\mu]$ and $\nu = \nu_\sigma \otimes [(P_f)_\#\nu]$ do not really "talk to each other"

---

[9]   The advantage of working with a generic set $G \subset \mathbb{R}^n \times \mathbb{R}^n$ rather than with $G = G(f)$ or with $G = \mathrm{spt}\,\gamma$ lies mainly in the flexibility this choice offers in taking care of the reduction to the case $\mathcal{ML}(f) = \emptyset$, see step two of the proof of Theorem 18.1.

since $(P_f)_\# \mu$ and $(P_f)_\# \nu$ are not the same measure! In other words, rather than the disintegration of ν with respect to P_f, we need a decomposition of the form

$$\nu = \lambda_\sigma \otimes [(P_f)_\# \mu], \tag{18.28}$$

for some Borel measurable family of probability measures $\{\lambda_\sigma\}_\sigma$. Interestingly, the validity of (18.28) has nothing to do with the issue of whether or not ν is concentrated on $\mathcal{T}_r(f) \cap F$: having proved that μ is concentrated on $\mathcal{T}(f)$, we see that γ is concentrated on $(\mathcal{T}(f) \times \mathbb{R}^n) \cap G(f)$, and thus we can consider its disintegration $\{\gamma_\tau\}_\tau$ with respect to the map

$$r_f : (\mathcal{T}(f) \times \mathbb{R}^n) \cap G(f) \to \mathcal{S}^c, \qquad r_f(x, y) = \rho^c(P_f(x)),$$

(which is Borel measurable by $r_f = \rho^c \circ P_f \circ \mathbf{p}$); as a consequence, we claim that (18.28) is established in the form

$$\nu = (\mathbf{q}_\# \gamma_{\rho^c(\sigma)}) \otimes [(P_f)_\# \mu], \tag{18.29}$$

since, indeed, if $\varphi \in C_c^0(\mathbb{R}^n)$, then

$$
\begin{aligned}
\int_{\mathbb{R}^n} \nu \, d\varphi &= \int_{\mathbb{R}^n \times \mathbb{R}^n} (\varphi \circ \mathbf{q}) \, d\gamma = \int_{\mathcal{S}^c} d[(r_f)_\# \gamma](\tau) \int_{\mathbb{R}^n \times \mathbb{R}^n} (\varphi \circ \mathbf{q}) \, d\gamma_\tau \\
&= \int_{\mathcal{S}^c} d[(\rho^c \circ P_f \circ \mathbf{p})_\# \gamma](\tau) \int_{\mathbb{R}^n \times \mathbb{R}^n} (\varphi \circ \mathbf{q}) \, d\gamma_\tau \\
&= \int_{\mathcal{S}^\circ} d[(P_f)_\# (\mathbf{p}_\# \gamma)](\sigma) \int_{\mathbb{R}^n \times \mathbb{R}^n} (\varphi \circ \mathbf{q}) \, d\gamma_{\rho^c(\sigma)} \tag{18.30} \\
&= \int_{\mathcal{S}^\circ} d[(P_f)_\# \mu](\sigma) \int_{\mathbb{R}^n \times \mathbb{R}^n} \varphi \, d[\mathbf{q}_\# (\gamma_{\rho^c(\sigma)})].
\end{aligned}
$$

By Theorem 16.1 (with $\mathbb{R}_\sigma$ in place of $\mathbb{R}$), we then find increasing transport maps $T_\sigma : \mathbb{R}_\sigma \to \mathbb{R}_\sigma$ with $(T_\sigma)_\# \mu_\sigma = \lambda_\sigma$, to be glued (by means of Theorem 18.7) into an optimal transport map T in $\mathbf{M}_1(\mu, \nu)$.

18.2 Construction of the Sudakov Maps

We now present a formalization of Sudakov's construction, see Theorem 18.7. Starting from a generic plan $\gamma \in \Gamma(\mu, \nu)$ concentrated on a set G satisfying the non-crossing condition (18.21) and such that the direction map τ_G has the countable Lipschitz property, we construct a transport map T that induces a plan $\gamma_T = (\mathbf{id} \times T)_\# \mu$ with lower (linear) transport cost than γ. Here, a Borel measurable map $S : E \to \mathbb{R}^n$ ($E \in \mathcal{B}(\mathbb{R}^n)$) is said to have the **countable Lipschitz property** on E, if there exists $\{E_j\}_{j \in \mathbb{N}}$ such that $E_j \subset E$, $\mathrm{Lip}(S; E_j) < j$, and

$$\mathcal{L}^n \left(E \setminus \bigcup_{j \in \mathbb{N}} E_j \right) = 0. \tag{18.31}$$

Theorem 18.7 (Sudakov transport maps)   *Let $G \subset \mathbb{R}^n \times \mathbb{R}^n$ be a Borel set satisfying the non-crossing condition* (18.21), *and such that $\mathcal{T}(G)$, $P_G : \mathcal{T}(G) \to S^\circ$ and $\tau_G : \mathcal{T}(G) \to \mathbb{S}^{n-1}$ are Borel measurable. Let $E \subset \mathbb{R}^n$ be a Borel set such that $\mathcal{T}(G) \subset E \subset \mathcal{T}_c(G)$, and,*

(a) *P_G can be uniquely extended from $\mathcal{T}(G)$ to E, in the sense that for every $x \in E$ there is a unique $\sigma \in \Theta_{\max}(G)$ with $x \in \Phi(\rho^c(\sigma))$; the extension $P_G : E \to S^\circ$ is Borel measurable, and so is the corresponding extension of τ_G;*

(b) *τ_G has countable Lipschitz property on E.*

Then, whenever $\mu \ll \mathcal{L}^n \llcorner E$ is a finite Borel measure and $\{\mu_\sigma\}_\sigma$ is the disintegration of μ with respect to P_G, we have

$$\mu_\sigma \ll \mathcal{H}^1 \llcorner P_G^{-1}(\sigma) \qquad for \ (P_G)_{\#}\mu\text{-a.e. } \sigma \in S^\circ. \tag{18.32}$$

Let us now assume, in addition to the above properties, that

(c) *$\mathcal{ML}(G) = \emptyset$;*

(d) *$\mathcal{T}_\ell(G) \setminus \mathcal{T}(G)$ is μ-negligible.*

Then, whenever $\mu, \nu \in \mathcal{P}_1(\mathbb{R}^n)$, $\gamma \in \Gamma(\mu, \nu)$ is concentrated on G, and $\{\gamma_\tau\}_\tau$ is the disintegration of γ with respect to $r_G = \rho^c \circ P_G \circ \mathbf{p} : (E \times \mathbb{R}^n) \cap G \to S^\circ$, we have

(i) *$\nu = \lambda_\sigma \otimes [(P_G)_{\#}\mu]$ where $\lambda_\sigma = \mathbf{q}_{\#}\gamma_{\rho^c(\sigma)}$;*

(ii) *for $(P_G)_{\#}\mu$-a.e. $\sigma \in S^\circ$, there is an increasing and left continuous map $T_\sigma : \mathbb{R}_\sigma \to \mathbb{R}_\sigma$ such that $(\mathrm{id} \times T_\sigma)_{\#}\mu_\sigma$ is an optimal transport plan in $\mathbf{K}_1(\mu_\sigma, \lambda_\sigma)$;*

(iii) *there is $E' \subset E$ with $\mathcal{L}^n(E \setminus E') = 0$ and a Borel measurable map $T : E' \to \mathbb{R}^n$ such that (for $(P_G)_{\#}\mu$-a.e. $\sigma \in S^\circ$) $T = T_\sigma$ μ_σ-a.e. on $\mathbb{R}_\sigma$, $T_{\#}\mu = \nu$, and*

$$\int_{\mathbb{R}^n} |x - T(x)| \, d\mu(x) \le \int_{\mathbb{R}^n \times \mathbb{R}^n} |x - y| \, d\gamma(x, y). \tag{18.33}$$

In particular, if γ is optimal in $\mathbf{K}_1(\mu, \nu)$, then $(\mathrm{id} \times T)_{\#}\mu$ is optimal in $\mathbf{K}_1(\mu, \nu)$ and T is optimal in $\mathbf{M}_1(\mu, \nu)$.

Remark 18.8   Notice that we are not assuming that G is closed (as the examples $G = \mathrm{spt}\,\gamma$ and $G = G(f)$ would suggest to do) so to be eventually able to address the reduction to the case of an empty motionless set in step two of the proof of Theorem 18.1. In a situation where $G = G(f)$, the extension domain E appearing in assumption (a) is of course $\mathcal{T}_r(f) \cap F$, F the set of differentiability points of f. We also recall that, without assumption (b), conclusion (18.32) may fail (see the Bibliographical Notes).

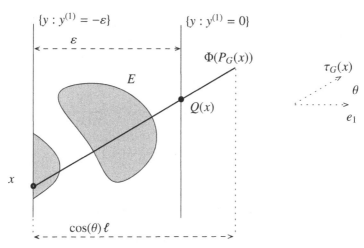

Figure 18.1 If $x \in E$, then the open segment $\Phi(P_G(x))$, whose direction is $\tau_G(x)$, makes an angle $\theta \le \pi/4$ with the direction e_1. Since $\mathcal{H}^1(\Phi(P_G(x))) > \ell > \sqrt{2}\,\varepsilon$ and x is at distance at most ε from $\{y : y^{(1)} = 0\}$, the inequality $\ell \cos \theta > \varepsilon$ ensures that $\Phi(P_G(x))$ intersects $\{y : y^{(1)} = 0\}$ at exactly one point, denoted by $Q(x)$. Thus, in the special case considered in step one of the proof of Theorem 18.7, we can use Remark 17.6 to reparametrize the disintegration of μ with respect to P_G in terms of the disintegration of μ with respect to $Q : E \rightarrow \{y : y^{(1)} = 0\} \equiv \mathbb{R}^{n-1}$, and then apply the coarea formula as in Remark 17.5 to conclude the required absolute continuity property.

Proof of Theorem 18.7 The first two steps are devoted to the proof of (18.32). We then focus on conclusions (i), (ii), and (iii).

Step one: We prove (18.32) under assumptions (a), (b), and
(b1): $\mathrm{Lip}(\tau_G, E) \le L$;
(b2): $\tau_G(x) \cdot e_1 \ge 1/\sqrt{2}$ for every $x \in E$;
(b3): $E \subset \{y \in \mathbb{R}^n : -\varepsilon \le y^{(1)} \le 0\}$, where $y^{(i)} = y \cdot e_i$, $1 \le i \le n$;
(b4): for each $x \in E$, $\mathcal{H}^1(\Phi(P_G(x))) > \ell$,
where $L \in (0, \infty)$, $\varepsilon > 0$ is small enough in terms of L and n, and $\ell > \sqrt{2}\,\varepsilon$. Indeed, by (b2–4), for every $x \in E$ there is a unique $Q(x)$ such that

$$\Phi(P_G(x)) \cap \{y : y^{(1)} = 0\} = \{Q(x)\}; \tag{18.34}$$

see Figure 18.1. We can prove that the resulting map $Q : E \rightarrow \{y : y^{(1)} = 0\}$ is Lipschitz continuous (and obtain a formula for it) by the following remark: if $x \in E$, then the half-line $\{x + t\,\tau_G(x) : t > 0\}$ intersects $\{y : y^{(1)} = 0\}$ when $t = t(x)$ is given by

$$t(x) = \frac{-x^{(1)}}{\tau_G(x)^{(1)}} \in [0, \sqrt{2}\,\varepsilon];$$

the corresponding map $t : E \to [0, \sqrt{2}\,\varepsilon]$ is Lipschitz continuous on E, since properties (b1–3) give

$$|t(x_1) - t(x_2)| = \left| \frac{x_1^{(1)}}{\tau_G(x_1)^{(1)}} - \frac{x_2^{(1)}}{\tau_G(x_2)^{(1)}} \right|$$

$$\leq \sqrt{2}\,|x_1^{(1)} - x_2^{(1)}| + \varepsilon \left| \frac{1}{\tau_G(x_1)^{(1)}} - \frac{1}{\tau_G(x_2)^{(1)}} \right|$$

$$\leq \sqrt{2}\,|x_1 - x_2| + \varepsilon \max_{i=1,2} \left(\frac{1}{\tau_G(x_i)^{(1)}} \right)^2 |\tau_G(x_1) - \tau_G(x_2)|$$

$$\leq (\sqrt{2} + 2\varepsilon L)\,|x_1 - x_2|,$$

and then Q is Lipschitz continuous on E since it satisfies

$$Q(x) = x + t(x)\,\tau_G(x) = x - \frac{x^{(1)}}{\tau_G(x)^{(1)}}\,\tau_G(x), \qquad \forall x \in E. \tag{18.35}$$

Let us now $\{\mu_y^*\}_y$ be the disintegration of μ with respect to Q. We claim that

$$\mu_y^* \ll \mathcal{H}^1 \llcorner Q^{-1}(y), \qquad \text{for } (Q_\# \mu)\text{-a.e. } y \in \{y : y^{(1)} = 0\}, \tag{18.36}$$

and that (18.36) implies (18.32) (under the additional assumptions (b1–4)). Let us first address the second point. Since $Q(x) \in \Phi(P_G(x))$ by construction, and since $Q(x) = Q(z)$, $x, z \in E$, implies $P_G(x) = P_G(z)$ (by assumption (a)), we can define a bijection $\eta : (P_G(E) \subset S^\circ) \to Q(E) \subset \{y : y^{(1)} = 0\}$ by setting

$$\eta(P_G(x)) = Q(x).$$

It is easily seen that η is continuous and, thus, Borel measurable. Since $Q = \eta \circ P_G$, by Remark 17.6 we conclude that, denoting by $\{\mu_\sigma\}_\sigma$ the disintegration of μ with respect to P_G, it holds

$$\mu_\sigma = \mu_{\eta(\sigma)}^* \qquad \text{for } (P_G)_\# \mu\text{-a.e. } \sigma \in S^\circ;$$

in particular, (18.36) implies (18.32), as claimed.

We are thus left to prove (18.36). Setting $x = (x^{(1)}, \hat{x}) \in \mathbb{R} \times \mathbb{R}^{n-1}$ for the generic point $x \in \mathbb{R}^n$, we see that $\{y : y^{(1)} = 0\}$ can be identified with $\mathbb{R}^{n-1}$, and Q with a Lipschitz map $\hat{Q} : E \subset \mathbb{R}^n \to \mathbb{R}^{n-1}$. In particular, we can deduce (18.36) by Remark 17.5 if we can show that $C\hat{Q} > 0$ $\mathcal{L}^n$-a.e. on E. Now, $(C\hat{Q})^2$ is the sum of the squares of the $(n-1) \times (n-1)$-minors of $\nabla \hat{Q}$, in particular

$$C\hat{Q} \geq \det(\nabla_{\hat{x}} \hat{Q}),$$

where $\nabla_{\hat{x}}$ denotes the gradient along the variables $(x^{(2)}, \ldots, x^{(n)})$. From the second identity in (18.35), since $\nabla_{\hat{x}} x^{(1)} = 0$, $\nabla_{\hat{x}} \hat{x} = \mathrm{Id}_{(n-1) \times (n-1)}$ and τ_G is Lipschitz continuous on E, we find that

$$\nabla_{\hat{x}} \hat{Q}(x) = \text{Id}_{(n-1) \times (n-1)} - \frac{x^{(1)}}{\tau_G(x)^{(1)}} \nabla_{\hat{x}} \widehat{\tau_G} + \frac{x^{(1)}}{[\tau_G(x)^{(1)}]^2} \widehat{\tau_G(x)} \otimes \nabla_{\hat{x}} \tau_G^{(1)},$$
(18.37)

for $\mathcal{L}^n$-a.e. on $x \in E$. Since, by assumptions (b1–3) we have $\text{Lip}(\tau_G; E) \leq L$ as well as $|x^{(1)}| \leq \varepsilon$ and $|\tau_G^{(1)}| \geq 1/\sqrt{2}$ on E, we conclude from (18.37) that

$$|\nabla_{\hat{x}} \hat{Q}(x) - \text{Id}_{(n-1) \times (n-1)}| \leq C(n, L)\, \varepsilon,$$

and thus that $\det (\nabla_{\hat{x}} \hat{Q}) \geq 1 - C(n, L)\, \varepsilon$, where $C(n, L)$ denotes a suitable constant depending on n and L only. Taking ε small enough in terms of n and L we complete the proof of (18.36), thus of step one.

Step two: We complete the proof of (18.32) (under the sole assumptions (a) and (b)) by reducing to the situation of step one. To this aim, we consider the following partitioning process: first, we use assumption (b) to introduce a countable Borel partition $\{E_j\}_j$ of E (modulo a Lebesgue negligible set) such that $\text{Lip}(\tau_G, E_j) \leq j \in \mathbb{N}$; second, we pick a finite set of directions $\{v_k\}_k \subset \mathbb{S}^{n-1}$ with the property that for every $\tau \in \mathbb{S}^{n-1}$ there is at least one k such that $\tau \cdot v_k \geq 1/\sqrt{2}$, and then we further subdivide each E_j into Borel sets E_{jk} such that

$$\tau_G(x) \cdot v_k \geq \frac{1}{2}, \qquad \forall x \in E_{jk};$$

obviously the sets E_{jk} are Borel sets since the map $x \mapsto \tau_G(x) \cdot v_k$ is Borel measurable (assumption (a)); third, given $\ell > 0$ and setting $\Sigma(\ell) = \{\sigma \in \mathcal{S}^{\circ} : \mathcal{H}^1(\Phi(\sigma)) > \ell\}$, we notice that $\Sigma(\ell)$ is open in $\mathcal{S}^{\circ}$, and thus that $P_G^{-1}(\Sigma(\ell))$ is a Borel subset of $\mathbb{R}^n$ (assumption (a) again): in particular, we can construct a Borel partition $\{E_{jkm}\}_m$ of E_{jk} so that

$$P_G(x) \in \Sigma(1/m), \qquad \forall x \in E_{jkm};$$

fourth, for a positive ε_j to be chosen later on in terms of n and $j (\geq \text{Lip}(\tau_G, E_j))$, we further subdivide each E_{jkm} into Borel sets of the form

$$E_{jkmh} = E_{jkm} \cap Z_{jkmh}, \quad Z_{jkmh} = \left\{ y \in \mathbb{R}^n : \alpha_h - \min\left\{\varepsilon_j, \frac{1}{\sqrt{2}\, m}\right\} \leq y \cdot v_k \leq \alpha_h \right\}$$

corresponding to a choice of $\{\alpha_h\}_h \subset \mathbb{R}$ such that the stripes Z_{jkmh} cover the whole $\mathbb{R}^n$. Finally, we notice that, up to a rotation taking v_k into e_1, and a translation taking $\{y : y^{(1)} = \alpha_h\}$ into $\{y : y^{(1)} = 0\}$, the set E_{jkmh} satisfies assumptions (b1–4) with $L = j$, $\varepsilon = \varepsilon_j$, and $\ell = 1/m$. We can thus apply step one to deduce that, if $\{\mu_\sigma^{jkmh}\}_\sigma$ denotes the disintegration of $\mu \llcorner E_{jkmh}$ with respect to $(P_G)|_{E_{jkmh}}$, then

$$\mu_\sigma^{jkmh} \ll \mathcal{H}^1 \llcorner [(P_G)^{-1}(\sigma) \cap E_{jkmh}] \qquad \text{for } (P_G)_{\#}\mu\text{-a.e. } \sigma \in \mathcal{S}^{\circ}.$$

By Remark 17.3, the disintegration $\{\mu_\sigma\}_\sigma$ of μ with respect to P_G is such that

$$\mu_\sigma \llcorner E_{jkmh} \ll \mu_\sigma^{jkmh}, \qquad \text{for } (P_G)_\#\mu\text{-a.e. } \sigma \in S^\circ.$$

We finally prove (18.32): by Remark 17.3, and thanks to $\mu \ll \mathcal{L}^n$ and

$$\mathcal{L}^n\left(E \setminus \bigcup_{j,k,m,h} E_{jkmh}\right) = 0,$$

so that for $(P_G)_\#\mu$-a.e. $\sigma \in S^\circ$ we have

$$\mu_\sigma = \sum_{j,k,m,h} \mu_\sigma \llcorner E_{jkmh} \ll \sum_{j,k,m,h} \mathcal{H}^1 \llcorner [(P_G)^{-1}(\sigma) \cap E_{jkmh}] \le \mathcal{H}^1 \llcorner (P_G)^{-1}(\sigma).$$

Step three: We now let $\mu, \nu \in \mathcal{P}_1(\mathbb{R}^n)$, $\gamma \in \Gamma(\mu, \nu)$ be concentrated on G, consider the disintegration $\{\gamma_\tau\}_\tau$ of γ with respect to

$$r_G = \rho^c \circ P_G \circ \mathbf{p} : (E \times \mathbb{R}^n) \cap G \to S^\circ,$$

and turn to prove conclusions (i), (ii), and (iii). Conclusion (i), namely

$$\nu = \lambda_\sigma \otimes [(P_G)_\#\mu], \qquad \lambda_\sigma = \mathbf{q}_\#\gamma_{\rho^c(\sigma)},$$

is proved by arguing exactly as in (18.30). Notice that $\mu, \nu \in \mathcal{P}_1(\mathbb{R}^n)$ implies $\mu_\sigma, \lambda_\sigma \in \mathcal{P}_1(\mathbb{R}_\sigma)$. Hence, thanks to (18.32), we can apply Theorem 16.1-(v) to find, for $(P_G)_\#\mu$-a.e. $\sigma \in S^\circ$, an increasing and left continuous map $T_\sigma : \mathbb{R}_\sigma \to \mathbb{R}_\sigma$ such that $(\mathbf{id} \times T_\sigma)_\#\mu_\sigma$ solves $\mathbf{K}_1(\mu_\sigma, \lambda_\sigma)$. In particular, thanks to $\mathbf{p}_\#\gamma_{\rho^c(\sigma)} = \mu_\sigma$, we have

$$\int_{\mathbb{R}^n} |T_\sigma(x) - x| \, d\mu_\sigma(x) \le \int_{\mathbb{R}^n \times \mathbb{R}^n} |x - y| \, d\gamma_{\rho^c(\sigma)}, \tag{18.38}$$

for $(P_G)_\#\mu$-a.e. $\sigma \in S^\circ$. All that remains to prove is that

$$\begin{aligned} &\exists E' \subset E, \ E' \in \mathcal{B}(\mathbb{R}^n), \text{ with } \mathcal{L}^n(E \setminus E') = 0 \\ &\text{and } T : E' \to \mathbb{R}^n \text{ Borel measurable and such that} \\ &T = T_\sigma \ \mu_\sigma\text{-a.e. on } \mathbb{R}_\sigma, \text{ for } (P_G)_\#\mu\text{-a.e. } \sigma \in S^\circ. \end{aligned} \tag{18.39}$$

Indeed, if (18.39) holds, we have $T_\#\mu = \nu$: indeed, for every $\varphi \in C_c^0(\mathbb{R}^n)$,

$$\begin{aligned} \int_{\mathbb{R}^n} \varphi \, d\nu &= \int_{S^\circ} d[(P_G)_\#\mu](\sigma) \int_{\mathbb{R}^n} \varphi \, d\lambda_\sigma = \int_{S^\circ} d[(P_G)_\#\mu](\sigma) \int_{\mathbb{R}^n} \varphi \circ T_\sigma \, d\mu_\sigma \\ &= \int_{S^\circ} d[(P_G)_\#\mu](\sigma) \int_{\mathbb{R}^n \times \mathbb{R}^n} (\varphi \circ \mathbf{q}) \, d[(\mathbf{id} \times T_\sigma)_\#\mu_\sigma] \\ &= \int_{S^\circ} d[(P_G)_\#\mu](\sigma) \int_{\mathbb{R}^n \times \mathbb{R}^n} (\varphi \circ \mathbf{q} \circ (\mathbf{id} \times T)) \, d\mu_\sigma \\ &= \int_{\mathbb{R}^n} (\varphi \circ \mathbf{q} \circ (\mathbf{id} \times T)) \, d\mu = \int_{\mathbb{R}^n} (\varphi \circ T) \, d\mu = \int_{\mathbb{R}^n} \varphi \, d[T_\#\mu]. \end{aligned}$$

Similarly, by (18.38) we see that

$$\int_{\mathbb{R}^n} |T - x| \, d\mu = \int_{S^\circ} d[(P_G)_\# \mu](\sigma) \int_{\mathbb{R}^n} |T_\sigma(x) - x| \, d\mu_\sigma(x)$$

$$\leq \int_{S^\circ} d[(P_G)_\# \mu](\sigma) \int_{\mathbb{R}^n \times \mathbb{R}^n} |x - y| \, d\gamma_{\rho^c(\sigma)}$$

$$\leq \int_{S^c} d[(r_G)_\# \gamma](\tau) \int_{\mathbb{R}^n \times \mathbb{R}^n} |x - y| \, d\gamma_\tau = \int_{\mathbb{R}^n \times \mathbb{R}^n} |x - y| \, d\gamma.$$

We are thus left to prove (18.39), the "gluing step" in Sudakov's strategy.

Proof of (18.39), *step one*: We prove that

$$\sigma \mapsto (\mathbf{id} \times T_\sigma)_\# \mu_\sigma \text{ is Borel measurable.} \tag{18.40}$$

Let us first of all notice that, since T_σ is an increasing map in the orientation of $\mathbb{R}_\sigma$ induced by σ, spt $[(\mathbf{id} \times T_\sigma)_\# \mu_\sigma]$ is an increasing subset of $\mathbb{R}_\sigma \times \mathbb{R}_\sigma$ (in that same orientation), and thus, by Theorem 16.1-(i) and thanks to $(T_\sigma)_\# \mu_\sigma = \lambda_\sigma$, it holds that

$$(\mathbf{id} \times T_\sigma)_\# \mu_\sigma \left([x, x^*] \times [y, y^*] \right) \tag{18.41}$$

$$= \min \{ \mu_\sigma ((-\infty, x^*]), \lambda_\sigma ((-\infty, y^*]) \} - \min \{ \mu_\sigma ((-\infty, x)), \lambda_\sigma ((-\infty, y)) \}.$$

Here, given $x \in \mathbb{R}_\sigma \subset \mathbb{R}^n$, $(-\infty, x)$ denotes the open half-line in $\mathbb{R}_\sigma$ with x as its *right* endpoint x (where what is "right" is defined by the orientation of σ itself), and $(-\infty, x]$ is similarly defined. This said, to prove (18.40) it is sufficient to fix two arbitrary closed cubes Q_1 and Q_2 in $\mathbb{R}^n$, and prove that

$$\sigma \mapsto f_{12}(\sigma) = (\mathbf{id} \times T_\sigma)_\# \mu_\sigma(Q_1 \times Q_2) \text{ is Borel measurable.} \tag{18.42}$$

Clearly, $\Sigma_0 = \{ \sigma \in S^\circ : \Phi(\rho^c(\sigma)) \text{ intersects both } Q_1 \text{ and } Q_2 \}$ is a closed subset of S°, and $f_{12} = 0$ on $S^\circ \setminus \Sigma_0$. Now, given $\sigma \in \Sigma_0$, let $x_k(\sigma)$ and $y_k(\sigma)$ denote, respectively, the left-most and right-most intersection points between $\Phi(\rho^c(\sigma))$ and ∂Q_k in the orientation induced by σ on $\Phi(\rho^c(\sigma))$. (Notice that, clearly, we could have $x_k(\sigma) = y_k(\sigma)$ if $\Phi(\rho^c(\sigma))$ intersect Q_k at one of its vertexes.) In this way, we have

$$\Phi(\rho^c(\sigma)) \cap Q_1 = [x_1(\sigma), y_1(\sigma)], \qquad \Phi(\rho^c(\sigma)) \cap Q_2 = [x_2(\sigma), y_2(\sigma)],$$

where we are slightly abusing the notation set in Section 18.1 by setting $[x, x] = \{x\}$. Since μ_σ and $\lambda_\sigma = (T_\sigma)_\# \mu_\sigma$ are both concentrated on $\Phi(\rho^c(\sigma))$, we conclude from (18.41) that

$$f_{12}(\sigma) = \min \{ \mu_\sigma ((-\infty, y_1]), \lambda_\sigma ((-\infty, y_2]) \}$$

$$- \min \{ \mu_\sigma ((-\infty, x_1)), \lambda_\sigma ((-\infty, x_2)) \},$$

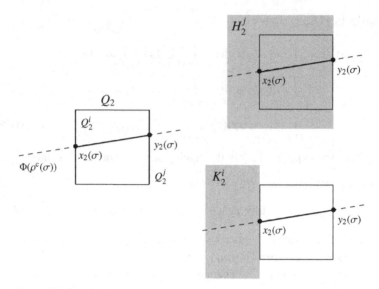

Figure 18.2 Proof of the measurability of $\sigma \mapsto (\mathbf{id} \times T_\sigma)_\# \mu_\sigma$. We have denoted by Q_2^i and Q_2^j the left and right sides of Q_2. The segment σ in the picture is such that $\sigma \in \Sigma_{0,\ell}^i$ (since $x_2(\sigma) \in Q_2^i$) and $\sigma \in \Sigma_{0,r}^j$ (since $y_2(\sigma) \in Q_2^j$). We have $\lambda_\sigma((-\infty, x_2(\sigma)]) = \lambda_\sigma(K_2^i)$, and since K_2^i is independent of the specific $\sigma \in \Sigma_{0,\ell}^i$ under consideration, and $\sigma \mapsto \lambda_\sigma(J)$ is Borel measurable for every fixed $J \in \mathcal{B}(\mathbb{R}^n)$, we conclude that $\sigma \mapsto \lambda_\sigma((-\infty, x_2(\sigma)])$ is Borel measurable. The case of $\lambda_\sigma((-\infty, y_2(\sigma)])$ is analogous, with $\lambda_\sigma((-\infty, y_2(\sigma)]) = \lambda_\sigma(H_2^j)$.

where $x_k = x_k(\sigma)$ and $y_k = y_k(\sigma)$. We claim that each of the four functions appearing in this formula is Borel measurable. Let us prove, for example, that

$$\sigma \in \Sigma_0 \mapsto \lambda_\sigma((-\infty, x_2(\sigma)]) \tag{18.43}$$

$$\sigma \in \Sigma_0 \mapsto \lambda_\sigma((-\infty, y_2(\sigma)]) \tag{18.44}$$

are Borel measurable. To this end (see Figure 18.2), let $\{Q_2^i\}_{i=1}^{2N}$ denote the collection of all the (closed) faces of Q_2, and let $\Sigma_{0,\ell}^i$ and $\Sigma_{0,r}^i$ denote the set of those $\sigma \in \Sigma_0$ such that, respectively, $x_2(\sigma) \in Q_2^i$ or $y_2(\sigma) \in Q_2^i$. Denoting by H_2^i the closed half-space containing Q_2^i and the whole Q_2^i, and by K_2^i the closed half-space containing Q_2^i but disjoint from $Q_2 \setminus Q_2^i$, we see that

$$\text{if } \sigma \in \Sigma_{0,\ell}^i, \text{ then } \lambda_\sigma((-\infty, x_2(\sigma)]) = \lambda_\sigma(K_2^i), \tag{18.45}$$

$$\text{if } \sigma \in \Sigma_{0,r}^i, \text{ then } \lambda_\sigma((-\infty, y_2(\sigma)]) = \lambda_\sigma(H_2^i). \tag{18.46}$$

Since $\sigma \mapsto \lambda_s$ is Borel measurable, and since K_2^i and H_2^i are independent, respectively, on the specific σ appearing (18.45) and (18.46), we conclude that the function in (18.43) is Borel measurable on each $\Sigma_{0,\ell}^i$, and that the one in

(18.44) is Borel measurable on each $\Sigma_{0,r}^i$, for $i = 1, \ldots, 2^N$. Since Σ_0 is covered by both $\{\Sigma_{0,\ell}^i\}_{i=1}^{2^N}$ and $\{\Sigma_{0,r}^i\}_{i=1}^{2^N}$, the proof of (18.40) is thus complete.

Proof of (18.39), *conclusion*: We complete the proof of (18.39) with a rather abstract measure-theoretic argument that can be safely omitted on a first reading. Let Σ be a Borel set of full $(P_G)_\#\mu$-measure in $\mathcal{S}^\circ$ such that $\sigma \mapsto \mu_\sigma$ and $\sigma \mapsto \lambda_\sigma$ are both defined and Borel measurable on Σ. For each $\sigma \in \Sigma$, we can consider an increasing sequence $\{T_\sigma^m\}_m$ of increasing maps $T_\sigma^m : \mathbb{R}_\sigma \to \mathbb{R}_\sigma$ such that $\mathrm{Lip}(T_\sigma^m) \leq m$ and $T_\sigma^m \to T_\sigma$ on $\mathbb{R}_\sigma$ as $m \to \infty$. We then make the following claim (which clearly implies (18.39)):

Claim: there exist increasing sequences of compact sets $\Sigma_j \subset \Sigma$ and $K_j \subset E$ with $(P_G)_\#\mu(\Sigma \setminus \Sigma_j) \to 0$ and $\mu(E \setminus K_j) \to 0$ as $j \to \infty$, and such that the maps $T^m : K_j \to \mathbb{R}^n$, defined by

$$T^m(x) = T_\sigma^m(x) \qquad \text{for } x \in \mathbb{R}_\sigma \cap K_j \text{ and } \sigma \in \Sigma_j \tag{18.47}$$

are well-defined and continuous on K_j.

To begin with, let us notice that $\mathcal{S}_k^\circ = \{]\!] x, y [\![: |x|, |y| \leq k, |x - y| \geq 1/k \}$ ($k \in \mathbb{N}$) are compact sets in $\mathcal{S}^\circ$, that $E_k = P_G^{-1}(\mathcal{S}_k^\circ)$ ($k \in \mathbb{N}$) are bounded Borel subsets of E, and that $\mathcal{S}_k^\circ$ and E_k are monotonically converging, respectively, to $\mathcal{S}^\circ$ and E. Therefore, in proving the claim, we can directly assume that E is a bounded Borel set, $\Sigma \subset \mathcal{S}_k^\circ$ for some $k \in \mathbb{N}$, as well as that $\Sigma = P_G(E)$.

The sets K_j and Σ_j are constructed by repeated applications of Lusin's theorem. In the first one, we infer from $\mu(E) < \infty$ the existence of compact sets $K_j^* \subset E$ such that $\mu(E \setminus K_j^*) \to 0$ as $j \to \infty$ and P_G is continuous on K_j^*. We now apply Lusin's theorem two more times to define the compact sets Σ_j:

Application of Lusin's theorem to the map $\sigma \in \Sigma \mapsto \mathrm{spt}\, \mu_\sigma$: Denote by $\mathbf{K}(\mathbb{R}^n)$ the family of all the compact subsets of $\mathbb{R}^n$. As well-known, this is a complete, separable metric space with the Hausdorff distance between compact sets

$$\mathrm{hd}(K_1, K_2) = \max \left\{ \sup_{x \in K_1} \mathrm{dist}(x, K_2), \sup_{x \in K_2} \mathrm{dist}(x, K_1) \right\},$$

whose σ-algebra of Borel sets is generated[10] by the family

$$\mathbf{K}(\mathbb{R}^n; A) = \left\{ K \in \mathbf{K}(\mathbb{R}^n) : K \cap A \neq \emptyset \right\} \qquad \text{for some } A \subset \mathbb{R}^n \text{ open.}$$

Since E is bounded and, for each $\sigma \in \Sigma$, μ_σ is concentrated on the bounded set $\mathbb{R}_\sigma \cap E$, we have $\mathrm{spt}\, \mu_\sigma \in \mathbf{K}(\mathbb{R}^n)$, and we can consider the map

$$\sigma \in \Sigma \mapsto \mathrm{spt}\, \mu_\sigma \in \mathbf{K}(\mathbb{R}^n).$$

---

[10] See, e.g., [CV77, Corollary II-9, Theorem II-10].

We claim that this map is Borel measurable. Indeed, for each open set $A \subset \mathbb{R}^n$ the map $(\sigma \in \Sigma) \mapsto \mu_\sigma(A)$ is Borel measurable, and therefore

$$\{\sigma \in \Sigma : \operatorname{spt} \mu_\sigma \cap A \neq \emptyset\} = \{\sigma \in \Sigma : \mu_\sigma(A) > 0\}$$

is a Borel subset of S°. Since $(\sigma \in \Sigma) \mapsto (\operatorname{spt} \mu_\sigma \in \mathbf{K}(\mathbb{R}^n))$ is a Borel map from a metric space into a separable metric space, and since $(P_G)_\# \mu(\Sigma)$ is finite, we can apply Lusin's theorem[11] and find an increasing sequence of S°-compact sets $\Sigma_j \subset \Sigma$ such that $(P_G)_\# \mu(S^\circ \setminus \Sigma_j) \to 0$ as $j \to \infty$ and such that

$$(\sigma \in \Sigma_j) \mapsto (\operatorname{spt} \mu_\sigma \in \mathbf{K}(\mathbb{R}^n)) \text{ is continuous}, \qquad \forall j \in \mathbb{N}. \quad (18.48)$$

We now further refine the sets Σ_j.

Application of Lusin's theorem to the maps $\sigma \in \Sigma \mapsto (\mathbf{id} \times T_\sigma^m)_\# \mu_\sigma$: Arguing as in the proof of (18.40) we see that each of these maps is Borel measurable. In particular, given $\mathcal{F}$ a countable subset of $C_c^0(\mathbb{R}^n \times \mathbb{R}^n)$ that is dense in the uniform topology, since $(P_G)_\# \mu(\Sigma)$ is finite, we can apply Lusin's theorem to each finite valued Borel function $\sigma \in \Sigma \mapsto \int_{\mathbb{R}^n} \psi(x, T_\sigma^m(x)) \, d\mu_\sigma(x)$ with $m \in \mathbb{N}$ and $\psi \in \mathcal{F}$. By the separability of $\mathcal{F}$ and up to further restricting the compact sets $\Sigma_j \subset \Sigma$, we can thus maintain that $(P_G)_\# \mu(S^\circ \setminus \Sigma_j) \to 0$ as $j \to \infty$ while adding that, for every $\psi \in C_c^0(\mathbb{R}^n \times \mathbb{R}^n)$ and $m \in \mathbb{N}$,

$$(\sigma \in \Sigma_j) \mapsto \int_{\mathbb{R}^n} \psi(x, T_\sigma^m(x)) \, d\mu_\sigma(x) \text{ is continuous}. \quad (18.49)$$

Conclusion of the proof: We are finally in the position to prove the claim. Let

$$K_j^{**} = \bigcup_{\sigma \in \Sigma_j} \operatorname{spt} \mu_\sigma.$$

Notice that is a compact subset of E since $\sigma \mapsto \operatorname{spt} \mu_\sigma$ is continuous on the S°-compact set Σ_j. Moreover $\mu(E \setminus K_j^{**}) \to 0$ as $j \to \infty$, since, by the disintegration formula $\mu = \mu_\sigma \otimes [(P_G)_\# \mu]$, we have

$$\mu(E \setminus K_j^{**}) = \int_\Sigma \mu_{\sigma^*}\left(E \setminus \bigcup_{\sigma \in \Sigma_j} \operatorname{spt} \mu_\sigma\right) d[(P_G)_\# \mu](\sigma^*)$$

$$\leq \int_{\Sigma \setminus \Sigma_j} \mu_{\sigma^*}(E) \, d[(P_G)_\# \mu](\sigma^*) = (P_G)_\# \mu(S^\circ \setminus \Sigma_j).$$

Now consider the Borel sets

$$F_j = P_G^{-1}(\Sigma_j) = \bigcup_{\sigma \in \Sigma_j} \Phi(\sigma),$$

---

[11] For example, in the general version stated in [Fed69, 2.3.5].

so that $T^m : F_j \to \mathbb{R}^n$ is well-defined by setting $T^m(x) = T^m_\sigma(x)$ if $x \in \Phi(\sigma)$, $\sigma \in \Sigma_j$. We have $\mu(E \setminus F_j) = (P_G)_\# \mu(S^\circ \setminus \Sigma_j) \to 0$ as $j \to \infty$, and given compact sets $K^{***}_j \subset F_j$ such that $\mu(F_j \setminus K^{***}_j) \to 0$ as $j \to \infty$, we see that T^m is well defined on the compact sets

$$K_j = K^*_j \cap K^{**}_j \cap K^{***}_j.$$

Since, clearly, we have $\mu(E \setminus K_j) \to \infty$ as $j \to \infty$, we are left to show that T^m is continuous on K_j. Indeed, let $x_h, x_0 \in K_j$ with $x_h \to x_0$ as $h \to \infty$. By the properties of F_j, K^*_j, K^{**}_j and K^{***}_j, $\sigma_h = P_G(x_h)$ and $\sigma_0 = P_G(x_0)$ are such that $\sigma_h, \sigma_0 \in \Sigma_j$ with

$$x_h \in \operatorname{spt} \mu_{\sigma_h} \subset \mathbb{R}_{\sigma_h}, \qquad x_0 \in \operatorname{spt} \mu_{\sigma_0} \subset \mathbb{R}_{\sigma_0}, \qquad \sigma_h \to \sigma_0 \text{ in } S^\circ. \quad (18.50)$$

Let us now assume by contradiction that there is $\delta > 0$ such that

$$|T^m(x_h) - T^m(x_0)| \geq \delta, \qquad \forall h.$$

By (18.50), and since $\operatorname{Lip}(T^m_\sigma; \mathbb{R}_\sigma) \leq m$ for every $\sigma \in \Sigma$, if $x \in B_r(x_0) \cap \operatorname{spt} \mu_{\sigma_h}$ with $r < \delta/2m$, then

$$|T^m_{\sigma_h}(x) - T^m_{\sigma_0}(x_0)| \geq \delta - |T^m_{\sigma_h}(x) - T^m_{\sigma_h}(x_h)| \geq \delta - m\,r > \frac{\delta}{2}.$$

Therefore, given non-negative functions

$$\psi_1 \in C^0_c(B_r(x_0)), \qquad \psi_2 \in C^0_c(B_{\delta/2}(T^m(x_0)))$$

with $\psi_1(x_0) = \psi_2(T^m(x_0)) = 1$, by (18.51) we find that

$$\int_{\mathbb{R}^n \times \mathbb{R}^n} \psi_1(x)\,\psi_2(y)\,d[(\mathrm{id} \times T^m_{\sigma_h})_\# \mu_{\sigma_h}]$$
$$= \int_{B_r(x_0) \cap \operatorname{spt} \mu_{\sigma_h}} \psi_1(x)\,\psi_2(T^m_{\sigma_h}(x))\,d\mu_{\sigma_h}(x) = 0,$$

while $|T^m_{\sigma_0}(x) - T^m_{\sigma_0}(x_0)| \leq m\,r < \delta/2$ for $x \in B_r(x_0) \cap \operatorname{spt} \mu_{\sigma_0}$ gives

$$\int_{\mathbb{R}^n \times \mathbb{R}^n} \psi_1(x)\,\psi_2(y)\,d[(\mathrm{id} \times T^m_{\sigma_0})_\# \mu_{\sigma_0}]$$
$$= \int_{B_r(x_0) \cap \operatorname{spt} \mu_{\sigma_0}} \psi_1(x)\,\psi_2(T^m_{\sigma_0}(x))\,d\mu_{\sigma_0}(x) > 0,$$

up to further decrease the value of r. Since $\sigma_h \to \sigma_0$ in S° as $h \to \infty$, we have reached a contradiction with (18.49). $\qquad\square$

18.3 Kantorovich Potentials Are Countably $C^{1,1}$-Regular

We now prove that, for every 1-Lipschitz function f, ∇f has the countable Lipschitz property on the transport set with endpoints $\mathcal{T}_c(f)$, and then use this information to discuss the validity of the assumptions of Theorem 18.7 when $G \subset G(f) = \{(x,y) : f(y) = f(x) + |x - y|\}$.

Theorem 18.9 (Countable $C^{1,1}$-regularity of the Kantorovich potentials) *If $f : \mathbb{R}^n \to \mathbb{R}$ is a 1-Lipschitz function with set of differentiability points F, then*

$$\nabla f \text{ has the countable Lipschitz property on } F \cap \mathcal{T}_c(f). \qquad (18.51)$$

Moreover, if $G \subset G(f)$ is such that $\mathcal{T}(G)$, $\mathcal{T}_c(G)$, and $P_G : \mathcal{T}(G) \to S^\circ$ are Borel measurable, then the following properties hold:

(i) *G satisfies the non-crossing condition;*

(ii) *for every $x \in E = F \cap \mathcal{T}_c(G)$, there is a unique $\sigma \in \Theta_{\max}(G)$ with $x \in \Phi(\rho^c(\sigma))$, the corresponding extensions of P_G and τ_G from $\mathcal{T}(G)$ to E are both Borel measurable, and $\tau_G = (\nabla f)|_E$ has the countable Lipschitz property on E;*

(iii) *$\mathcal{L}^n(\mathcal{T}_c(G) \setminus \mathcal{T}(G)) = 0$.*

Proof We first prove that ∇f has the countable Lipschitz property on $\mathcal{T}_\ell(f) \cap F$; then, by a symmetric argument, the same will hold on $\mathcal{T}_r(f) \cap F$, and thus on $\mathcal{T}_c(f) \cap F$. Given $e \in \mathbb{S}^{n-1}$ and $\alpha \in \mathbb{R}$, we consider the open half-space

$$H_{e,\alpha} = \{x \in \mathbb{R}^n : x \cdot e < \alpha\},$$

the set $Y_{e,\alpha}$ of right endpoints of transport rays of f that lie *outside* of $H_{e,\alpha}$,

$$Y_{e,\alpha} = \Big\{y \in \mathbb{R}^n : y \cdot e \geq \alpha, \exists\, x \neq y \text{ s.t. } (x,y) \in G(f)\Big\},$$

and the set

$$E_{e,\alpha} = F \cap H_{e,\alpha} \cap \bigcup \Big\{[\![x,y[\![: (x,y) \in G(f), y \in Y_{e,\alpha}\Big\},$$

obtained by intersecting $F \cap H_{e,\alpha}$ with the union of the left-closed/right-open segments defined by $G(f)$ having their right-end point in $Y_{e,\alpha}$. Since $G' = \{(x,y) \in G(f) : y \in Y_{e,\alpha}\}$ is closed in $\mathbb{R}^n \times \mathbb{R}^n$, by Proposition 18.5 we have that $E_{e,\alpha} = F \cap H_{e,\alpha} \cap \mathcal{T}_\ell(G')$ is a Borel set for every choice of e and α. Moreover, by using countably many choices of e and α, we can cover $F \cap \mathcal{T}_\ell(f)$ with sets of the form $E_{e,\alpha}$. We have thus reduced to prove that ∇f has the countable Lipschitz property on $E_{e,\alpha}$ for fixed $e \in \mathbb{S}^{n-1}$ and $\alpha \in \mathbb{R}$.

To this end, let us define $g_{e,\alpha} : \mathbb{R}^n \to \mathbb{R}$ by setting

$$g_{e,\alpha}(z) = \sup\Big\{f(y) - |z - y| : y \in Y_{e,\alpha}\Big\}, \qquad z \in \mathbb{R}^n.$$

Clearly, $g_{e,\alpha}$ is a Lipschitz function with $\mathrm{Lip}(g_{e,\alpha}) \le 1$: in addition to that,

$$g_{e,\alpha} \le f \text{ on } \mathbb{R}^n \text{ with } g_{e,\alpha} = f \text{ on } E_{e,\alpha}, \tag{18.52}$$

$$\begin{cases} \forall \beta < \alpha \text{ there is } C(\alpha,\beta) > 0 \text{ s.t.} \\ z \mapsto g_{e,\alpha}(z) + C(\alpha,\beta)\,|z|^2 \text{ is convex on } H_{e,\beta}. \end{cases} \tag{18.53}$$

To prove (18.52): since $f(y) - |z - y| \le f(z)$ for every $z, y \in \mathbb{R}^n$ we obviously have $g_{e,\alpha} \le f$ on $\mathbb{R}^n$; while if $z \in E_{e,\alpha}$, then $z \in [\![x, y]\!]$ for some $(x, y) \in G(f)$ with $x \ne y$ and $y \in Y_{e,\alpha}$, so that $z = y + t(x - y)$ for some $t \in (0, 1)$ and, and thanks to $(x, y) \in G(f)$ we find

$$f(z) = f(y + t(x - y)) = f(y) - t\,|x - y| = f(y) - |z - y| \le g_{e,\alpha}(z) \le f(z).$$

To prove (18.53): Given $y \in Y_{e,\alpha}$, we have that $|z - y| \ge \alpha - \beta$ for every $z \in H_{e,\beta}$: hence, the second derivatives of $z \mapsto |z - y|$ are uniformly bounded by $C(n)/(\alpha - \beta)$ on $z \in H_{e,\beta}$, and therefore

$$z \in H_{e,\beta} \mapsto f(y) - |z - y| + C\,|z|^2,$$

is convex on $H_{e,\beta}$ provided C is large enough depending on n, α and β only. We conclude since a supremum of convex functions is a convex function.

By Theorem 7.1 (gradients of convex functions have locally bounded variation), by (18.53), and with C large enough depending on n, α and β, we have

$$\nabla g_{e,\alpha} + 2C\,\mathrm{Id} \in BV_{\mathrm{loc}}(H_{e,\beta}; \mathbb{R}^n) \qquad \forall \beta < \alpha,$$

so that $\nabla g_{e,\alpha} \in BV_{\mathrm{loc}}(H_{e,\alpha}; \mathbb{R}^n)$. A classical result about maximal functions (see Theorem A.2 in the Appendix) then implies that $\nabla g_{e,\alpha}$ has the countable Lipschitz property on $H_{e,\alpha}$. Since, by (18.52), it holds that $\nabla g_{e,\alpha} = \nabla f$ a.e. on $E_{e,\alpha} \subset H_{e,\alpha}$, we conclude that ∇f has the countable Lipschitz property on $E_{e,\alpha}$.

We now consider $G \subset G(f)$ such that $\mathcal{T}(G)$ and $P_G : \mathcal{T}(G) \to \mathcal{S}^\circ$ are Borel measurable, and prove conclusions (i), (ii), and (iii).

Proof of (i): By Proposition 18.6, $G(f)$ satisfies the non-crossing condition (18.21), which is thus satisfied by G thanks to $G \subset G(f)$.

Proof of (ii): The inclusion $G \subset G(f)$ implies that $\Theta(G) \subset \Theta(f)$, and thus that $\mathcal{T}(G) \subset \mathcal{T}(f)$ and $\mathcal{T}_c(G) \subset \mathcal{T}_c(f)$. Moreover, every $\sigma \in \Theta^{\max}(G)$ is a (possibly strict) sub-segment of some $\sigma' \subset \Theta^{\max}(f)$. We conclude by combining these properties with the possibility of extending P_f and ∇f from $\mathcal{T}(f)$ to $F \cap \mathcal{T}_c(f)$ (Proposition 18.6) with the fact that ∇f has the countable Lipschitz property on $F \cap \mathcal{T}_c(f)$.

Proof of (iii): By conclusion (ii), G satisfies assumptions (a) and (b) of Theorem 18.7 with the Borel set $E = F \cap \mathcal{T}_c(G)$. Let us assume for a moment that $\mathcal{L}^n(E) < \infty$, and apply conclusion (18.32) in Theorem 18.7 to the disintegration $\{\mu_\sigma\}_\sigma$ of the finite measure $\mu = \mathcal{L}^n \llcorner E$ with respect to P_G : $E \to S°$: we find, in particular, that $\mu_\sigma \ll \mathcal{H}^1 \llcorner \mathbb{R}_\sigma$. Since, by assumption, $Z = F \cap [\mathcal{T}_c(G) \setminus \mathcal{T}(G)]$ is a Borel set, we can apply the disintegration formula to deduce that

$$\mathcal{L}^n(Z) = \int_{S°} d\,[(P_G)_\# \mu](\sigma) \int_{\mathbb{R}_\sigma \cap Z} d\,\mu_\sigma,$$

where (for $(P_G)_\# \mu$-a.e. $\sigma \in S°$) $\mu_\sigma(\mathbb{R}_\sigma \cap Z) = 0$ since $\mathbb{R}_\sigma \cap Z$ consists of at most two points and $\mu_\sigma \ll \mathcal{H}^1 \llcorner \mathbb{R}_\sigma$. Considering that $\mathcal{L}^n(\mathbb{R}^n \setminus F) = 0$ by Rademacher's theorem, we have proved $\mathcal{L}^n(\mathcal{T}_c(G) \setminus \mathcal{T}(G)) = 0$. When $\mathcal{L}^n(E) = +\infty$ we can of course repeat these considerations with $\mu = \mathcal{L}^n \llcorner (E \cap B_R)$ $(R > 0)$, and then send $R \to \infty$, to obtain the same conclusion. □

18.4 Proof of the Sudakov Theorem

Proof of Theorem 18.1   We are given $\mu, \nu \in \mathcal{P}_1(\mathbb{R}^n)$ with $\mu \ll \mathcal{L}^n$, and want to construct an optimal transport map T in $\mathbf{M}_1(\mu, \nu)$. We let f be a Kantorovich potential from μ to ν (with F the set of differentiability of f), γ an optimal plan in $\mathbf{K}_1(\mu, \nu)$, and set $G = \operatorname{spt} \gamma$, so that $G \subset G(f) = \{(x,y) : f(y) = f(x) + |y - x|\}$ by (18.4). By Proposition 18.5, since G is closed, $\mathcal{T}(G)$, $\mathcal{T}_c(G)$, $\mathcal{T}_r(G)$, $\mathcal{T}_\ell(G)$, P_G and r_G are all Borel measurable. Hence, by Theorem 18.9, G satisfies the non-crossing condition, P_G and τ_G can be properly extended from $\mathcal{T}(G)$ to the Borel set $F \cap \mathcal{T}_c(G)$, and $\tau_G = (\nabla f)|_{F \cap \mathcal{T}_c(G)}$ has the countable Lipschitz property on $F \cap \mathcal{T}_c(G)$; moreover, $\mathcal{L}^n(\mathcal{T}_c(G) \setminus \mathcal{T}(G)) = 0$.

Step one: If $\mathcal{ML}(G) = \emptyset$, then Theorem 18.7 can be applied to find an optimal transport map T in $\mathbf{M}_1(\mu, \nu)$ such that $\gamma_T = (\mathrm{id} \times T)_\# \mu$ is optimal in $\mathbf{K}_1(\mu, \nu)$, and with T monotone increasing along the transport rays of G.

Step two: Let us now assume that $M = \mathcal{ML}(G) \ne \emptyset$, and let

$$G^* = G \setminus \Delta(M), \qquad \text{where } \Delta(E) = \{(x,y) : x = y \in E\} \text{ for } E \subset \mathbb{R}^n.$$

If $\gamma(G^*) = 0$, then we conclude by taking $T = \mathrm{id}$. Assuming that $\gamma(G^*) > 0$, let

$$\gamma^* = \frac{\gamma \llcorner G^*}{\gamma(G^*)}, \qquad \mu^* = \mathbf{p}_\# \gamma^*, \qquad \nu^* = \mathbf{q}_\# \gamma^*.$$

We claim that, for every Borel set $E \subset \mathbb{R}^n$,

$$\mu^* = \frac{\mu \llcorner (\mathbb{R}^n \setminus M)}{\gamma(G^*)}, \qquad \nu^* = \frac{\nu - \mu \llcorner M}{\gamma(G^*)}. \tag{18.54}$$

Indeed, since $\gamma(\Delta(E \cap M)) = \mu(E \cap M)$ for every $E \subset \mathbb{R}^n$, we easily find that

$$\mathbf{p}_\#[\gamma \llcorner G^*](E) = \gamma(G \cap (E \times \mathbb{R}^n)) - \gamma(\Delta(M) \cap (E \times \mathbb{R}^n))$$
$$= \mu(E) - \mu(E \cap M),$$
$$\mathbf{q}_\#[\gamma \llcorner G^*](E) = \gamma(G \cap (\mathbb{R}^n \times E)) - \gamma(\Delta(M) \cap (\mathbb{R}^n \times E))$$
$$= \nu(E) - \mu(E \cap M),$$

where we have also used $\Delta(M) \cap (\mathbb{R}^n \times E) = \Delta(M) \cap (E \times \mathbb{R}^n)$.

We now check that we can apply Theorem 18.7 to G^*, μ^*, ν^* and γ^*. Clearly γ^* is concentrated on G^*, and G^* is a Borel set (G closed implies that M, and thus $\Delta(M)$, is a Borel set, see (18.63)) that satisfies the non-crossing condition (by $G^* \subset G \subset G(f)$) and $\Theta(G^*) = \Theta(G)$ (by construction). In particular, $\mathcal{T}(G) = \mathcal{T}(G^*)$, $\mathcal{T}_\star(G) = \mathcal{T}_\star(G^*)$ ($\star \in \{c,r,\ell\}$), $P_G = P_{G^*}$, $r_G = r_{G^*}$, and $\tau_G = \tau_{G^*}$ are all Borel regular, $\mathcal{T}_\ell(G) \setminus \mathcal{T}(G) = \mathcal{T}_\ell(G^*) \setminus \mathcal{T}(G^*)$ is Lebesgue negligible, and τ_{G^*} has the countable Lipschitz property on $\mathcal{T}(G) = \mathcal{T}(G^*)$. Finally, by construction, $\mathcal{ML}(G^*) = \emptyset$.

By Theorem 18.7 we find $T^* : \mathcal{T}(G) \to \mathbb{R}^n$ such that T^* transports μ^* into ν^* and (18.33) holds (with T^*, μ^* and γ^* in place of T, μ and ν). Taking into account (18.54) we thus conclude

$$\int_{\mathcal{T}(G) \setminus M} |T^*(x) - x| \, d\mu(x) \le \int_{G \setminus \Delta(M)} |x - y| \, d\gamma(x,y) = \int_{\mathbb{R}^n \times \mathbb{R}^n} |x - y| \, d\gamma(x,y). \tag{18.55}$$

Now let us consider the Borel map $T : \mathcal{T}(G) \cup M \to \mathbb{R}^n$ defined by setting $T = T^*$ on $\mathcal{T}(G)$ and $T = \mathbf{id}$ on M. By (18.17) and $\mathcal{L}^n(\mathcal{T}_\ell(G) \setminus \mathcal{T}(G)) = 0$ we see that μ is concentrated on $\mathcal{T}(G) \cup M$, so that T transports μ, with cost

$$\int_{\mathbb{R}^n} |T(x) - x| \, d\mu(x) = \int_{\mathcal{T}(G) \setminus M} |T^*(x) - x| \, d\mu(x). \tag{18.56}$$

By (18.55) and (18.56), we are going to deduce that T is optimal in the Monge problem as soon as we show that $T_\#\mu = \nu$. Indeed, $T^*_\#\mu^* = \nu^*$ and (18.54) imply $\mu((T^*)^{-1}(E)) = \nu(E) - \mu(E \cap M)$ for every $E \in \mathcal{B}(\mathbb{R}^n)$, so that

$$T^{-1}(E) = [T^{-1}(E) \setminus M] \cup [T^{-1}(E) \cap M] = (T^*)^{-1}(E) \cup [E \cap M],$$

gives $\mu(T^{-1}(E)) = \nu(E) - \mu(E \cap M) + \mu(E \cap M) = \nu(E)$, as required.

Step three: When $\mu = \rho \, d\mathcal{L}^n$ and $\nu = \sigma \, d\mathcal{L}^n$ we can also apply Sudakov's argument inverting the roles of μ and ν. If f is a Kantorovich potential from μ to ν, then $-f$ is a Kantorovich potential from ν to μ, and denoting by S the map resulting from Sudakov's argument applied to $(-f, \nu, \mu)$, and taking into account Theorem 16.1-(vi) and the way T and S have been constructed, we easily infer that $T \circ S = \mathbf{id}$ ν-a.e. on $\mathbb{R}^n$ and $S \circ T = \mathbf{id}$ μ-a.e. on $\mathbb{R}^n$. $\qquad \square$

18.5 Some Technical Measure-Theoretic Arguments

We finally present the proofs of Proposition 18.3, of Proposition 18.5.

Proof of Proposition 18.3   We want prove that if Σ is a countable union of closed subset of $S^\star$, then $\Phi(\Sigma)$, the image of Σ through the realization map Φ (recall 18.11), is a Borel set in $\mathbb{R}^n$.

Step one: Given $\Sigma \subset S^\star$, define the **set of bounded sub-segments of Σ** as

$$\text{Sub}(\Sigma) = \left\{ \sigma \in S^\star : \sigma \text{ is a bounded sub-segment of some } \sigma^* \in \Sigma \right\}, \quad (18.57)$$

and notice that
$$\Phi(\Sigma) = \Phi(\text{Sub}(\Sigma)), \qquad \forall \Sigma \subset S^\star. \qquad (18.58)$$

We claim that

$$\Sigma \text{ is closed in } S^\star \quad \Rightarrow \quad \text{Sub}(\Sigma) \text{ is closed in } S^\star. \qquad (18.59)$$

Indeed, let us first of all notice that

$$\begin{cases} \text{if } \sigma_j \text{ is a sub-segment of } \sigma_j^*, \text{ and } \sigma_j \to \sigma \text{ in } S^\star \text{ as } j \to \infty, \\ \text{then } \exists \sigma^* \in S^\star \text{ s.t., up to extracting subsequences,} \\ \sigma_j^* \to \sigma^* \text{ in } S^\star \text{ as } j \to \infty, \text{ and } \sigma \text{ is a sub-segment of } \sigma^*. \end{cases} \qquad (18.60)$$

Then, let σ_j be a $S^\star$-convergent sequence in $\text{Sub}(\Sigma)$, with limit σ, and let $\sigma_j^* \in \Sigma$ be such that σ_j is a sub-segment of σ_j^*. By applying (18.60), and by noticing that Σ being closed implies $\sigma^* \in \Sigma$, we find that $\sigma \in \text{Sub}(\Sigma)$, thus proving (18.59).

Step two: We now complete the proof in the case $S^\star = S^\circ$. Let Σ be as in the statement, so that

$$\Sigma = \bigcup_{i=1}^{\infty} \Sigma_i, \qquad \Sigma_i \text{ is closed in } S^\circ,$$

and let

$$\Sigma_i^k = \left\{ \sigma \in \text{Sub}(\Sigma_i) : \sigma =]\!]x, y[\![, |x|, |y| \le k, |x - y| \ge \frac{1}{k} \right\}.$$

Denoting by $I_\varepsilon(A)$ the open ε-neighborhood of $A \subset \mathbb{R}^n$, we obtain the desired conclusion by showing that

$$\Phi(\text{Sub}(\Sigma)) = \bigcup_{i=1}^{\infty} \bigcup_{k=1}^{\infty} \bigcup_{h=3}^{\infty} \bigcap_{j=h}^{\infty} \bigcup_{]\!]x, y[\![\in \Sigma_i^k} I_{|x-y|/2j}\left(\left]x + \frac{y-x}{h}, y - \frac{y-x}{h}\right[\right),$$
$$(18.61)$$

since, indeed, (18.58) and (18.61) characterize $\Phi(\Sigma)$ as a countable union of countable intersections of open sets, and thus prove that $\Phi(\Sigma)$ is a Borel set.

Let us prove (18.61). On the one hand, if $z \in \Phi(\mathrm{Sub}(\Sigma))$, then $z \in \,]x,y[$ for some $]\!]x,y[\![\,\in \mathrm{Sub}(\Sigma_i)$, thus there exist integers $k \geq 1$ and $h \geq 3$ such that

$$|x|,|y| \leq k, \qquad |x - y| \geq \frac{1}{k}, \qquad z \in \,\Big]x + \frac{y-x}{h}, y - \frac{y-x}{h}\Big[$$

and the inclusion $\subset$ in (18.61) holds. On the other hand let $z \in \mathbb{R}^n$ be such that there exist integers $i, k \geq 1$ and $h \geq 3$ such that, for every $j \geq h$, one can find $\sigma_j =]\!]x_j, y_j[\![\,\in \Sigma_i^k$ such that

$$z \in I_{|x_j - y_j|/2j}\left(\Big]x_j + \frac{y_j - x_j}{h}, y_j - \frac{y_j - x_j}{h}\Big[\right). \tag{18.62}$$

We now notice that Σ_i^k is *compact* in $\mathcal{S}^\circ$: indeed, Σ_i^k is the intersection of the $\mathcal{S}^\circ$-closed set $\mathrm{Sub}(\Sigma_i)$ (recall (18.59)) with the $\mathcal{S}^\circ$-compact set $\{]\!]x, y[\![\,: |x|,|y| \leq k, |x - y| \geq 1/k\}$ (the uniform lower bound on $|x - y|$ is used here). By $\mathcal{S}^\circ$-compactness of Σ_i^k we find that, up to extracting a not relabeled subsequence in j, $\sigma_j \to \sigma$ in $\mathcal{S}^\circ$ as $j \to \infty$, for some $\sigma =]\!]x, y[\![\,\in \Sigma_i^k$ corresponding to $x, y \in \mathbb{R}^n$ with $|x|,|y| \leq k$ and $|x - y| \geq 1/k$. We claim that $z \in \,]x,y[$. Indeed, thanks to $|x_j - y_j| \leq 2k$, (18.62) implies the existence, for each $j \geq 3$, of $t_j \in (0,1)$ such that

$$\Big|z - \big(x_j + t_j(y_j - x_j)\big)\Big| < \frac{k}{j}, \qquad \frac{1}{h} < t_j < 1 - \frac{1}{h}.$$

Up to extracting a final subsequence in j, we can assume that $t_j \to t \in [1/h, 1 - (1/h)]$, and therefore conclude that $z = x + t\,(y - x) \in \,]x,y[$, as claimed. We have thus proved that $z \in \Phi(\sigma)$ where $\sigma \in \Sigma_i^k \subset \mathrm{Sub}(\Sigma_i) \subset \mathrm{Sub}(\Sigma)$.

Step three: The argument of step two can be adapted to cover the cases of $\mathcal{S}^\star$ with $\star \in \{\mathrm{c}, \ell, \mathrm{r}\}$. For example, in the case when $\mathcal{S}^\star = \mathcal{S}^\ell$, is sufficient to replace

$$\Big]x + \frac{y-x}{h}, y - \frac{y-x}{h}\Big[\qquad \text{by} \qquad \Big[x, y - \frac{y-x}{h}\Big[$$

in (18.61), while in the case $\mathcal{S}^\star = \mathcal{S}^c$ we do not need the index h and just work with $[x,y]$ in (18.61). □

Proof of Proposition 18.5 We want to prove that if G is a closed subset of $\mathbb{R}^n \times \mathbb{R}^n$, then $\mathcal{T}(G)$, $\mathcal{T}_c(G)$, $\mathcal{T}_\ell(G)$, $\mathcal{T}_r(G)$ and $\mathcal{ML}(G)$ are Borel sets in $\mathbb{R}^n$. Moreover, if the non-crossing condition (18.21) holds, then the map $P_G : \mathcal{T}(G) \to \mathcal{S}^\circ$ is well-defined and Borel measurable. Of course, the identity

$$\mathcal{ML}(G) = \mathbf{p}(G \cap \Delta) \setminus \mathbf{p}(G \setminus \Delta), \tag{18.63}$$

(where $\Delta = \{(x,y) \in \mathbb{R}^n \times \mathbb{R}^n : x = y\}$) shows immediately that $\mathcal{ML}(G)$ is a Borel set if G is closed. The other conclusions require more work.

Step one: Given $\Sigma \subset S^\circ$ and $r > 0$, we let

$$\text{Join}(\Sigma, r) \subset S^\circ,$$

denote the family of those open oriented segments $]\!]x, y[\![$ for which there exist $z, w \in \mathbb{R}^n$ such that (i) $]\!]x, z[\![$ and $]\!]w, y[\![$ belong to Σ; (ii) the intersection between $]x, z[$ and $]w, y[$ is the open segment $]w, z[$; and (iii) $|z - w| \geq r$. In plain terms, *segments in* $\text{Join}(\Sigma, r)$ *are obtained by joining segments in* Σ *whose realizations have an oriented overlapping of length at least* r: in particular, all the segments in Σ of length at least r are contained in $\text{Join}(\Sigma, r)$ (although Σ may contain shorter segments, and thus not be contained in $\text{Join}(\Sigma, r)$), and every segment in $\text{Join}(\Sigma, r)$ has length at least r. If we extend Σ by adding all the segments in $\text{Join}(\Sigma, r)$, and thus consider

$$\text{Ext}(\Sigma, r) = \Sigma \cup \text{Join}(\Sigma, r),$$

then

$$\begin{cases} \Sigma \text{ is closed in } S^\circ \\ \Phi(\Sigma) \text{ bounded} \end{cases} \Rightarrow \begin{cases} \text{Ext}(\Sigma, r) \text{ is closed in } S^\circ \\ \text{with } \Phi(\text{Ext}(\Sigma, r)) \text{ bounded in } \mathbb{R}^n. \end{cases}$$
$$(18.64)$$

We just need to prove that $\text{Join}(\Sigma, r)$ is closed in S°. Indeed, let $\sigma_j =]\!]x_j, y_j[\![$ be a sequence in $\text{Join}(\Sigma, r)$ associated to points z_j, w_j satisfying (i), (ii) and (iii), and assume that $\sigma_j \to \sigma$ in S°. Since $x_j, y_j \in \Phi(\Sigma)$, which is bounded, it must be $\sigma =]\!]x, y[\![$ for some $x, y \in \mathbb{R}^n$, $x \neq y$. The fact that $]\!]x_j, z_j[\![$ and $]\!]w_j, y_j[\![$ lie on a same oriented line implies that, up to extracting subsequences, $z_j \to z$ and $w_j \to w$ in $\mathbb{R}^n$, with $|z - w| \geq r$ and with $x < w < z < y$ in the orientation of the line generated by σ, and with $]\!]x, z[\![$ and $]\!]w, y[\![$ in Σ thanks to the facts that $]\!]x_j, z_j[\![$ and $]\!]w_j, y_j[\![$ are in Σ and that Σ is closed. This shows that $\sigma \in \text{Join}(\Sigma, r)$, and thus (18.64) is proved.

Step two: We notice that if $\Sigma \subset \text{Sub}(\Theta^{\max}(G))$, then

$$\text{Sub}(\Sigma) \subset \text{Sub}(\Theta^{\max}(G)), \quad \text{Ext}(\Sigma, r) \subset \text{Sub}(\Theta^{\max}(G)) \quad \forall r > 0.$$
$$(18.65)$$

The simple proof is omitted.

Step three: We claim that

$$\text{Sub}(\Theta^{\max}(G)) = \bigcup_{k=1}^{\infty} \Theta_{k,k},$$
$$(18.66)$$

where $\Theta_{k,k}$ is the sequence of closed subsets of S° defined as follows: we set

$$\Theta_k = \left\{]\!]x, y[\![: (x, y) \in G, x \neq y, |x|, |y| \leq k \right\} \subset \Theta(G), \quad k \in \mathbb{N},$$

for the set of all the transport rays of G with end-points in $\text{Cl}(B_k)$, and then define

$$\Theta_{k,0} = \text{Sub}(\Theta_k),$$
$$\Theta_{k,h} = \text{Sub}\big(\text{Ext}(\Theta_{k,h-1}, 1/h)\big), \qquad 1 \le h \le k.$$

Notice that, at step h, we define $\Theta_{k,h}$ by first adding to $\Theta_{k,h-1}$ all the segments obtained by joining segments in $\Theta_{k,h-1}$ with oriented overlapping of length at least $1/h$, and then by taking all the possible sub-segments of the resulting family. For this reason, it is easily seen that $\Theta_{k,h}$ is monotone increasing both in k and in h with respect to set inclusion. The fact that each $\Theta_{k,h}$ is closed in $\mathcal{S}^\circ$ is immediate from (18.59) and (18.64), and the inclusion in $\text{Sub}(\Theta^{\max}(G))$ follows from (18.65). We are thus left to prove that if σ is a sub-segment of some maximal transport ray $\sigma^* \in \Theta^{\max}(G)$, then $\sigma \in \Theta_{k,k}$ for k large enough. To this end, let us notice that by definition of maximal transport ray we can find countably many transport rays σ_j of G, such that $\Phi(\sigma^*)$ is covered by $\{\Phi(\sigma_j)\}_{j\in\mathbb{N}}$. Since $\Phi(\sigma)$ is bounded, we can find an integer N such that $\Phi(\sigma)$ is covered by $\{\Phi(\sigma_j)\}_{j=1}^N$; moreover, by connectedness of $\Phi(\sigma)$ and up to rearranging in the index j, we find that for each $j = 1,\ldots,N-1$, $\Phi(\sigma_j)$ has a positive overlapping with $\Phi(\sigma_{j+1})$; we denote by ε the corresponding infimum overlapping length, which, evidently, is positive. We now take k_0 large enough so that $1/k_0 < \varepsilon$ and $\sigma_j \in \Theta_{k_0}$ for every $j = 1,\ldots,N$, and then pick $k = k_0 + N - 1$. Since σ is a sub-segment of the segment obtained by joining $(N-1)$-segments in Θ_{k_0} with overlapping $1/k_0$, the choice of k ensures that $\sigma \in \Theta_{k,k}$.

Step four: Let Σ be a closed subset of $\mathcal{S}^\circ$, and notice that, by (18.66),

$$\Phi(\Sigma \cap \Theta^{\max}(G)) = \Phi\big(\Sigma \cap \text{Sub}(\Theta^{\max}(G))\big)$$
$$= \Phi\Big(\Sigma \cap \bigcup_{k=1}^\infty \Theta_{k,k}\Big) = \Phi\Big(\bigcup_{k=1}^\infty [\Sigma \cap \Theta_{k,k}]\Big).$$

Since $\Sigma \cap \Theta_{k,k}$ is $\mathcal{S}^0$-closed, by Proposition 18.3 we see that

$$\Phi(\Sigma \cap \Theta^{\max}(G)) \in \mathcal{B}(\mathbb{R}^n), \qquad \forall \Sigma \subset \mathcal{S}^\circ, \Sigma \text{ closed.} \qquad (18.67)$$

Taking $\Sigma = \mathcal{S}^\circ$ in (18.67), we see that $\mathcal{T}(G) = \Phi(\Theta^{\max}(G))$ is a Borel set. When the non-crossing condition (18.21) holds, and thus P_G is well-defined, we have of course

$$(P_G)^{-1}(\Sigma) = \Phi(\Sigma \cap \Theta^{\max}(G)), \qquad \forall \Sigma \subset \mathcal{S}^\circ,$$

and the Borel measurability of P_G follows by arbitrariness of Σ closed in (18.67).

Step five: The Borel measurability of $\mathcal{T}_\star(G)$ with $\star \in \{c, \ell, r\}$ is discussed similarly. Indeed, the definitions of Join and Ext are easily adapted in $\mathcal{S}^\star$ (compare to what done with the definition of Sub given in (18.57)). The assertions in step two, three and four are easily seen to hold with identical arguments if $\Theta^{\max}(G)$ is systematically replaced with $\Theta^{\max}_\star(G) = \{\rho^\star(\sigma) : \sigma \in \Theta^{\max}(G)\} \subset \mathcal{S}^\star$. In particular, the argument in step three shows that $\Theta^{\max}_\star(G)$ can always be characterized as countable union of $\mathcal{S}^\star$-closed sets $\Theta^\star_{k,k}$.   □

19

An Introduction to the Needle
Decomposition Method

In this final chapter we relate OMT to the "needle decomposition method" originating in the classical work by Payne and Weinberger [PW60] on sharp forms of the Poincaré inequality on convex domains, and subsequently reintroduced as a "localization theorem" in Convex Geometry by various authors [GM87, LS93, KLS95]. The idea of framing the needle decomposition technique as a disintegration procedure along transport rays [Kla17] and the concurrent development of OMT in (weighted) Riemannian manifolds first and in metric measure spaces with curvature/dimension conditions later have led to realize the (proper formulation and) validity of the localization theorem in those very general settings and opened the way to several striking advancements on geometric and functional inequalities, some of them previously unknown even in the Riemannian setting. Although a reasonably complete account on these developments is evidently beyond the scope of our Euclidean-centric discussion, we can still take advantage of the results presented in Chapter 18 to illustrate some of the key ideas involved in the method, thus offering a particularly gentle introduction to this important topic; and this will indeed be the goal of this final chapter.

We start, in Section 19.1, with an account on the original proof of the Payne–Weinberger comparison theorem (Theorem 19.1). This is instructive for at least two reasons: first, it shows that the basic idea behind the needle decomposition method arises from completely elementary and direct considerations (no need to solve the Monge problem to use this method – at least in $\mathbb{R}^n$, and in other highly symmetric spaces); second, it introduces readers to the subject of comparison theorems, a class of statements of fundamental importance in Riemannian geometry, and the typical "target" of proofs based on the method. In Section 19.2 we briefly introduce the localization theorem (Theorem 19.4) as the most efficient formalization of the ideas behind the needle decomposition method, and we apply it to deduce, quite immediately, the Payne–Weinberger comparison theorem. Although the localization proof requires, on aggregate, more work than the original proof, it offers an undeniably simpler approach from the *conceptual* viewpoint. This simplicity is reflected, on a more pragmatic level, in the fact that the localization theorem can be generalized, *mutatis mutandis*, to much more general settings than Euclidean spaces, where the simple geometric arguments presented in Section 19.1 are not even conceivable; and, once such generalizations are obtained, they allow the far-reaching extension of Theorem 19.1 to those settings by repeating the simple argument presented in Section 19.2. The chapter is closed by a proof of the localization theorem, which is achieved by first (re-)discussing, in Section 19.3, the $C^{1,1}$-differentiability properties of Kantorovich potentials, and by then proving a key concavity property of disintegration densities in Section 19.4.

19.1 The Payne–Weinberger Comparison Theorem

Given a bounded, open, *connected* set $\Omega \subset \mathbb{R}^n$ with Lipschitz boundary, the Poincaré inequality states the existence of a positive constant $\lambda(\Omega)$ such that

$$\int_\Omega |\nabla u|^2 \geq \lambda(\Omega) \int_\Omega u^2, \qquad \forall u \in C^1(\Omega), \int_\Omega u = 0. \qquad (19.1)$$

The constant $\lambda(\Omega)$ appearing in (19.1) is understood to be the largest possible constant such that (19.1) holds, i.e., $\lambda(\Omega)$ is by definition

$$\lambda(\Omega) = \inf \left\{ \int_\Omega |\nabla u|^2 : u \in C^1(\Omega) \quad \int_\Omega u^2 = 1, \int_\Omega u = 0 \right\}.$$

Notice that $\lambda(\Omega)$ has the dimensions of a squared inverse length and could have been equivalently defined as the first positive eigenvalue of the Laplace operator on Ω with Neumann boundary condition, i.e., as the smallest number[1] λ such that there is a nonconstant function u with the property that $-\Delta u = \lambda u$ in Ω and $\nabla u \cdot \nu_\Omega = 0$ on $\partial\Omega$. We call[2] $\lambda(\Omega)$ **the Poincaré constant of** Ω. Since the Poincaré inequality itself does not assert anything but the positivity of $\lambda(\Omega)$, and given that one can hope to explicitly compute $\lambda(\Omega)$ only in very specific situations (e.g., if $n = 1$ and Ω is an interval of length d, then $\lambda(\Omega) = (\pi/d)^2$ by basic Fourier analysis), a natural question is bounding $\lambda(\Omega)$ in terms of other (hopefully more directly accessible) geometric/functional quantities depending on Ω. The Payne–Weinberger comparison theorem takes on this question in the class of **convex domains** (i.e., bounded open convex sets).

---

[1] Any such λ is necessarily positive, as seen if we integrate by parts $\lambda u^2 = -u \Delta u$.
[2] The same term is routinely used also for $1/\lambda(\Omega)$ in the literature.

Theorem 19.1 (Payne–Weinberger comparison theorem) *If Ω is a convex domain in $\mathbb{R}^n$ and $u \in C^1(\Omega)$ with $\int_\Omega u = 0$, then*

$$\int_\Omega |\nabla u|^2 \geq \left(\frac{\pi}{\text{diam}(\Omega)}\right)^2 \int_\Omega u^2. \tag{19.2}$$

In other words, at a fixed diameter, one-dimensional intervals have the lowest possible Poincaré constant among convex domains in any dimension, i.e.,

$$\lambda(\Omega) \geq \lambda((0, \text{diam}(\Omega))), \qquad \forall \Omega \text{ convex domain.}$$

The original proof of Theorem 19.1 by Payne and Weinberger introduces (*ante litteram*) the needle decomposition method and provides us with a key insight into the localization theorem (Theorem 19.4). The method allows to reduce the proof of Theorem 19.1 to the discussion of the family of one-dimensional variational problems addressed in the following statement:

Lemma 19.2 *If $d > 0$, $h : (0, d) \to \mathbb{R}$ is concave, and $\ell = e^h$, then*

$$\int_0^d (u')^2 \ell \geq \left(\frac{\pi}{d}\right)^2 \int_0^d u^2 \ell, \tag{19.3}$$

whenever $u \in W^{1,2}(0, d)$ is such that $\int_0^d u \ell = 0$.

Remark 19.3 Notice that (19.3) is sharp, since in the case $h = $ constant we have indeed $\lambda((0, d)) = (\pi/d)^2$. It is also useful to notice that Lemma 19.2 applies whenever $\ell = k^m$ for some $m > 0$ and $k : (0, d) \to \mathbb{R}$ is positive and concave, since in this case $\ell = e^h$ with $h = m \log k$, which is (trivially) concave on $(0, d)$. The interest of this case will become apparent in the proof of Theorem 19.1, see, in particular, (19.13) and (19.14).

Proof of Lemma 19.2 For $\varepsilon \in (0, d/2)$, the ε-regularization by convolution of h, $h_\varepsilon(x) = \int h(y) \rho_\varepsilon(y-x)\, dx$, defines a smooth concave function on $(\varepsilon, d-\varepsilon)$ so that $h_\varepsilon \uparrow h$ locally in $(0, d)$ as $\varepsilon \to 0^+$. Exploiting this approximation, we reduce to the case when h is smooth and convex on $(0, d)$ (and, in particular, ℓ is smooth, positive, and bounded from above) and consider the problem

$$\gamma = \inf\left\{\int_0^d (u')^2 \ell : \int_0^d u^2 \ell = 1, \int_0^d u \ell = 0\right\},$$

which is easily seen to admit a minimizer $u \in W^{1,2}(0, d)$ by the Direct Method. Given $\zeta \in W^{1,2}(0, d)$ with $\int_0^d \zeta \ell = 0$, we notice that

$$\varphi = \zeta - \left(\int_0^d \zeta u \ell\right) u \tag{19.4}$$

is such that

$$\int_0^d \varphi \ell = \int_0^d \varphi u \, \ell = 0. \tag{19.5}$$

Picking $\psi \in W^{1,2}(0,d)$ with

$$\int_0^d \psi \, \ell = 0, \qquad \int_0^d \psi u \, \ell = 1, \tag{19.6}$$

(e.g., pick $\psi = u$), we consider

$$f(t,s) = -1 + \int_0^d (u + t\,\varphi + s\,\psi)^2 \, \ell.$$

Clearly, $f(0,0) = 0$, and (19.5) and (19.6) give

$$\frac{\partial f}{\partial t}(0,0) = 0, \qquad \frac{\partial f}{\partial s}(0,0) = 2,$$

so that we can find $\varepsilon > 0$ and $s : (-\varepsilon, \varepsilon) \to \mathbb{R}$ such that $s(0) = 0$, $s'(0) = 0$ and $f(t, s(t)) = 0$ for every $|t| < \varepsilon$. Since, by construction, $(u + t\,\varphi + s(t)\,\psi)\,\ell$ has zero average on $(0,d)$, we see that $u + t\,\varphi + s(t)\,\psi$ is admissible in γ for every $|t| < \varepsilon$, and exploiting the minimality of u in γ, $s'(0) = 0$ and (19.4), we conclude that

$$0 = \int_0^d \varphi' u' \, \ell = \int_0^d \zeta' u' \, \ell - \gamma \int_0^d \zeta u \, \ell, \tag{19.7}$$

for every $\zeta \in W^{1,2}(0,d)$ with $\int_0^d \zeta \ell = 0$. Given $\eta \in C^\infty[0,d]$ with $\int_0^d \eta = 0$, and applying (19.7) to $\zeta = \eta/\ell$, we find $0 = \int_0^d \eta' u' - \eta\,[(\ell'/\ell)\,u' + \gamma u]$, that is (counting that $\ell'/\ell = h'$)

$$\eta u' \Big|_0^d = \int_0^d \eta\,(u'' + h'\,u' + \gamma u), \qquad \forall \eta \in C^\infty[0,d] \text{ with } \int_0^d \eta = 0. \tag{19.8}$$

By testing (19.8) with η compactly supported in $(0,d)$ and by a classical regularity argument, $u \in C^\infty(0,d)$, and there exists $\lambda \in \mathbb{R}$ such that

$$u'' + h'\,u' + \gamma u = \lambda, \qquad \text{on } (0,d), \tag{19.9}$$

which can then be combined again into (19.8) with choices of η that are nonzero at $\{0,d\}$ to conclude that $u \in C^\infty[0,d]$ with $u'(0) = u'(d) = 0$. Differentiating (19.9) we get

$$u''' + h'\,u'' + (h'' + \gamma)\,u' = 0, \qquad \text{on } (0,d); \tag{19.10}$$

at the same time, setting $v = \sqrt{\ell}\,u'$, by $(\sqrt{\ell})' = (h'/2)\,\sqrt{\ell}$ and (19.10), we find that

$$v' = \sqrt{\ell}\left\{u'' + \frac{h'}{2}u'\right\}$$

$$v'' = \sqrt{\ell}\,u''' + 2\sqrt{\ell}\,\frac{h'}{2}u'' + \left(\sqrt{\ell}\,\frac{h'}{2}\right)'u' = \sqrt{\ell}\left\{u''' + h'\,u'' + \left(\frac{h'}{2}\right)^2 u' + \frac{h''}{2}u'\right\}$$

$$= \left\{-\frac{h''}{2} + \frac{(h')^2}{4} - \gamma\right\}v \ge -\gamma\,v,$$

where in the las step we have used $h'' \le 0$. Multiplying by v and integrating by parts with the help of $v(0) = v(d) = 0$, we find $-\int_0^d (v')^2 \ge -\gamma \int_0^d v^2$, which gives

$$\gamma \ge \frac{\int_0^d (v')^2}{\int_0^d v^2} \ge \inf\left\{\frac{\int_0^d (w')^2}{\int_0^d w^2} : w \in W_0^{1,2}(0,d)\right\} = \left(\frac{\pi}{d}\right)^2,$$

if v is not identically zero on $(0,d)$. It is easily that it cannot be $v \equiv 0$; indeed, this would give u constant, which combined with $\int_0^d u\,\ell = 0$ and $\ell > 0$ on $(0,d)$ would give $u \equiv 0$, in contradiction with $\int_0^\ell u^2\,\ell = 1$. □

We now present the original proof of Theorem 19.1.

Proof of Theorem 19.1   *Step one*: Given positive d and δ, let us say that a convex domain Ω is a **needle of length d and waist δ** if, up to rigid motions,

$$\Omega \subset \{x : x_1 \in (0,d), |x_i| < \delta \,\forall i \ne 1\}. \tag{19.11}$$

(It is implicitly intended that the values of d and δ cannot be further decreased.) In this step we prove that if Ω is a needle of length d and waist δ, then

$$\int_\Omega |\nabla u|^2 \ge \left(\frac{\pi}{d}\right)^2 \left\{\int_\Omega u^2 - C(u)\,\mathcal{L}^n(\Omega)\,\delta\right\} - C(u)\,\mathcal{L}^n(\Omega)\,\delta, \tag{19.12}$$

for every $u \in C^2(\Omega)$ with $\int_\Omega u = 0$, and for a constant $C(u)$ depending only on the C^2-norm of u in Ω. Indeed, after a rigid motion that takes Ω in a position such that (19.11) holds, for $t \in (0,d)$ let us set

$$\Omega(t) = \Omega \cap \{x_1 = t\}, \qquad \ell(t) = \mathcal{H}^{n-1}(\Omega(t)), \tag{19.13}$$

and notice that, for every Borel function $g : \Omega \to \mathbb{R}$ with $g(x) = G(x_1)$, one has

$$\int_\Omega g = \int_0^d G(t)\,\ell(t)\,dt, \tag{19.14}$$

thanks to Fubini's theorem. By the Brunn–Minkowski inequality, if $t_1, t_2 \in [0,d]$ and $s \in (0,1)$, then

$$(1-s)\,\ell(t_1)^{1/(n-1)} + s\,\ell(t_2)^{1/(n-1)} \le \mathcal{H}^{n-1}((1-s)\,\Omega(t_1) + s\,\Omega(t_2))^{1/(n-1)},$$

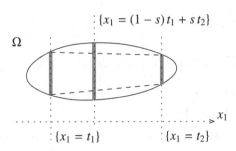

Figure 19.1 If Ω is convex, then the section of Ω by $\{x_1 = (1 - s)\,t_1 + s\,t_2\}$ contains the convex combination $(1 - s)\,\Omega(t_1) + s\,\Omega(t_2)$ of the sections $\Omega(t_i) = \Omega \cap \{x_1 = t_i\}$, $i = 1, 2$. These various sections are depicted in gray.

while, by convexity of Ω,

$$(1 - s)\,\Omega(t_1) + s\,\Omega(t_2) \subset \Omega((1 - s)\,t_1 + s\,t_2),$$

see Figure 19.1, so that $\ell^{1/(n-1)}$ is concave on $[0, d]$ (and positive on $(0, d)$ since the value of d cannot be further decreased). We can thus exploit Lemma 19.2 (as explained in Remark 19.3 with $m = n-1$) to the function $x_1 \in (0, d) \mapsto u(x_1, 0)$ and deduce by (19.14) that

$$\int_\Omega \left(\frac{\partial u}{\partial x_1}(x_1, 0)\right)^2 dx \geq \left(\frac{\pi}{d}\right)^2 \int_\Omega \left(u(x_1, 0) - \frac{1}{\mathcal{L}^n(\Omega)} \int_\Omega u(y_1, 0)\,dy\right)^2 dx.$$

Since $\int_\Omega u = 0$, for a constant $C(u)$ depending only on the C^2-norm of u in Ω we have

$$\left| \int_\Omega \left(\frac{\partial u}{\partial x_1}(x_1, 0)\right)^2 dx - \int_\Omega \left(\frac{\partial u}{\partial x_1}\right)^2 \right| \leq C(u)\,\delta\,\mathcal{L}^n(\Omega),$$

$$\left| \int_\Omega \left(u(x_1, 0) - \frac{1}{\mathcal{L}^n(\Omega)} \int_\Omega u(y_1, 0)\,dy\right)^2 dx - \int_\Omega u^2 \right| \leq C(u)\,\delta\,\mathcal{L}^n(\Omega),$$

and therefore we deduce

$$\int_\Omega |\nabla u|^2 \geq \int_\Omega \left(\frac{\partial u}{\partial x_1}\right)^2 \geq -C(u)\,\delta\,\mathcal{L}^n(\Omega) + \left(\frac{\pi}{d}\right)^2 \left\{ \int_\Omega u^2 - C(u)\,\delta\,\mathcal{L}^n(\Omega)\right\},$$

which is (19.12).

Step two: We conclude the proof. Let Ω be a convex domain and let $u \in C^2(\Omega)$ with $\int_\Omega u = 0$: we *claim* that, for every $\delta > 0$, we can find a finite family of convex domains $\{\Omega_i\}_{i=1}^N$ such that $\mathcal{L}^n(\Omega \setminus \bigcup_{i=1}^N \Omega_i) = 0$, $\Omega_i \cap \Omega_j = \emptyset$, each Ω_i is a needle with length $d_i \leq \mathrm{diam}(\Omega)$ and waist δ, and

$$\int_{\Omega_i} u = 0, \qquad \forall i = 1, \dots, N. \tag{19.15}$$

Notice that the claim implies the theorem: indeed, by (19.15) we can apply step two to u on each Ω_i (notice that the C^2-norm of u on Ω_i is bounded by the C^2-norm of u on Ω) to find

$$\left(\frac{d_i}{\pi}\right)^2 \left\{ \int_{\Omega_i} |\nabla u|^2 + C(u) \, \mathcal{L}^n(\Omega_i) \, \delta \right\} \geq \int_{\Omega_i} u^2 - C(u) \, \mathcal{L}^n(\Omega_i) \, \delta.$$

Exploiting $d_i \leq \mathrm{diam}(\Omega)$ and adding up over i we find

$$\left(\frac{\mathrm{diam}(\Omega)}{\pi}\right)^2 \left\{ \int_{\Omega} |\nabla u|^2 + C(u) \, \mathcal{L}^n(\Omega) \, \delta \right\} \geq \int_{\Omega} u^2 - C(u) \, \mathcal{L}^n(\Omega) \, \delta.$$

Letting $\delta \to 0^+$ we obtain (19.2).

To prove the claim: We first consider the case $n = 2$. Let $\mathrm{waist}(\Omega)$ denote the infimum over $v \in \mathbb{S}^1$ of $\max\{|(x - y) \cdot v| : x, y \in \Omega\}$ so that $\mathrm{waist}(\Omega)$ is the size of the narrowest strip containing Ω, and let us notice that

$$\mathrm{waist}(\Omega) \leq \sqrt{2 \, \mathcal{L}^2(\Omega)}. \tag{19.16}$$

(Indeed, Ω always contains two intersecting orthogonal segments, one of length $a = \mathrm{waist}(\Omega)$ and one of length $b \geq a$; thus, it contains their convex envelope, whose area is at least $a^2/2$.) Next, for every $v \in \mathbb{S}^1$ we can find two complementary half-planes $\{H_v^+, H_v^-\}$, v pointing outward H_v^+ and inward H_v^-, such that, setting $\Omega_v^\pm = \Omega \cap H_v^\pm$, we have $\mathcal{L}^2(\Omega_v^\pm) = \mathcal{L}^2(\Omega)/2$. Since $\int_{\Omega_v^+} u = - \int_{\Omega_v^-} u$, by a continuity argument we can find $v \in \mathbb{S}^1$ such that $\int_{\Omega_v^+} u = - \int_{\Omega_v^-} u = 0$. Iterating this construction, after m steps we have defined a family $\{\Omega_i^{(m)}\}_{i=1}^{2^m}$ of mutually disjoint convex domains such that, for every $i = 1, \ldots, 2^m$,

$$\mathcal{L}^2\left(\Omega \setminus \bigcup_{i=1}^{2^m} \Omega_i^{(m)}\right) = 0, \qquad \mathcal{L}^2(\Omega_i^{(m)}) = \frac{\mathcal{L}^2(\Omega)}{2^m}, \qquad \int_{\Omega_i^{(m)}} u = 0.$$

Thanks to (19.16), $\Omega_i^{(m)}$ is a needle with length $d \leq \mathrm{diam}(\Omega)$ and waist $\delta \leq C \, (\mathcal{L}^2(\Omega)/2^m)^{1/2}$, and the claim is proved taking m large with respect to δ. In the case $n \geq 3$ we argue as follows. Given an $(n - 2)$-dimensional plane J in $\mathbb{R}^n$, we denote by $\{v_J, w_J\}$ an orthonormal basis to $J^\perp$. Given $v \in \mathbb{S}^2_{J^\perp} = \{\cos \alpha \, v_J + \sin \alpha \, w_J : \alpha \in [0, 2\pi]\}$, we now define $H_v^\pm$ and $\Omega_v^\pm$ by requiring

$$\mathcal{H}^2(\mathbf{p}_J^\perp(\Omega_v^\pm)) = \frac{\mathcal{H}^2(\mathbf{p}_J^\perp(\Omega))}{2},$$

where $\mathbf{p}_J^\perp$ denotes the projection of $\mathbb{R}^n$ over $J^\perp$. Again by a continuity argument, a specific $v \in \mathbb{S}^2_{J^\perp}$ can be selected to ensure $\int_{\Omega_v^+} u = - \int_{\Omega_v^-} u = 0$. The iteration procedure is then repeated, this time finding

$$\mathcal{H}^2(\mathbf{p}_J^\perp(\Omega_i^{(m)})) = \frac{\mathcal{H}^2(\mathbf{p}_J^\perp(\Omega))}{2^m}$$

(in place of $\mathcal{L}^2(\Omega_i^{(m)}) = \mathcal{L}^2(\Omega)/2^m$), thus proving that

$$\lim_{m \to \infty} \sup_{1 \le i \le 2^m} \text{waist}(\mathbf{p}_J^\perp(\Omega_i^{(m)})) = 0.$$

We thus obtain a partition of Ω into convex domains $\Omega_i^{(m)}$, such that u has zero average on each $\Omega_i^{(m)}$ and such that each $\Omega_i^{(m)}$ is δ-thin with respect to a direction contained in $J^\perp$. Let us take any one of these pieces, say, $\Omega_1^{(m)}$, and assume, as it can be done without loss of generality up to a rotation, that $\Omega_1^{(m)}$ is δ-thin in the direction e_n: by repeating the above argument starting from $\Omega_1^{(m)}$ in place of Ω, and with $J^\perp = \text{span}\{e_{n-2}, e_{n-1}\}$, we further subdivide $\Omega_1^{(m)}$ into convex domains on which u has zero average and which are δ-thin in the direction e_n *and* in a direction contained in $\text{span}\{e_{n-2}, e_{n-1}\}$. By continuing this procedure, we eventually prove the claim and, thus, the theorem. □

19.2 The Localization Theorem

We now introduce (the Euclidean version of) the localization theorem. Here and in the rest of this chapter we use the notation $\mathcal{T}(f)$, $\mathcal{T}_c(f)$, P_f, and so on for transport set, transport set with endpoints, maximal transport ray map, and others associated to a 1-Lipschitz function f. This notation was introduced in Section 18.1; see, in particular, Proposition 18.6.

Theorem 19.4 (Localization theorem)   *Let Ω be a convex domain in $\mathbb{R}^n$, $u :$ $\Omega \to \mathbb{R}$ a Borel measurable function with $\int_\Omega u = 0$ and $u \not\equiv 0$, and let $f :$ $\mathbb{R}^n \to \mathbb{R}$ be a Kantorovich potential from μ to ν, where*

$$\mu = \frac{u^+ \, d\mathcal{L}^n \llcorner \Omega}{\int_\Omega u^+}, \qquad \nu = \frac{u^- \, d\mathcal{L}^n \llcorner \Omega}{\int_\Omega u^-}.$$

Then,[3]

$$u = 0, \qquad \mathcal{L}^n\text{-a.e. on } \Omega \setminus \mathcal{T}(f), \qquad (19.17)$$

and for $(P_f)_\#(\mathcal{L}^n \llcorner (\Omega \cap \mathcal{T}(f)))$-a.e. $\sigma \in \mathcal{S}^\circ$, the set $I_\sigma := P_f^{-1}(\sigma) \cap \Omega$ is an open segment and there is a smooth function $\ell_\sigma : I_\sigma \to \mathbb{R}$ such that

(i) *$\ell_\sigma^{1/(n-1)}$ is concave on I_σ and $\int_{I_\sigma} u \, \ell_\sigma \, d\mathcal{H}^1 = 0$;*

(ii) *$\{\ell_\sigma \mathcal{H}^1 \llcorner I_\sigma\}_\sigma$ is the disintegration of $\mathcal{L}^n \llcorner (\Omega \cap \mathcal{T}(f))$ with respect to P_f.*

---

[3]   It could be that $u = 0$ on a set of positive Lebesgue measure inside $\mathcal{T}(f)$. For example, if $\Omega = (0, 3) \times (0, 1) \subset \mathbb{R}^2$ with $u = 1$ on $Q_1 = (0, 1) \times (0, 1)$, $u = 0$ on $Q_2 = (1, 2) \times (0, 1)$ and $u = -1$ on $Q_3 = (2, 3) \times (0, 1)$, then $f(x_1, x_2) = x_1$ is a Kantorovich potential from $\mu = \mathcal{L}^n \llcorner Q_1$ to $\nu = \mathcal{L}^n \llcorner Q_3$ with $\mathcal{T}(f) = \Omega$ and $P_f(x_1, x_2) =]\!](x_1, 0), (x_2, 3)[\![$ for every $(x_1, x_2) \in \Omega$.

Remark 19.5 (Uniqueness of Kantorovich potentials)   As a consequence of (19.17), if $\mu, \nu \in \mathcal{P}_1(\mathbb{R}^n)$, $\mu = \rho \, d\mathcal{L}^n$, $\nu = \sigma \, d\mathcal{L}^n$ and

$$\{\rho > 0\} \cup \{\sigma > 0\} \text{ is } \mathcal{L}^n\text{-equivalent to a convex domain } \Omega, \qquad (19.18)$$

then *all Kantorovich potentials from μ to ν agree up to an additive constant.* Indeed, if $u = \rho - \sigma$, then $u \neq 0$ $\mathcal{L}^n$-a.e. on Ω by (19.18), and thus (19.17) implies $\mathcal{L}^n(\Omega \setminus \mathcal{T}(f)) = 0$. In particular, if f is a Kantorovich potential from μ to ν, then $|\nabla f| = 1$ $\mathcal{L}^n$-a.e. on Ω. Now, if f_1 and f_2 are Kantorovich potentials from μ to ν, then $(f_1 + f_2)/2$ is a Kantorovich potential from μ to ν too (since in the dual Kantorovich problem we are maximizing the linear function $g \mapsto \int_{\mathbb{R}^n} g \, d\nu - \int_{\mathbb{R}^n} g \, d\mu$ on the convex set $\{g : \text{Lip}(g) \leq 1\}$, see (3.38)). Hence, $|\nabla(f_1 + f_2)/2| = 1$ $\mathcal{L}^n$-a.e. on Ω, and, by strict convexity of the Euclidean norm, $\nabla f_1 = \nabla f_2$ $\mathcal{L}^n$-a.e. on Ω. Since Ω is, in particular, open and connected, this implies that $f_1 = f_2 + \text{constant}$ in Ω, as claimed.

We now comment on the relation between Theorem 19.4 and the "needle decomposition" argument used to prove Theorem 19.1. There, given $\delta > 0$, we have constructed a partition of Ω into convex (solid) "needles" $\{\Omega_i^\delta\}_i$ with waist of size δ, so that $\int_{\Omega_i^\delta} u = 0$ for each i. Further assuming $u \in C^2(\Omega)$, we have deduced, by Lemma 19.2, that (19.2) holds on each needle Ω_i^δ up to an error of size $\|u\|_{C^2(\Omega)} \delta \, \mathcal{L}^n(\Omega_i^\delta)$, and then, by adding up the resulting approximate inequalities over i before sending $\delta \to 0^+$, we have deduced (19.2) on Ω. A natural question is thus if this approximation step is necessary at all – in other words, if one could directly work, so to say, with $\delta = 0$. Theorem 19.4 answers affirmatively to that question. By the properties of disintegration (see Theorem 17.1), the decomposition[4] of Ω into one-dimensional (rather than solid) "needles" $\{I_\sigma\}_\sigma$, or, better, the disintegration of $\mathcal{L}^n \llcorner \Omega$ into measures $\{\mu_\sigma = \ell_\sigma \mathcal{H}^1 \llcorner I_\sigma\}_\sigma$ allows us to decompose "bulk" integrals on Ω as superposition of one-dimensional integrals such that u has still zero average with respect to each μ_σ. Lemma 19.2 applies thanks to the concavity of $\ell_\sigma^{1/(n-1)}$, which descends, in the limit as $\delta \to 0^+$, from the concavity of the slice function (19.13) (indeed, each μ_σ can be seen as a weak-star limit of probability measures of the form $\mathcal{L}^n(\Omega_{i(j)}^{\delta_j})^{-1} \mathcal{L}^n \llcorner \Omega_{i(j)}^{\delta_j}$ with $\delta_j \to 0^+$ as $j \to \infty$).

We close this section by showing how easily the proof of Theorem 19.1 is reduced to that of Lemma 19.2 by using Theorem 19.4. The proof of Theorem 19.4 will then take the rest of the chapter.

---

[4]   In this sentence we are really thinking to the situation where $\mathcal{L}^n(\Omega \setminus \mathcal{T}(f)) = 0$, and thus we obtain a foliation by segments of the whole Ω, and not just of $\Omega \setminus \mathcal{T}(f)$ (which is a subset of $\{u = 0\}$ modulo $\mathcal{L}^n$-null sets by (19.17), and thus is irrelevant for the typical applications of the localization theorem, see for example the proof of Theorem 19.1 via Theorem 19.4).

Proof of Theorem 19.1 via Theorem 19.4   Given $u \in C^1(\Omega)$, let f be as in Theorem 19.4. By (19.17) and by Theorem 19.4-(ii), setting for brevity $\lambda = (P_f)_\#(\mathcal{L}^n \llcorner (\Omega \cap \mathcal{T}(f)))$, we have

$$
\int_\Omega |\nabla u|^2 = \int_{\Omega \cap \mathcal{T}(f)} |\nabla u|^2 = \int_{S^\circ} d\lambda(\sigma) \int_{I_\sigma} |\nabla u|^2 \, \ell_\sigma \, d\mathcal{H}^1
$$

$$
\geq \int_{S^\circ} d\lambda(\sigma) \int_{I_\sigma} (u'_\sigma)^2 \, \ell_\sigma \, d\mathcal{H}^1,
$$

where $u_\sigma \in C^1(I_\sigma)$ denotes the restriction of u to the open segment I_σ so that, trivially, $|\nabla u| \geq |u'_\sigma|$ on I_σ. By Theorem 19.4-(i) and Lemma 19.2,

$$
\int_{I_\sigma} (u'_\sigma)^2 \, \ell_\sigma \geq \left(\frac{\pi}{\mathrm{diam}(I_\sigma)} \right)^2 \int_{I_\sigma} u_\sigma^2 \, \ell_\sigma,
$$

where $\mathrm{diam}(I_\sigma) \leq \mathrm{diam}(\Omega)$, and thus, using again Theorem 19.4-(ii), we find

$$
\int_\Omega |\nabla u|^2 \geq \left(\frac{\pi}{\mathrm{diam}(\Omega)} \right)^2 \int_{S^\circ} d\lambda(\sigma) \int_{I_\sigma} u_\sigma^2 \, \ell_\sigma = \left(\frac{\pi}{\mathrm{diam}(\Omega)} \right)^2 \int_{\Omega \cap \mathcal{T}(f)} u^2.
$$

Using again (19.17), we conclude the proof.   □

19.3 $C^{1,1}$-Extensions of Kantorovich Potentials

In a first step toward the proof of Theorem 19.4 we revisit in more geometric terms the analysis from Section 18.3. There we proved that if f is a Kantorovich potential, then ∇f has the countable Lipschitz property on $\mathcal{T}_e(f)$, the transport set with endpoints of f (see Section 18.1 for all the relevant notation). We now show that ∇f is actually $O(1/\varepsilon)$-Lipschitz continuous on the transport-type sets $\Sigma_\varepsilon(f)$ of those $x \in \mathcal{T}(f)$ that are midpoints of intervals of length 2ε contained in some transport ray of f, i.e., on

$$
\Sigma_\varepsilon(f) = \left\{ x \in \mathcal{T}(f) : \,]x - \varepsilon \nabla f(x), x + \varepsilon \nabla f(x)[\, \subset \Phi(P_f(x)) \right\}. \tag{19.19}
$$

In addition, we construct a $C^{1,1}$-extension of f from $\Sigma_\varepsilon(f)$ to $\mathbb{R}^n$ with $C^{1,1}$-norm of order $O(1/\varepsilon)$ by using the Whitney extension theorem.

Theorem 19.6   *If $f : \mathbb{R}^n \to \mathbb{R}$ is 1-Lipschitz and $\varepsilon > 0$, then*

$$
|\nabla f(x) - \nabla f(y)| \leq \frac{48}{\varepsilon} |x - y|, \tag{19.20}
$$

$$
|f(y) - f(x) - \nabla f(x) \cdot (y - x)| \leq \frac{16}{\varepsilon} |x - y|^2, \qquad \forall x, y \in \Sigma_\varepsilon(f). \tag{19.21}
$$

In particular, there exists $f_\varepsilon \in C^{1,1}(\mathbb{R}^n)$ such that

$$
f = f_\varepsilon \text{ and } \nabla f_\varepsilon = \nabla f \text{ on } \Sigma_\varepsilon(f). \tag{19.22}
$$

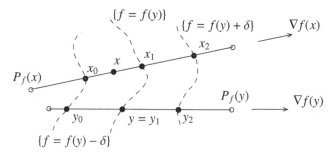

Figure 19.2 The situation in the proof of Theorem 19.6.

Proof Thanks to the Whitney extension theorem for $C^{1,1}$-functions (see Appendix A.16) (19.22) is a consequence of (19.20) and (19.21). Since $|\nabla f| \leq 1$, if $|x-y| \geq \varepsilon/8$, then (19.20) and (19.21) hold, respectively, with $(16/\varepsilon)|x-y|$ and with $(16/\varepsilon)|x - y|^2$ on their right-hand sides. Therefore, we can focus on proving (19.20) and (19.21) when $x, y \in \Sigma_\varepsilon(f)$ are such that

$$|x - y| \leq \frac{\varepsilon}{8}. \tag{19.23}$$

We set

$$x_1 = x + (f(x) - f(y)) \nabla f(x), \qquad y_1 = y.$$

By $\mathrm{Lip}(f) \leq 1$,

$$|x_1 - x| \leq |x - y|, \tag{19.24}$$

so (19.23) and $x \in \Sigma_\varepsilon(f)$ give $x_1 \in \Phi(P_f(x))$. Since f grows linearly with unit slope along $P_f(x)$, we have $f(x_1) = f(x) + (f(y) - f(x)) = f(y)$, and, setting

$$x_0 = x_1 - \delta \nabla f(x), \qquad x_2 = x_1 + \delta \nabla f(x),$$
$$y_0 = y_1 - \delta \nabla f(y), \qquad y_2 = y_1 + \delta \nabla f(y), \qquad \delta = \frac{\varepsilon}{8},$$

we deduce from $x, y \in \Sigma_\varepsilon(f)$ that $x_i \in \Phi(P_f(x))$, $y_i \in \Phi(P_f(y))$ $(i = 0, 1, 2)$, and

$$x_0, y_0 \in \{f = f(y) - \delta\}, \qquad x_1, y_1 \in \{f = f(y)\}, \qquad x_2, y_2 \in \{f = f(y) + \delta\}; \tag{19.25}$$

see Figure 19.2. We now *claim* that

$$\max \{|x_i - x_j|, |y_i - y_j|\} \leq |x_i - y_j|, \qquad \forall i, j = 0, 1, 2, \tag{19.26}$$
$$|x_1 - y_1| \leq 2 |x - y|, \tag{19.27}$$
$$\max\{|x_0 - y_0|, |x_2 - y_2|\} \leq 2 |x_1 - y_1|. \tag{19.28}$$

To prove (19.26): Thanks to $x_i, x_j \in \Phi(P_f(x))$ and to (19.25) we have

$$|x_i - x_j| = |f(x_i) - f(x_j)| = |f(x_i) - f(y_j)| \leq |x_i - y_j|.$$

To prove (19.27): By (19.24),

$$|x_1 - y_1| \leq |x_1 - x| + |x - y_1| \leq 2|x - y|.$$

To prove (19.28): The symmetries of the problem allow us to assume that $|x_0 - y_0| \leq |x_2 - y_2|$, and thus focus on proving $|x_2 - y_2| \leq 2|x_1 - y_1|$. By (19.26),

$$|x_0 - x_2| \leq |x_0 - y_2| = |(y_2 - x_2) + (x_2 - x_0)|$$

so that, taking squares, expanding them, and noticing that $x_0 - x_2 = 2(x_1 - x_2)$, we find

$$|x_2 - y_2|^2 \geq 2(y_2 - x_2) \cdot (x_0 - x_2) = 4(y_2 - x_2) \cdot (x_1 - x_2). \quad (19.29)$$

Similarly, again by (19.26),

$$|y_0 - y_2| \leq |y_0 - x_2| = |(y_2 - x_2) + (y_0 - y_2)|,$$

so $y_2 - y_0 = 2(y_2 - y_1)$ implies

$$|x_2 - y_2|^2 \geq 2(y_2 - x_2) \cdot (y_2 - y_0) = 4(y_2 - x_2) \cdot (y_2 - y_1). \quad (19.30)$$

By (19.29) and (19.30),

$$
\begin{aligned}
|y_2 - x_2||y_1 - x_1| &\geq (y_2 - x_2) \cdot (y_1 - x_1) \\
&= (y_2 - x_2) \cdot ((y_2 - x_2) - (y_2 - y_1) - (x_1 - x_2)) \\
&= |y_2 - x_2|^2 - (y_2 - x_2) \cdot (y_2 - y_1) - (y_2 - x_2) \cdot (x_1 - x_2) \\
&\geq \frac{|x_2 - y_2|^2}{2},
\end{aligned}
$$

thus completing the proof of (19.28). Finally:

To prove (19.20): By (19.28) and (19.27),

$$
\begin{aligned}
\delta |\nabla f(x) - \nabla f(y)| &= |(x_2 - x_1) - (y_2 - y_1)| \leq |x_2 - y_2| + |x_1 - y_1| \\
&\leq 3|x_1 - y_1| \leq 6|x - y|,
\end{aligned}
$$

and (19.20) follows since $\delta = \varepsilon/8$.

To prove (19.21): By $y = y_1$ we have

$$
\begin{aligned}
|f(y) - f(x) &- \nabla f(x) \cdot (y - x)| \\
&= |f(y_1) - (f(x) + \nabla f(x) \cdot (x_1 - x)) - \nabla f(x) \cdot (y - x_1)| \\
&= |\nabla f(x) \cdot (y - x_1)|,
\end{aligned}
$$

where we have used the identity $f(y_1) = f(x_1) = f(x) + \nabla f(x) \cdot (x_1 - x)$. Now, by (19.26),

$$|x_0 - x_1|^2 \le |x_0 - y_1|^2 = |x_0 - x_1|^2 + |x_1 - y_1|^2 + 2(x_0 - x_1) \cdot (x_1 - y_1),$$

which combined with $x_0 - x_1 = -\delta \nabla f(x)$ and with (19.27) gives

$$2\delta \nabla f(x) \cdot (x_1 - y_1) \le |x_1 - y_1|^2 \le 4|x - y|^2.$$

Similarly,

$$|x_2 - x_1|^2 \le |x_2 - y_1|^2 \le |x_2 - x_1|^2 + |x_1 - y_1|^2 + 2(x_2 - x_1) \cdot (x_1 - y_1),$$

so that $x_2 - x_1 = \delta \nabla f(x)$ and (19.27) give

$$-2\delta \nabla f(x) \cdot (x_1 - y_1) \le |x_1 - y_1|^2 \le 4|x - y|^2,$$

and hence $|\nabla f(x) \cdot (y - x_1)| \le (2/\delta)|x - y|^2 \le (16/\varepsilon)|x - y|^2$. $\qquad\square$

19.4 Concave Needles and Proof of the Localization Theorem

In Theorem 19.7, given a 1-Lipschitz function f, we consider the disintegration $\{\mu_\sigma\}_\sigma$ of $\mu = \mathcal{L}^n \llcorner \mathcal{T}_c(f)$ with respect to P_f and prove a crucial concavity property of the densities of μ_σ with respect to $\mathcal{H}^1 \llcorner P_f^{-1}(\sigma)$. Here, P_f has been extended as usual from $\mathcal{T}(f)$ to $\mathcal{T}_c(f) \cap F$ by Proposition 18.6, where F is the differentiability set of f, and thus $\mathcal{L}^n(\mathbb{R}^n \setminus F) = 0$: in particular, it makes sense to disintegrate μ with respect to P_f. With Theorem 19.7 proved, we will be finally ready to prove Theorem 19.4 and conclude our discussion.

Theorem 19.7 (Concave needles) *If $n \ge 2$, $f : \mathbb{R}^n \to \mathbb{R}$ is 1-Lipschitz and $\mu = \mathcal{L}^n \llcorner \mathcal{T}_c(f)$, then the disintegration $\{\mu_\sigma\}_\sigma$ of μ with respect to P_f is such that*

$$\mu_\sigma = \ell_\sigma \, d\mathcal{H}^1 \llcorner P_f^{-1}(\sigma) \qquad \text{for } (P_f)_\# \mu\text{-a.e. } \sigma \in \mathcal{S}^\circ, \qquad (19.31)$$

where ℓ_σ is a smooth function on $P_f^{-1}(\sigma)$ such that $(\ell_\sigma)^{1/(n-1)}$ is concave along $P_f^{-1}(\sigma)$.

Proof By Theorem 18.9 we can apply Theorem 18.7 to $G = G(f)$. Hence, for $(P_f)_\# \mu$-a.e. $\sigma \in \mathcal{S}^\circ$ there is $\ell_\sigma : P_f^{-1}(\sigma) \to \mathbb{R}$ Borel measurable such that

$$\mu_\sigma = \ell_\sigma \, d\mathcal{H}^1 \llcorner P_f^{-1}(\sigma). \qquad (19.32)$$

Therefore, our goal will be proving that ℓ_σ is smooth and $(\ell_\sigma)^{1/(n-1)}$ is concave on the open segment $P_f^{-1}(\sigma)$. In doing so we shall use the coarea formula for rectifiable sets, which works analogously to the most basic version already used in Remark 17.5, and which is recalled here in Appendix A.14 and A.15.

Step one: Given $k > 0$ large enough (any $k > 96$ will do), let $\varepsilon > 0$ and set

$$\delta = \left(1 - \frac{1}{k}\right)\varepsilon.$$

Denoting by f_δ the $C^{1,1}$-extension of f from $\Sigma_\delta(f)$ to $\mathbb{R}^n$ constructed in Theorem 19.6, since ∇f_δ is Lipschitz continuous, $\nabla f = \nabla f_\delta$ on $\Sigma_\delta(f) \subset \mathcal{T}(f)$ and $|\nabla f| = 1$ on $\mathcal{T}(f)$, by the coarea formula, if G_δ is the set of non-differentiability points of f_δ and A is an arbitrary open set containing G_δ, we have

$$\mathcal{L}^n(A \cap \Sigma_\delta(f)) = \int_{A \cap \Sigma_\delta(f)} |\nabla f_\delta| = \int_{\mathbb{R}} \mathcal{H}^{n-1}(A \cap \Sigma_\delta(f) \cap \{f_\delta = s\})\, ds.$$

Applying this to $A = A_j$ with $G_\delta \subset A_j \subset A_{j-1}$ and $\mathcal{L}^n(A_j) \to \mathcal{L}^n(G_\delta) = 0$ as $j \to \infty$, we find that

$$0 = \lim_{j \to \infty} \int_{\mathbb{R}} \mathcal{H}^{n-1}(A_j \cap \Sigma_\delta(f) \cap \{f_\delta = s\})\, ds$$

$$\geq \int_{\mathbb{R}} \mathcal{H}^{n-1}(G_\delta \cap \Sigma_\delta(f) \cap \{f_\delta = s\})\, ds \geq 0.$$

In particular, there is $J_0 \subset \mathbb{R}$ such that $\mathcal{L}^1(\mathbb{R} \setminus J_0) = 0$ and

if $s \in J_0$, then ∇f_δ is differentiable $\mathcal{H}^{n-1}$-a.e. on $\{f_\delta = s\} \cap \Sigma_\delta(f)$.   (19.33)

Given $s \in J_0$, we now define

$$M = \Sigma_\varepsilon(f) \cap \{f = s\},$$
$$T : M \times I_{\varepsilon/k} \to \mathbb{R}^n, \qquad T(y,t) = y + t\,\nabla f(y),$$
$$E = T(M \times I_{\varepsilon/k}) = \left\{y + t\,\nabla f(y) : y \in M, |t| < \frac{\varepsilon}{k}\right\} \subset \mathcal{T}(f);$$

see Figure 19.3, where $I_a = (-a, a)$ and where we have used the facts that $\Sigma_\varepsilon(f) \subset \mathcal{T}(f)$ and that f is differentiable on $\mathcal{T}(f)$ to define T. We notice that, setting $E_{\varepsilon,s}$ in place of E to momentarily stress the dependency of E on the choices of ε and $s \in J_0$, we have

$$\mathcal{T}(f) = \bigcup_{\varepsilon > 0} \bigcup_{s \in J_0} E_{\varepsilon,s}.$$ (19.34)

Indeed, if $x \in \mathcal{T}(f)$, then $x \in \Sigma_{2\varepsilon}(f)$ for some $\varepsilon > 0$, and, by $\mathcal{L}^1(\mathbb{R} \setminus J_0) = 0$, we can find $s \in J_0$ with $|s - f(x)| < \varepsilon/k$. Then $y = x - (f(x) - s)\nabla f(x)$ is such that $f(y) = s$, $\nabla f(y) = \nabla f(x)$, $x = y + t\,\nabla f(y)$ for $t = f(x) - s \in I_{\varepsilon/k}$, and $y \in \Sigma_{2\varepsilon - |t|}(f)$, where $2\varepsilon - |t| > \varepsilon$.

We claim that if k is large enough, then M is locally $\mathcal{H}^{n-1}$-rectifiable, E is a Borel set contained in $\Sigma_\delta(f)$, and T is Lipschitz continuous and invertible on $M \times I_{\varepsilon/k}$, with Lipschitz continuous inverse. To prove the claim, we start

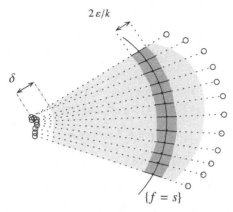

Figure 19.3 Construction of M, T, and E. The dotted segments with open end-points represent the maximal transport rays making up $\mathcal{T}(f)$. The light grey region is $\Sigma_\delta(f)$, which is obtained by cutting off segments of length δ from each end of every maximal transport ray. In particular, $\Sigma_\varepsilon(f)$ is obtained from $\Sigma_\delta(f)$ by an additional cutting off segments of length ε/k. We define M as the intersection of some level set $\{f = s\}$ of f with $\Sigma_\varepsilon(f)$. In this way, the set E (depicted in dark grey) obtained by taking the union of the segments of length $2\varepsilon/k$ centered on points in M, is contained in $\Sigma_\delta(f)$, and thus $f = f_\delta$ on E. The map T simply takes $M \times I_{\varepsilon/k}$ into E by the formula $T(y, t) = y + t\,\nabla f(y)$.

noticing that, since $f = f_\varepsilon$ and $|\nabla f_\varepsilon| = |\nabla f| = 1$ on $\Sigma_\varepsilon(f)$, by the implicit function theorem $\{f_\varepsilon = f(x_0)\}$ is a C^1-hypersurface in a neighborhood of M, and since $M \subset \{f_\varepsilon = f(x_0)\}$, we conclude that M is locally $\mathcal{H}^{n-1}$-rectifiable. Next, if $k > 48$, then, $\mathrm{Lip}(T; M \times I_{\varepsilon/k}) \le C(n)$ thanks to (19.20) and

$$|T(y_1,t_1) - T(y_2,t_2)| \le |y_1 - y_2| + |t_1 - t_2| + \frac{\varepsilon}{k}\,|\nabla f(y_1) - \nabla f(y_2)|$$
$$\le 2\,|y_1 - y_2| + |t_1 - t_2|.$$

In particular, T is Lipschitz continuous on $M \times I_{\varepsilon/k}$, and E is a Borel set (since it is the Lipschitz image of a Borel set). To prove that T is invertible with Lipschitz inverse, let $x_1, x_2 \in E$, $x_i = T(y_i, t_i)$: by $y_i \in \Sigma_\varepsilon(f)$ we find

$$f(x_i) = f(y_i + t_i\,\nabla f(y_i)) = f(y_i) + t_i = f(x_0) + t_i,$$

so that $|t_1 - t_2| \le |f(x_1) - f(x_2)| \le |x_1 - x_2|$; in particular,

$$|y_1 - y_2| \le |x_1 - x_2| + |t_1\,\nabla f(y_1) - t_2\,\nabla f(y_2)|$$

$$\le |x_1 - x_2| + |t_1 - t_2| + |t_1||\nabla f(y_1) - \nabla f(y_2)| \le 2|x_1 - x_2| + \frac{\varepsilon}{k}\frac{48}{\varepsilon}\,|y_1 - y_2|,$$

so that $(1/2)\,|y_1 - y_2| \le 2\,|x_1 - x_2|$ if $k \ge 96$, and $\mathrm{Lip}(T^{-1}; E) \le C(n)$. Finally, to prove $E \subset \Sigma_\delta(f)$, we notice that if $x \in E$, then $x = y + t\,\nabla f(y)$ for

$y \in M \subset \Sigma_\varepsilon(f)$ and $|t| < \varepsilon/k$, so that by $\nabla f(x) = \nabla f(y)$ we find

$$\left]x - \delta\,\nabla f(x), x + \delta\,\nabla f(x)\right[\ \subset \ \left]y - \varepsilon\,\nabla f(y), y + \varepsilon\,\nabla f(y)\right[\ \subset \Phi(P_f(y)) = \Phi(P_f(x))$$

and thus $x \in \Sigma_\delta(E)$.

Having proved the claim, and for reasons that will become apparent in step two, we set $\mathbf{p} : \mathbb{R}^n \times \mathbb{R} \to \mathbb{R}^n$, $\mathbf{p}(x,t) = x$, and consider the Lipschitz map

$$Q = \mathbf{p} \circ T^{-1} : E \to M, \tag{19.35}$$

and make a second claim: for $\mathcal{L}^n$-a.e. $x \in E$, setting

$$\mathbf{C}Q(x) = \det(\nabla Q(x)\,(\nabla Q(x))^*)^{1/2}$$

for $\nabla Q(x) \in (T_{Q(x)}M) \otimes \mathbb{R}^n$, there is $M_0 \subset M$ with $\mathcal{H}^{n-1}(M \setminus M_0)$ such that, for every $y \in M_0$,

> along each segment $\{y + t\nabla f(y) : t \in I_{\varepsilon/k}\}$ (19.36)
>
> $1/\mathbf{C}Q$ is $\mathcal{H}^1$-a.e. equal to a smooth function
>
> whose $(1/(n-1))$-power is concave.

To prove this second claim, since ∇f_δ is Lipschitz continuous on $\Sigma_\delta(f)$ and $M \subset E \subset \Sigma_\delta(f)$, we can find $M_0 \subset M$ with $\mathcal{H}^{n-1}(M \setminus M_0) = 0$ such that

> at every $y \in M_0$, ∇f_δ is tangentially differentiable along M.   (19.37)

Since $s_0 \in J_0$ and $M \subset \{f_\delta = s_0\}$, thanks to (19.33) and up to shrinking M_0, we can still retain $\mathcal{H}^{n-1}(M \setminus M_0) = 0$ while achieving, in addition to (19.37)

> at every $y \in M_0$, ∇f_δ is differentiable.   (19.38)

By (19.37) and (19.38) (see (A.31)), we thus find that

$$\nabla^M (\nabla f_\delta)(y)[\tau] = \nabla^2 f_\delta(x)[\tau] \qquad \forall \tau \in T_y M, \forall y \in M_0. \tag{19.39}$$

Since $\nabla f = \nabla f_\delta$ on $\Sigma_\delta(f)$ and ∇f is constant on the segment $\{y + t\,\nabla f(y) : t \in I_{\varepsilon/k}\} \subset \Sigma_\delta(f)$, we see that $\nabla^2 f_\delta[\nabla f(y)] = 0$, where $T_y M = (\nabla f(y))^\perp$. By the symmetry of $\nabla^2 f(y)$ we thus conclude that for every $y \in M_0$ we can find an orthonormal basis $\{\tau_i(y)\}_{i=1}^{n-1}$ of $T_y M \equiv \nabla f(y)^\perp$ and $\{\kappa_i(y)\}_{i=1}^{n-1}$ such that

$$\nabla^2 f_\delta(y) = \sum_{i=1}^{n-1} \kappa_i(y)\,\tau_i(y) \otimes \tau_i(y),$$

which combined with (19.39) gives

$$\nabla^M (\nabla f_\delta)(y) = \sum_{i=1}^{n-1} \kappa_i(y)\,\tau_i(y) \otimes \tau_i(y), \qquad \forall y \in M_0. \tag{19.40}$$

Now, since $M \subset E \subset \Sigma_\delta(f)$, we have $T(y,t) = y + t \nabla f_\delta(y)$ for $(y,t) \in M \times I_{\varepsilon/k}$, and thus we deduce from (19.37) and (19.40) that

for every $(y,t) \in M_0 \times I_{\varepsilon/k}$,

T is tangentially differentiable along $M \times I$ (19.41)

with $\quad \nabla^{M \times I} T(y,t) = \sum_{i=1}^{n-1} (1 + t \kappa_i(y)) \, \tau_i(y) \otimes \tau_i(y) + \nabla f_\delta(y) \otimes e,$

if e is such that $\{\tau_1(y), \ldots, \tau_{n-1}(y), e\}$ is an orthonormal basis of $T_y M \times \mathbb{R}$. We now notice that, since $\mathcal{H}^{n-1}(M \setminus M_0) = 0$, we have

$$\mathcal{L}^n \left(T((M \setminus M_0) \times I_{\varepsilon/k}) \right) = \int_{(M \setminus M_0) \times I_{\varepsilon/k}} J^{M \times I_{\varepsilon/k}} T \, d \left(\mathcal{H}^{n-1} \times \mathcal{L}^1 \right) = 0,$$

so that E is $\mathcal{L}^n$-equivalent to $T(M_0 \times I_{\varepsilon/k})$ and

there is $E_0 \subset E$ with $\mathcal{L}^n(E \setminus E_0) = 0$,

and, for every $x \in E_0$, Q is differentiable at x, (19.42)

and $x = T(y,t)$ for some $(y,t) \in M_0 \times I_{\varepsilon/k}$.

Now, by construction, $y = Q(y + t \nabla f(y))$ for every $(y,t) \in M \times I_{\varepsilon/k}$, and since $M \subset \Sigma_\delta(f) \subset \{\nabla f = \nabla f_\delta\}$, we have

$$y = Q(y + t \nabla f_\delta(y)) \qquad \forall (y,t) \in M \times I_{\varepsilon/k}. \tag{19.43}$$

For every $x = T(y,t)$ as in (19.42), we can differentiate (19.43) along $\tau_i(y)$ and exploit (19.41) to find

$$\tau_i(y) = \nabla Q(x) \big[(1 + t \kappa_i(y)) \, \tau_i(y) \big] = (1 + t \kappa_i(y)) \, \nabla Q(x)[\tau_i(y)];$$

similarly, we can differentiate along e to find that

$$0 = \nabla Q(x)[\nabla f(y)].$$

We have thus proved that for every $x \in E_0$ there is $(y,t) \in M_0 \times I_{\varepsilon/k}$ such that

$$\nabla Q(x) = \sum_{i=1}^{n-1} \frac{\tau_i(y) \otimes \tau_i(y)}{(1 + t \kappa_i(x))}, \qquad x = T(y,t), \tag{19.44}$$

and hence, by definition of $\mathbf{C}Q$, such that

$$\frac{1}{\mathbf{C}Q(x)} = \prod_{i=1}^{n-1} (1 + t \kappa_i(y)). \tag{19.45}$$

Now, the function on the right-hand side of this last identity is smooth, with concave $[1/(n-1)]$-power, along the segment $\{y + t\nabla f(y) : t \in I_{\varepsilon/k}\} \subset E$. To conclude, we notice that for every $y \in M_0$, the set

$$I_0(y) = \left\{ t \in I_{\varepsilon/k} : T(y,t) \in E_0 \right\}$$

is such that $\mathcal{L}^1(I_{\varepsilon/k} \setminus L_0(y)) = 0$: indeed, $J^{M \times I}T > 0$ on $M_0 \times I_{\varepsilon/k}$, and, by the area formula,

$$0 = \mathcal{L}^n(E \setminus E_0) = \int_{M_0} d\mathcal{H}^{n-1}(y) \int_{I_{\varepsilon/k} \setminus I_0(y)} J^{M \times I}T \, d\mathcal{L}^1.$$

Therefore, if $y \in M_0$, then for $\mathcal{L}^1$-a.e. $t \in I_{\varepsilon/k}$ we have $x = T(y,t) \in E_0$, and thus (19.45) holds for $\mathcal{H}^1$-a.e. $x \in \{y + t\nabla f(y) : t \in I_{\varepsilon/k}\}$. This proves (19.36).

Step two: To conclude the proof, we notice that since $P_f(M) = P_f(E)$ and $(P_f)|_M$ is Borel measurable and injective, it turns out that

$$\eta = [(P_f)|_M]^{-1} : P_f(E) \to M$$

is Borel measurable and satisfies

$$Q = \eta \circ (P_f)|_E. \tag{19.46}$$

By Remark 17.3, $\{\mu_\sigma \llcorner E/\mu_\sigma(E)\}_\sigma$ is the disintegration of $\mu\llcorner E = \mathcal{L}^n\llcorner(E \cap \mathcal{T}_c(f)) = \mathcal{L}^n\llcorner E$ with respect to $(P_f)|_E$; and thanks to (19.46), we can apply Remark 17.6 to find that if $\{\mu_y^Q\}_y$ is the disintegration of $\mu\llcorner E$ with respect to Q (so that, for $Q_\#\mu$-a.e. $y \in M$, μ_y^Q is concentrated on the open segment $Q^{-1}(y) = E \cap \Phi(P_f(y)))$, then

$$\frac{\mu_\sigma \llcorner E}{\mu_\sigma(E)} = \mu_{\eta(\sigma)}^Q \qquad \text{for } [(P_f)|_E]_\#\mu\text{-a.e. } \sigma \in P_f(E) \subset S^\circ.$$

In particular, thanks to (19.32),

$$\ell_\sigma \, d\mathcal{H}^1\llcorner(P_f^{-1}(\sigma) \cap E) = \mu_\sigma(E)\, \mu_{\eta(\sigma)}^Q \qquad \text{for } (P_f)_\#\mu\text{-a.e. } \sigma \in P_f(E) \subset S^\circ. \tag{19.47}$$

(where also $\mu_{\eta(\sigma)}^Q$ is concentrated on the open segment $P_f^{-1}(\sigma) \cap E$). Notice that Q is a Lipschitz map from a Borel set $E \subset \mathbb{R}^n$ to a locally $\mathcal{H}^{n-1}$-rectifiable set $M \subset \mathbb{R}^n$ so that the coarea formula

$$\int_F CQ = \int_M \mathcal{H}^1(F \cap Q^{-1}(y))\, d\mathcal{H}^{n-1}(y) \qquad \forall F \subset E \text{ Borel} \tag{19.48}$$

holds (see, in particular, Remark A.3 and (A.32)), provided

$$(CQ)^2 = \det[(\nabla Q)(\nabla Q)^*]$$

and $\nabla Q(x) \in (T_{Q(x)}M) \otimes \mathbb{R}^n$ (as explained in the appendix, for $\mathcal{L}^n$-a.e. $x \in E$, Q is differentiable at E, M has an approximate tangent plane at $Q(x)$, and $\nabla Q(x)$ is well defined as an element of $(T_{Q(x)}M) \otimes \mathbb{R}^n$). Based on (19.48), assuming that $\mathbf{C}^M Q > 0$ $\mathcal{L}^n$-a.e. on E (as we are going to prove in a moment) we can repeat the argument of Remark 17.5 to find $M^* \subset M$ with

$$\mu_y^Q = \frac{1}{CQ} d\mathcal{H}^1 \llcorner Q^{-1}(y) \qquad \forall y \in M^*, \qquad Q_\# \mu(M \setminus M^*) = 0.$$

In particular, if $\sigma \in \eta^{-1}(M^*)$, then $Q^{-1}(\eta(\sigma)) = [(P_f)|_E]^{-1}(\sigma) = E \cap P_f^{-1}(\sigma)$ gives

$$\mu_{\eta(\sigma)}^Q = \frac{1}{CQ} d\mathcal{H}^1 \llcorner [E \cap P_f^{-1}(\sigma)], \tag{19.49}$$

where

$$[(P_f)_E]_\# \mu \ (\eta^{-1}(M \setminus M^*)) = Q_\# \mu(M \setminus M^*) = 0.$$

Having proved that (19.49) holds for $(P_f)_\# \mu$-a.e. $\sigma \in P_f(E)$, and recalling (19.47), we conclude that, for $(P_f)_\# \mu$-a.e. $\sigma \in P_f(E)$,

$$\ell_\sigma = \frac{\mu_\sigma(E)}{CQ} \qquad \mathcal{H}^1\text{-a.e. on } E \cap P_f^{-1}(\sigma), \tag{19.50}$$

and thus that ℓ_σ is $\mathcal{H}^1$-equivalent to a smooth function with concave $1/(n-1)$-power thanks to (19.36). Thanks to (19.34), resorting back to the notation where $E = E_{\varepsilon,s}$, each open interval $P_f^{-1}(\sigma)$ is covered by the open intervals $\{E_{\varepsilon,s} \cap P_f^{-1}(\sigma)\}_{\varepsilon > 0, s \in I_0}$, and the proof is complete. $\qquad\square$

We conclude with the proof of the localization theorem.

Proof of Theorem 19.4 Let Ω, u, μ, v and f be as in the statement of the theorem. By Theorem 19.7 we only need to prove (19.17) and the second stated property in Theorem 19.4-(i). To this end, we divide the proof in a few steps.

Step one: We claim that if K is compact in $\mathbb{R}^n$, then

$$\left| \int_K u \right| \leq \int_{\mathbb{R}^n \setminus [K \cup \text{Dis}(K;f)]} |u|, \tag{19.51}$$

where we have denoted by

$$\text{Dis}(K; f) = \left\{ x \in \mathbb{R}^n : |f(x) - f(y)| < |x - y| \text{ for every } y \in K \right\}$$

the set of those $x \in \mathbb{R}^n$ that are "not connected to K" by the transport ray geometry of f. To prove (19.51), let us define $g_\delta : \mathbb{R}^n \to \mathbb{R}$ by setting

$$g_\delta(x) = \inf_{y \in \mathbb{R}^n} f(y) + |x - y| - \delta \, 1_K(y), \qquad x \in \mathbb{R}^n.$$

Evidently, g_δ is 1-Lipschitz (so that

$$\int_{\mathbb{R}^n} u f \geq \int_{\mathbb{R}^n} u g_\delta \qquad (19.52)$$

by Theorem 3.17), with

$$0 \leq f - g_\delta \leq \delta \text{ on } \mathbb{R}^n, \qquad g_\delta + \delta = f \text{ on } K. \qquad (19.53)$$

Now, if we set $v_\delta = (f - g_\delta)/\delta$ ($\delta > 0$), then by (19.53) we have $v_\delta : \mathbb{R}^n \to [0,1]$; moreover, we easily see that, for every $x \in \mathbb{R}^n$, $\delta \in (0,\infty) \mapsto v_\delta(x)$ is increasing, and, in particular, $v = \lim_{\delta \to 0^+} v_\delta : \mathbb{R}^n \to [0,1]$ defines a Borel function such that (thanks to (19.52))

$$\int_{\mathbb{R}^n} u v \geq 0. \qquad (19.54)$$

Now $v = 1$ on K by (19.53), while we claim that $v = 0$ on $\mathrm{Dis}(K; f)$: indeed, if $x \in \mathrm{Dis}(K; f)$, then $|f(x) - f(y)| < |x - y|$ for every $y \in K$ and, in particular,

$$\inf_{y \in K} f(y) + |x - y| = f(y_x) + |x - y_x| \geq f(x) + \delta_x,$$

for some $\delta_x > 0$; and hence, for every $\delta \in (0, \delta_x)$,

$$f(y) + |x - y| - \delta 1_K(y) - f(x) \geq (\delta_x - \delta) 1_K(y) \geq 0$$

so that $g_\delta \geq f$ on $\mathrm{Dis}(K; f)$, and hence $g_\delta = f$ therein by (19.53). Having proved $v = 1$ on K and $v = 0$ on $\mathrm{Dis}(K; f)$ we conclude from (19.54)

$$\int_K u \geq - \int_{\mathbb{R}^n \setminus [K \cup \mathrm{Dis}(K;f)]} |u|. \qquad (19.55)$$

Considering that $-f$ is a Kantorovich potential from v to μ, that $(-u)^+ = u^-$ and $(-u)^- = u^+$, and that $\mathrm{Dis}(K; f) = \mathrm{Dis}(K; -f)$, the above argument also entails

$$\int_K (-u) \geq - \int_{\mathbb{R}^n \setminus [K \cup \mathrm{Dis}(K;f)]} |u|,$$

which combined with (19.55) gives (19.51).

Step two: We now claim that

$$\int_A u = 0, \qquad (19.56)$$

whenever A is a **partial transport set for** f, that is, whenever $A \subset \mathcal{T}(f)$ and for every $x \in A$, it holds that $\Phi(P_f(x)) \subset A$. Indeed, if $K \subset A$ is compact and

$$x \in \mathbb{R}^n \setminus (K \cup \mathrm{Dis}(K; f)),$$

then there is $y \in K$ such that $|f(x) - f(y)| = |x - y|$; since $x \notin K$, it must be $x \neq y$, and

$$]x, y[\subset \Phi(P_f(y)) \subset A.$$

In particular, either $x \in \Phi(P_f(y))$, and thus $x \in A$; or x is an end-point of $P_f(y)$, and thus $x \in \mathcal{T}_c(f) \setminus \mathcal{T}(f)$; this shows that

$$\mathbb{R}^n \setminus (K \cup \text{Dis}(K; f)) \subset A \cup (\mathcal{T}_c(f) \setminus \mathcal{T}(f)).$$

Since $\mathcal{L}^n(\mathcal{T}_c(f) \setminus \mathcal{T}(f)) = 0$, we conclude from (19.51)

$$\left| \int_K u \right| \le \int_{A \setminus K} |u|. \tag{19.57}$$

By the arbitrariness of $K \subset A$ and by the fact that $u \in L^1(\mathbb{R}^n)$, we find (19.56).

Step four: We now prove (19.17), that is,

$$u = 0 \text{ a.e. on } \Omega \setminus \mathcal{T}(f). \tag{19.58}$$

Indeed, let us consider the **isolated set of** f,

$$\text{Iso}(f) = \left\{ x \in \mathbb{R}^n : |f(x) - f(y)| < |x - y| \quad \forall y \in \mathbb{R}^n \right\} = \text{Dis}(\mathbb{R}^n; f),$$

i.e., those points that are "totally isolated" in the transport geometry of f, and notice that

$$\mathbb{R}^n \setminus \mathcal{T}(f) = \text{Iso}(f) \cup \left(\mathcal{T}_c(f) \setminus \mathcal{T}(f) \right). \tag{19.59}$$

If $K \subset \text{Iso}(f)$ is compact, then

$$\text{Dis}(K; f) \supset \text{Dis}(\text{Iso}(f); f) = \mathbb{R}^n,$$

so that (19.51) implies $\int_K u = 0$; by arbitrariness of K, $u = 0$ $\mathcal{L}^n$-a.e. on $\text{Iso}(f)$ and thus on $\mathcal{T}(f)$, thanks to $\mathcal{L}^n(\mathcal{T}_c(f) \setminus \mathcal{T}(f)) = 0$ and (19.59).

Conclusion: Let $\{\lambda_\sigma\}_\sigma$ be the disintegration of $\lambda = \mathcal{L}^n \llcorner (\Omega \cap \mathcal{T}(f))$ with respect to P_f. For every $\Sigma \subset S^\circ$ we have, thanks to (19.56) applied to $A = P_f^{-1}(\Sigma)$,

$$\int_\Sigma d[(P_f)_\# \lambda](\sigma) \int_{P_f^{-1}(\sigma)} u \, d\lambda_\sigma = \int_{P_f^{-1}(\Sigma)} u = 0.$$

By the arbitrariness of Σ we conclude that

$$\int_{P_f^{-1}(\sigma)} u \, d\lambda_\sigma = 0 \qquad \text{for } (P_f)_\# \lambda\text{-a.e. } \sigma \in S^\circ, \tag{19.60}$$

which proves the second stated property in Theorem 19.4-(i). This completes the proof of the localization theorem. $\qquad \square$

Appendix A
Radon Measures on $\mathbb{R}^n$ and Related Topics

Here we summarize a few aspects of the theory of Radon measures on $\mathbb{R}^n$ that provide the basic language for the rest of the notes. We will follow the presentation given in [Mag12, Part I]. We assume readers have a certain familiarity with abstract measure theory, including topics like measurability of functions, integration, limit theorems, and Lebesgue spaces.

A.1 Borel and Radon Measures, Main Examples

A **Borel measure** μ is a measure on the σ-algebra of Borel sets $\mathcal{B}(\mathbb{R}^n)$. A **Radon measure** on $\mathbb{R}^n$ is a Borel measure on $\mathbb{R}^n$, which is locally finite (i.e., finite on compact sets). A **probability measure** (for the purpose of these notes!) is a Radon measure μ on $\mathbb{R}^n$ such that $\mu(\mathbb{R}^n) = 1$. The **Lebesgue measure on** $\mathbb{R}^n$ is defined on an arbitrary set $E \subset \mathbb{R}^n$ as

$$\mathcal{L}^n(E) = |E| = \inf_{\mathcal{F}} \sum_{Q \in \mathcal{F}} r(Q)^n,$$

where $\mathcal{F}$ ranges among the countable coverings of E by cubes Q with sides parallel to the coordinate axes, and $r(Q)$ denotes the side length of Q. Notice that $\mathcal{L}^n$ is countably additive on $\mathcal{B}(\mathbb{R}^n)$ (and actually on a larger σ-algebra than $\mathcal{B}(\mathbb{R}^n)$, but not on the whole $2^{\mathbb{R}^n}$) and that it is locally finite, so $\mathcal{L}^n$ is a Radon measure. Given $k \in \mathbb{N}$ and $\delta \in (0, \infty]$, the k-**dimensional Hausdorff measure of step** δ is defined on $E \subset \mathbb{R}^n$ by

$$\mathcal{H}^k_\delta(E) = \inf_{\mathcal{F}} \sum_{F \in \mathcal{F}} \omega_k \left(\frac{\mathrm{diam}(F)}{2} \right)^k,$$

where $\omega_k = \mathcal{L}^k(\{x \in \mathbb{R}^k : |x| < 1\})$ and $\mathcal{F}$ ranges among all the countable coverings of E by sets F with $\mathrm{diam}(F) < \delta$; the k-**dimensional Hausdorff measure** of $E \subset \mathbb{R}^n$ is then given by

$$\mathcal{H}^k(E) = \lim_{\delta \to 0^+} \mathcal{H}^k_\delta(E) = \sup_{\delta > 0} \mathcal{H}^k_\delta(E).$$

It turns out that $\mathcal{H}^k$ defines a Borel measure with the following three important properties: (i) $\mathcal{H}^0$ is the **counting measure**, it measures the number of elements of subsets of $\mathbb{R}^n$: for this reason we sometimes write #(E) in place of $\mathcal{H}^0(E)$; (ii) $\mathcal{H}^n = \mathcal{L}^n$ on Borel sets of $\mathbb{R}^n$; (iii) if $1 \le k \le n-1$, M is a k-dimensional C^1-surface in $\mathbb{R}^n$, and Area(M) denotes the area of M (as defined by using parameterizations and partitions of unity as in basic Differential Geometry), then

$$\mathcal{H}^k(M) = \text{Area}(M).$$

Given a Borel measure ν on $\mathbb{R}^n$ and $E \in \mathcal{B}(\mathbb{R}^n)$, the **restriction** $\mu = \nu \llcorner E$ **of** ν **to** E, defined by $\mu(F) = \nu(E \cap F)$ for every $F \in \mathcal{B}(\mathbb{R}^n)$, is still a Borel measure; if $\nu(E \cap B_R) < \infty$ for every $R > 0$, then $\mu = \nu \llcorner E$ is a Radon measure on $\mathbb{R}^n$. For example, if $x \in \mathbb{R}^n$, then **Dirac's delta** $\delta_x = \mathcal{H}^0 \llcorner \{x\}$ is a Radon measure on $\mathbb{R}^n$; if M is a k-dimensional C^1-surface in $\mathbb{R}^n$ with $\mathcal{H}^k(M \cap B_R) < \infty$ for every $R > 0$, then $\mathcal{H}^k \llcorner M$ is a Radon measure on $\mathbb{R}^n$ (so that the convergence of sequences of surfaces can be studied in terms of the convergence of their associated Radon measures).

A.2 Support and Concentration Set of a Measure

A Borel measure μ is **concentrated** on a Borel set E if $\mu(\mathbb{R}^n \setminus E) = 0$; in this case, if a property holds at every $x \in E$, we say that it holds μ-a.e. The intersection of all the closed sets of concentration of μ defines the **support of** μ, which satisfies

$$\text{spt}\,\mu = \{x \in \mathbb{R}^n : \mu(B_r(x)) > 0 \quad \forall r > 0\}. \tag{A.1}$$

So if a property holds at every $x \in \text{spt}\mu$, then it holds μ-a.e. A measure may be concentrated on a strict subset of its support, e.g., $\mu = \sum_{j=1}^{\infty} 2^{-j} \delta_{1/j}$ is a Radon measure on $\mathbb{R}$ with $\text{spt}\mu = \{1/j : j \ge 1\} \cup \{0\}$, and μ is concentrated on $\text{spt}\mu \setminus \{0\}$.

A.3 Fubini's Theorem

Given that Radon measures on Euclidean spaces are automatically σ-finite, if μ and ν are Radon measures, then we can always apply **Fubini's theorem** to write integrals in the product measure $\mu \times \nu$ as iterated integrals in μ and ν. Of particular interest is the **layer cake formula**

$$\int_{\mathbb{R}^n} g \, d\mu = \int_0^\infty \mu(\{g > t\}) \, dt, \qquad \forall g : \mathbb{R}^n \to [0, \infty] \text{ Borel function, (A.2)}$$

which follows by applying Fubini's theorem to μ and $\mathcal{L}^1$,

$$\int_{\mathbb{R}^n} g \, d\mu = \int_{\mathbb{R}^n} d\mu(x) \int_{\mathbb{R}} 1_{[0,g(x))}(t) \, dt$$

$$= \int_{\mathbb{R}} dt \int_{\mathbb{R}^n} 1_{[0,g(x))}(t) \, d\mu(x) = \int_0^\infty \mu(\{g > t\}) \, dt,$$

where in the last identity one uses $1_{[0,g(x))}(t) = 1_{[0,\infty)}(t) \, 1_{\{g > t\}}(x)$. Notice that, thanks to (A.11), and the fact that the Borel sets $\{g = t\}$, $t > 0$, are disjoint, identity (A.2) also holds with $\{g \geq t\}$ replacing $\{g > t\}$.

A.4 Push-Forward of a Measure

A **Borel function** is a function $T : F \to \mathbb{R}^m$ (F a Borel set in $\mathbb{R}^n$) such that $T^{-1}(A) = \{x \in F : T(x) \in A\} \in \mathcal{B}(\mathbb{R}^n)$ for every open set $A \subset \mathbb{R}^m$. If μ is a Borel measure on $\mathbb{R}^n$, concentrated on F, then the **push-forward measure** $T_\# \mu$ is the Borel measure on $\mathbb{R}^m$ defined by

$$T_\# \mu(E) = \mu(T^{-1}(E)) \qquad \forall E \in \mathcal{B}(\mathbb{R}^m), \tag{A.3}$$

or, alternatively, by

$$\int_{\mathbb{R}^m} \varphi \, d[T_\# \mu] = \int_{\mathbb{R}^n} (\varphi \circ T) \, d\mu, \qquad \forall \varphi \in C_c^0(\mathbb{R}^n). \tag{A.4}$$

(By standard approximation arguments, (A.4) holds also for every Borel function $\varphi : \mathbb{R}^n \to [0, \infty]$, and for every bounded Borel function $\varphi : \mathbb{R}^m \to \mathbb{R}$.) If T is **proper** (i.e., $T^{-1}(K)$ is compact in $\mathbb{R}^n$ for every compact set K in $\mathbb{R}^m$) or if μ is **finite** (i.e., $\mu(\mathbb{R}^n) < \infty$), then $T_\# \mu$ is a Radon measure on $\mathbb{R}^m$. Notice that since μ is concentrated on F, we have that $T_\# \mu$ is concentrated on $T(F)$. Moreover,

$$\mathrm{spt}(T_\# \mu) = T(\mathrm{spt}\mu) \tag{A.5}$$

whenever T can be extended to a continuous function on the whole $\mathrm{spt}\mu$.

A.5 Approximation Properties

Countable additivity on Borel sets ties Radon measures to the Euclidean topology, and leads to a crucial **approximation property**: if μ is a Radon measure, then for every Borel set $E \subset \mathbb{R}^n$ we have

$$\mu(E) = \inf\{\mu(A) : E \subset A \text{ open}\} = \sup\{\mu(K) : K \subset E, K \text{ compact}\}. \tag{A.6}$$

As a consequence, one has the classical **Lusin's theorem**, from which we readily deduce that $C_c^0(\mathbb{R}^n)$ is dense in $L^p(\mu)$ whenever $1 \leq p < \infty$ and μ is a Radon measure on $\mathbb{R}^n$.

A.6 Weak-Star Convergence

A sequence $\{\mu_j\}_j$ of Radon measures is **weakly-star converging** as $j \to \infty$ to a Radon measure μ, and we write $\mu_j \stackrel{*}{\rightharpoonup} \mu$, if

$$\lim_{j\to\infty} \int_{\mathbb{R}^n} \varphi \, d\mu_j = \int_{\mathbb{R}^n} \varphi \, d\mu, \qquad \forall \varphi \in C_c^0(\mathbb{R}^n). \tag{A.7}$$

Recall that (A.7) is equivalent to

$$\mu(A) \le \liminf_{j\to\infty} \mu_j(A) \qquad \text{for all open sets } A \subset \mathbb{R}^n \tag{A.8}$$

$$\mu(K) \ge \limsup_{j\to\infty} \mu_j(K) \qquad \text{for all compact sets } K \subset \mathbb{R}^n \tag{A.9}$$

and that in turn (A.8)+(A.9) is equivalent to

$$\mu(E) = \lim_{j\to\infty} \mu_j(E) \qquad \text{for all bounded } E \in \mathcal{B}(\mathbb{R}^n) \text{ s.t. } \mu(\partial E) = 0. \tag{A.10}$$

Condition (A.10) is often used in conjunction with the following fact: If $\{S_t\}_{t\in T}$ is a disjoint family of Borel sets in $\mathbb{R}^n$, indexed over an arbitrary set T, and if μ is a Borel measure on $\mathbb{R}^n$ which is finite on $S = \bigcup_{t\in T} S_t$, there are at most countably many values of t such that $\mu(S_t) > 0$: indeed,

$$\# \left\{ t \in T : \mu(S_t) > \varepsilon \right\} \le \frac{\mu(E)}{\varepsilon}, \tag{A.11}$$

whenever $\varepsilon > 0$. For example, if μ is a Radon measure on $\mathbb{R}^n$ and $x \in \mathbb{R}^n$, then we have $\mu(\partial B_r(x)) = 0$ for a.e. $r > 0$ (actually, for all but countably many values of r), so that if $\mu_j \stackrel{*}{\rightharpoonup} \mu$ on $\mathbb{R}^n$, then $\mu_j(B_r(x)) \to \mu(B_r(x))$ for a.e. $r > 0$ thanks to (A.10). Finally, we notice that $\mu_j \stackrel{*}{\rightharpoonup} \mu$ implies that

$$\forall x_0 \in \mathrm{spt}\mu \text{ there exist } x_j \in \mathrm{spt}\mu_j \text{ s.t. } x_j \to x_0 \text{ as } j \to \infty. \tag{A.12}$$

The converse property is in general false, that is to say, one can have $x_j \in \mathrm{spt}\,\mu_j$, $x_j \to x_0$, and $x_0 \notin \mathrm{spt}\mu$. For example, consider $\mu_j = (1 - (1/j))\,\delta_1 + (1/j)\delta_{1/j}$ on $\mathbb{R}$ with $x_j = 1/j$.

A.7 Weak-Star Compactness

The paramount importance of weak-star convergence lies in two facts: first, many different mathematical objects (functions, surfaces, etc.) can be seen as Radon measures; second, every sequence of Radon measures with locally bounded mass has a weak-star subsequential limit. This is the **Compactness Theorem for Radon measures**: if $\{\mu_j\}_j$ is a sequence of Radon measures on $\mathbb{R}^n$ with

$$\sup_j \mu_j(B_R) < \infty \qquad \forall R > 0,$$

then there exists a Radon measure μ on $\mathbb{R}^n$ such that, up to extract subsequences, $\mu_j \overset{*}{\rightharpoonup} \mu$ on $\mathbb{R}^n$.

A.8 Narrow Convergence

A sequence $\{\mu_h\}_h$ of Radon measures is **narrowly converging** as $h \to \infty$ to a Radon measure μ, and we write $\mu_h \overset{n}{\rightharpoonup} \mu$, if

$$\lim_{h\to\infty} \int_{\mathbb{R}^n} \varphi \, d\mu_h = \int_{\mathbb{R}^n} \varphi \, d\mu, \qquad \forall \varphi \in C_b^0(\mathbb{R}^n); \qquad (A.13)$$

where $C_b^0(\mathbb{R}^n)$ is the space of all bounded continuous functions on $\mathbb{R}^n$. Notice that in general $\mu_h \overset{*}{\rightharpoonup} \mu$ on $\mathbb{R}^n$ does not imply $\mu_h(\mathbb{R}^n) \to \mu(\mathbb{R}^n)$, since some mass contained in μ_h may be "lost at infinity": for example, $\mu_h = \delta_{x_h} \overset{*}{\rightharpoonup} \mu = 0$ if the points $x_h \in \mathbb{R}^n$ are such that $|x_h| \to \infty$. The importance of narrow convergence is that is exactly equal to weak-star convergence plus conservation of total mass (notice indeed that $\mu_h = \delta_{x_h}$ does not admit a narrow limit as soon as $|x_h| \to \infty$).

Proposition A.1 (Narrow convergence criterion) *Assume that μ_h and μ are finite Radon measures on $\mathbb{R}^n$ with $\mu_h \overset{*}{\rightharpoonup} \mu$ on $\mathbb{R}^n$ as $h \to \infty$. Then the following statements are equivalent:*
(i): $\mu_h(\mathbb{R}^n) \to \mu(\mathbb{R}^n)$ as $h \to \infty$;
(ii): μ_h is narrowly converging to μ on $\mathbb{R}^n$;
(iii): for all $\varepsilon > 0$ there exists $K_\varepsilon \subset \mathbb{R}^n$ compact such that $\mu_h(\mathbb{R}^n \setminus K_\varepsilon) < \varepsilon$ for every h.

Proof   Proof that (i) is equivalent to (iii): Assume (iii), fix $\varepsilon > 0$ and consider K_ε such that $\mu_h(\mathbb{R}^n \setminus K_\varepsilon) < \varepsilon$ for every h: then, by (A.9), we have

$$\mu(\mathbb{R}^n) \geq \mu(K_\varepsilon) \geq \limsup_{h\to\infty} \mu_h(K_\varepsilon) \geq \limsup_{h\to\infty} \mu_h(\mathbb{R}^n) - \varepsilon \geq \mu(\mathbb{R}^n) - \varepsilon,$$

where in the last inequality we have used (A.8). Viceversa, we just need to prove that (i) implies

$$\lim_{R\to 0^+} \sup_{h \in \mathbb{N}} \mu_h(\mathbb{R}^n \setminus B_R) = 0.$$

Indeed, by $\mu(\mathbb{R}^n) < \infty$, given $\varepsilon > 0$ we can find $R > 0$ such that $\mu(\mathbb{R}^n \setminus \mathrm{Cl}(B_R)) < \varepsilon$. Moreover, for all but countably many values of r we have $\mu(\partial B_r) = 0$ so that, up to increase the value of R we can also assume that $\mu(\partial B_R) = 0$. By applying (A.10) with $E = \mathrm{Cl}(B_R)$ we find that $\mu_h(\mathrm{Cl}(B_R)) \to \mu(\mathrm{Cl}(B_R))$, and thus, thanks to (i), that $\mu_h(\mathbb{R}^n \setminus \mathrm{Cl}(B_R)) \to \mu(\mathbb{R}^n \setminus \mathrm{Cl}(B_R))$ as $h \to \infty$, which gives $\mu_h(\mathbb{R}^n \setminus \mathrm{Cl}(B_R)) < \varepsilon$ for every h large enough.

Proof that (i) is equivalent to (ii): If (ii) holds then we can test (A.7) at $\varphi \equiv 1$ and deduce (i). Conversely, let $\varphi \in C_b^0(\mathbb{R}^n)$ and let $\zeta_R \in C_c^0(B_R)$ with $0 \leq \zeta_R \leq 1$ and $\zeta_R \to 1$ on $\mathbb{R}^n$ as $R \to \infty$. We have that

$$\left| \int_{\mathbb{R}^n} \varphi \, d\mu_h - \int_{\mathbb{R}^n} \varphi \, \zeta_R \, d\mu_h \right| \leq \|\varphi\|_{C^0(\mathbb{R}^n)} \, \mu_h(\mathbb{R}^n \setminus B_R),$$

$$\left| \int_{\mathbb{R}^n} \varphi \, d\mu - \int_{\mathbb{R}^n} \varphi \, \zeta_R \, d\mu \right| \leq \|\varphi\|_{C^0(\mathbb{R}^n)} \, \mu(\mathbb{R}^n \setminus B_R),$$

and that, for every $R > 0$, $\int_{\mathbb{R}^n} \varphi \, \zeta_R \, d\mu_h \to \int_{\mathbb{R}^n} \varphi \, \zeta_R \, d\mu$ as $h \to \infty$. Therefore, to show that $\int_{\mathbb{R}^n} \varphi \, d\mu_h \to \int_{\mathbb{R}^n} \varphi \, d\mu$ as $h \to \infty$,

$\square$

A.9 Differentiation Theory of Radon Measures

Given two Borel measures μ and ν, we say that μ **is absolutely continuous with respect to** ν, and write $\mu \ll \nu$, if $\nu(E) = 0$, $E \in \mathcal{B}(\mathbb{R}^n)$ implies $\mu(E) = 0$. We say that μ **and** ν **are mutually singular**, and write $\mu \perp \nu$, if there exists $E \in \mathcal{B}(\mathbb{R}^n)$ such that $\mu(E) = \nu(\mathbb{R}^n \setminus E) = 0$. The two concepts are complementary, since $\mu \ll \nu$ and $\mu \perp \nu$ imply $\mu = 0$. When μ is a Radon measure, we define Borel functions $D_\mu^+ \nu, D_\mu^- \nu : \mathrm{spt}\mu \to [0, \infty]$ by setting

$$D_\mu^+ \nu(x) = \limsup_{r \to 0^+} \frac{\nu(B_r(x))}{\mu(B_r(x))}, \qquad D_\mu^- \nu(x) = \liminf_{r \to 0^+} \frac{\nu(B_r(x))}{\mu(B_r(x))}, \qquad \forall x \in \mathrm{spt}\mu.$$

On the Borel set $E = \{x \in \mathrm{spt}\mu : D_\mu^+ \nu(x) = D_\mu^- \nu(x)\}$ we define the μ-**density** $D_\mu \nu(x)$ **of** ν **at** x as the common value of the two limits, and notice that $D_\mu \nu : E \to [0, \infty]$ is a Borel function. The **Lebesgue-Besicovitch covering theorem** states that if μ and ν are Radon measures, then $D_\mu \nu$ is defined μ-a.e. on $\mathbb{R}^n$, $D_\mu \nu \in L^1_{\mathrm{loc}}(\mu)$, and

$$\nu = D_\mu \nu \, d\mu + \nu_\mu^s \qquad \text{on } \mathcal{B}(\mathbb{R}^n),$$

where the Radon measure ν_μ^s is concentrated on the Borel set

$$Y = (\mathbb{R}^n \setminus \mathrm{spt}\mu) \cup \left\{ x \in \mathrm{spt}\mu : D_\mu^+ \nu(x) = +\infty \right\}, \tag{A.14}$$

and is thus mutually singular with μ. In the above formula we are using the notation $f \, d\mu$ to denote the integral measure $f \, d\mu(E) = \int_E f \, d\mu$. When $\mu = \mathcal{L}^n$, we typically write $f(x) \, dx$ in place of $f \, d\mathcal{L}^n$. The classical theory of Lebesgue differentiation of monotone functions is recovered (for example) by first identifying a monotone function $u : \mathbb{R} \to \mathbb{R}$ with the Radon measure μ defined by $\mu((a,b)) = u(b^-) - u(a^+)$ for $a < b$, and then by differentiating μ with respect to $\mathcal{L}^1$ in the sense of Radon measures. Another classical consequence of the differentiation theory is that if $f \in L^1_{\mathrm{loc}}(\mu)$, then for μ-a.e. $x \in \mathbb{R}^n$ one has

$$\lim_{r \to 0^+} \frac{1}{\mu(B_r(x))} \int_{B_r(x)} |f - f(x)| \, d\mu = 0. \tag{A.15}$$

Every such point x is called a **Lebesgue point of** f **with respect to** μ, and the set of Lebesgue points of a function is always a Borel set. The following variant of (A.15) is often useful: we say that a sequence of Borel sets $\{E_j\}_j$ shrinks nicely at $x \in \mathbb{R}^n$ (with respect to μ) if there exist $r_j \to 0^+$ such that

$$E_j \subset B_{r_j}(x), \qquad \liminf_{j \to \infty} \frac{\mu(E_j)}{\mu(B_{r_j}(x))} = c > 0. \tag{A.16}$$

Clearly, (A.15) and (A.16) imply that

$$\lim_{j \to \infty} \frac{1}{\mu(E_j)} \int_{E_j} |f - f(x)| \, d\mu = 0 \tag{A.17}$$

whenever x is a Lebesgue point of μ and $\{E_j\}_j$ shrinks nicely at $x \in \mathbb{R}^n$ with respect to μ.

A.10 Lipschitz Functions and Area Formula

Given $E \subset \mathbb{R}^n$ and $f : E \to \mathbb{R}^m$, the Lipschitz constant of f on E, $\mathrm{Lip}(f; E)$, is the supremum of $|f(x) - f(y)|/|x - y|$ over $x, y \in E$, $x \neq y$. When $E = \mathbb{R}^n$ we set $\mathrm{Lip}(f) = \mathrm{Lip}(f; \mathbb{R}^n)$. By **Kirszbraun's theorem**, if $\mathrm{Lip}(f; E) < \infty$ then there exists $g : \mathbb{R}^n \to \mathbb{R}^m$ with $\mathrm{Lip}(g) = \mathrm{Lip}(f; E)$ and $f = g$ on E. For this reason one often considers Lipschitz functions as defined on the whole space.

If f is a Lipschitz function over an open set $\Omega \subset \mathbb{R}^n$, then f admits a distributional gradient $\nabla f \in L^\infty_{\mathrm{loc}}(\Omega; \mathbb{R}^n)$, [Mag12, Proposition 7.7]. Moreover, f is differentiable at every Lebesgue point of ∇f [Mag12, Theorem 7.7], and thus is a.e. differentiable in Ω (**Rademacher's theorem**).

If $n \leq m$ and $f : \mathbb{R}^n \to \mathbb{R}^m$ is Lipschitz continuous, then the Jacobian Jf of f is defined by

$$Jf(x) = \sqrt{\det((\nabla f)^*(x)\nabla f(x))}$$

(where A^* denotes the transpose of the matrix A) at every x point of differentiability of f. By the **area formula**, for every Borel function $g : \mathbb{R}^n \to [0, \infty]$, we have

$$\int_{\mathbb{R}^m} dy \int_{\{f=y\}} g(x) \, d\mathcal{H}^0(x) = \int_{\mathbb{R}^n} g(x) \, Jf(x) \, dx. \tag{A.18}$$

Of particular importance is the case when f is injective, $\{f = y\}$ consists of exactly one point for every $y \in f(\mathbb{R}^n)$, and thus (A.18) boils down to

$$\int_{\mathbb{R}^m} g(f^{-1}(y)) \, dy = \int_{\mathbb{R}^n} g(x) \, Jf(x) \, dx, \tag{A.19}$$

for every Borel function $g : \mathbb{R}^n \to [0, \infty]$, or equivalently to

$$\int_{\mathbb{R}^m} h(y) \, dy = \int_{\mathbb{R}^n} h(f(x)) \, Jf(x) \, dx, \tag{A.20}$$

for every Borel function $h : \mathbb{R}^m \to [0, \infty]$.

A.11 Vector-valued Radon Measures

An $\mathbb{R}^m$-**valued Radon measure** μ in an open set $\mathbb{R}^n$ of $\mathbb{R}^n$ is a linear functional $\mu : C_c^0(\mathbb{R}^n) \to \mathbb{R}^m$ such that for every $\Omega \subset\subset \mathbb{R}^n$ we can find a constant $C(\Omega)$ such that $|\mu[\varphi]| \le C(\Omega) \, \|\varphi\|_{C^0}$ for every $\varphi \in C_c^0(\Omega)$. When $m = 1$ we say that μ is a **signed Radon measure**. The **total variation** of μ is the Radon measure $|\mu|$ defined by

$$|\mu|(A) = \sup \left\{ \mu[\varphi] : |\varphi| \le 1, \varphi \in C_c^0(A) \right\} \qquad \forall A \subset \mathbb{R}^n \text{ open}$$

and by $|\mu|(E) = \inf\{|\mu|(A) : E \subset A, A \text{ open}\}$ if $E \in \mathcal{B}(\mathbb{R}^n)$. The concepts of absolute continuity and mutual singularity with respect to a Borel measure ν on $\mathbb{R}^n$ are introduced by requiring, respectively, $|\mu| \ll \nu$ or $|\mu| \perp \nu$.

When $m = 1$ and μ is nonnegative, in the sense that $\mu[\varphi] \ge 0$ whenever $\varphi \in C_c^0(\Omega)$ is nonnegative, then by the **Riesz theorem** μ can be identified as the integration with respect to a Radon measure on $\mathbb{R}^n$, still denoted by μ, so that $\mu[\varphi] = \int \varphi \, d\mu$ and $|\mu| = \mu$. Coming back to the general case, if μ is an $\mathbb{R}^m$-valued Radon measure, then there exists a Borel map $g : \mathbb{R}^n \to \mathbb{R}^m$ such that $|g(x)| = 1$ for $|\mu|$-a.e. $x \in \mathbb{R}^n$ and

$$\mu[\varphi] = \int_{\mathbb{R}^n} \varphi g \, d\nu, \qquad \forall \varphi \in C_c^0(\mathbb{R}^n).$$

For $i = 1, \ldots, m$, the components of μ are the signed Radon measures $\mu_i = (e_i \cdot g) \, d|\mu|$. When $m = 1$, i.e., when μ is a signed Radon measure, then g take values in $\{\pm 1\}$ $|\mu|$-a.e., and the **positive and negative parts of** μ are defined by $\mu^+ = g^+ \, d|\mu|$ and $\mu^- = g^- \, d|\mu|$, so that $\mu = \mu^+ - \mu^-$, $\mu^+ \perp \mu^-$, and $\mu^\pm$ are Radon measures. More generally, if μ is an $\mathbb{R}^m$-valued Radon measure and ν is a Radon measure, then we have the decomposition $\mu = D_\nu \mu \, d\nu + \mu^s$ where $\mu^s \perp \nu$ and where $D_\nu \mu$ is a Borel function defined ν-a.e. on $\mathbb{R}^n$ and with values in $\mathbb{R}^m$, defined by noticing that

$$D_\nu \mu(x) = \lim_{r \to 0^+} \frac{\mu(B_r(x))}{\nu(B_r(x))}$$

exists for ν-a.e. $x \in \mathbb{R}^n$. The function g such that $\mu = g \, d|\mu|$ can be indeed identified with $D_{|\mu|}\mu$.

A.12 Regularization by Convolution

Given $\rho \geq 0$, $\rho \in C_c^\infty(B_1)$, with $\rho(x) = \rho(-x)$ and $\int_{\mathbb{R}^n} \rho = 1$, we set $\rho_\varepsilon(x) = \varepsilon^{-n} \rho(x/\varepsilon)$ and define, for every $\mathbb{R}^m$-valued Radon measure μ on $\mathbb{R}^n$, the functions

$$\mu_\varepsilon(x) = (\mu \star \rho_\varepsilon)(x) = \int_{\mathbb{R}^n} \rho_\varepsilon(y - x) \, d\mu(y) \qquad x \in \mathbb{R}^n, \varepsilon > 0.$$

In this way $\mu_\varepsilon \in C_c^\infty(\mathbb{R}^n; \mathbb{R}^m)$ and

$$\mu_\varepsilon \, d\mathcal{L}^n \overset{*}{\rightharpoonup} \mu \qquad \text{as } \varepsilon \to 0^+, \tag{A.21}$$

with $|\mu_\varepsilon| \, d\mathcal{L}^n \overset{*}{\rightharpoonup} |\mu|$ and $\int_{\mathbb{R}^n} |\mu_\varepsilon| \to |\mu|(\mathbb{R}^n)$ thanks to the basic estimate

$$\int_{B_r(x)} |\mu_\varepsilon| \leq |\mu|(B_{r+\varepsilon}(x)), \qquad \forall x \in \mathbb{R}^n, r > 0, \varepsilon > 0. \tag{A.22}$$

It is useful to notice that (A.22) can be improved in those (not so frequently met) situations where r is substantially smaller than ε, as expressed in the following estimate:

$$\int_{B_r(x)} |\mu_\varepsilon| \leq C(n) \frac{\min\{r^n, \varepsilon^n\}}{\varepsilon^n} |\mu|(B_{r+\varepsilon}(x)), \qquad \forall x \in \mathbb{R}^n, r > 0, \varepsilon > 0. \tag{A.23}$$

Indeed, by $\|\rho\|_{C^0} \leq C(n)/\varepsilon^n$ and

$$\int_{B_r(x)} |\mu_\varepsilon| \leq \int_{B_r(x)} dy \int_{B_\varepsilon(y)} \rho_\varepsilon(z - y) \, d|\mu|(z)$$

$$= \int_{B_{r+\varepsilon}(x)} d|\mu|(z) \int_{B_r(x) \cap B_\varepsilon(z)} \rho_\varepsilon(z - y) \, dy,$$

which gives (A.23).

A.13 Lipschitz Approximation of Functions of Bounded Variation

As an immediate consequence of (A.15) we see that, if $1 \leq p < \infty$ and $f \in L_{loc}^p(\mathbb{R}^n)$, then, for $\mathcal{L}^n$-a.e. $x \in \mathbb{R}^n$,

$$\lim_{r \to 0^+} \left(\frac{1}{r^n} \int_{B_r(x)} |f - (f)_{x,r}|^p \right)^{1/p} = 0, \qquad \text{where } (f)_{x,r} = \frac{1}{|B_r(x)|} \int_{B_r(x)} f. \tag{A.24}$$

For an arbitrary measurable function, the rate of convergence of the limit in (A.24) will drastically change depending on the point x. However, it is easily seen that as soon as f agrees $\mathcal{L}^n$-a.e. with an α-Hölder continuous function,

$\alpha \in (0, 1]$, then the rate of convergence is $O(r^{\alpha})$ as $r \to 0^{+}$, uniformly in x. The converse is true, and this is the content of **Campanato's criterion**, see [Mag12, Section 6.1]. This basic result, combined with the Poincaré inequality and a covering argument, is sufficient to conclude that every BV-function is countably Lipschitz.

Theorem A.2 *If Ω is an open set in $\mathbb{R}^{n}$, $f \in L^{1}_{\text{loc}}(\Omega)$ and Df is a Radon measure in Ω, then there exists an increasing family of Borel sets $\{E_{j}\}_{j}$ such that*

$$\mathcal{L}^{n}\Big(\Omega \setminus \bigcup_{j} E_{j}\Big) = 0, \qquad \text{Lip}(f; E_{j}) \leq j.$$

Proof If $\Omega(t) = \{x \in \Omega \cap B_{1/t} : \text{dist}(x, \partial\Omega) > t\}$, then, for a.e. $t > 0$, $\Omega(t)$ is an open bounded set of finite perimeter in $\mathbb{R}^{n}$ (see [Mag12, Remark 18.2]), and, in particular, $g = 1_{\Omega(t)} f \in L^{1}(\mathbb{R}^{n})$ and Dg is a finite Borel measure on $\mathbb{R}^{n}$. We can thus reduce to prove the theorem in the case $\Omega = \mathbb{R}^{n}$. Now, given $\ell > 0$, it turns out that f is $C(n)$ ℓ-Lipschitz on the Borel set

$$E_{\ell} = \Big\{ x \in \mathbb{R}^{n} : \frac{|Df|(B_{r}(x))}{r^{n}} \leq \ell, \quad \forall r > 0 \Big\}. \tag{A.25}$$

Indeed, by the Poincaré inequality for smooth functions, we have

$$\frac{1}{|B_{r}|} \int_{B_{r}(x)} |f - (f)_{x,r}| \leq C(n) \frac{|Df|(B_{r}(x))}{r^{n-1}}$$

(indeed, when f is smooth $|Df|(B_{r}(x)) = \int_{B_{r}(x)} |\nabla f|$, so that the general case follows by approximation), and thus

$$\frac{1}{|B_{r}|} \int_{B_{r}(x)} |f - (f)_{x,r}| \leq C(n) \ell r, \qquad \forall x \in E_{\ell}.$$

Arguing as in the proof of Campanato's criterion [Mag12, Section 6.1], we deduce that $|f(x) - f(y)| \leq C(n) \ell$ for every $x, y \in E_{\ell}$, so that $\text{Lip}(f; E_{\ell}) \leq C(n) \ell$. At the same time, by Vitali's covering theorem, we can find disjoint balls $\{B_{r_{j}}(x_{j})\}_{j=1}^{\infty}$ such that

$$\mathbb{R}^{n} \setminus E_{\ell} \subset \bigcup_{j=1}^{\infty} B_{5r_{j}}(x_{j}), \qquad \frac{|Df|(B_{r_{j}}(x_{j}))}{r_{j}^{n}} > \ell.$$

Therefore

$$\mathcal{L}^{n}(\mathbb{R}^{n} \setminus E_{\ell}) \leq 5^{n} \omega_{n} \sum_{j=1}^{\infty} r_{j}^{n} \leq \frac{C(n)}{\ell} \sum_{j=1}^{\infty} |Df|(B_{r_{j}}(x_{j})) \leq \frac{C(n)}{\ell} |Df|(\mathbb{R}^{n}),$$

which immediately implies $\mathcal{L}^{n}(\mathbb{R}^{n} \setminus E_{\ell}) \to 0$ as $\ell \to +\infty$. $\qquad\square$

A.14 Coarea Formula

The coarea formula is a generalization of Fubini's theorem to "curvilinear coordinates." In its simplest instance, the coarea formula pertains to Lipschitz functions $f : \mathbb{R}^n \to \mathbb{R}$ and states that

$$\int_E |\nabla f| = \int_{\mathbb{R}} \mathcal{H}^{n-1}(E \cap \{f = t\}) \, dt, \qquad \forall E \in \mathcal{B}(\mathbb{R}^n); \qquad (A.26)$$

or, equivalently $\int_{\mathbb{R}^n} g \, |\nabla f| = \int_{\mathbb{R}} dt \int_{\{f=t\}} g \, d\mathcal{H}^{n-1}$ for every Borel function $g : \mathbb{R}^n \to \mathbb{R}$, which is either bounded or nonnegative; see [Mag12, Theorem 18.1]. The case when f is affine is equivalent to Fubini's theorem on $\mathbb{R}^n \equiv \mathbb{R}^{n-1} \times \mathbb{R}$. For a Lipschitz function $f : \mathbb{R}^n \to \mathbb{R}^k$ with $1 \le k \le n$, at every differentiability point x of f we can define the **coarea factor of** f as

$$\mathbf{C}f(x) = \det \left(\nabla f(x) \, \nabla f(x)^* \right)^{1/2},$$

and correspondingly find that

$$\int_E \mathbf{C}f = \int_{\mathbb{R}^k} \mathcal{H}^{n-k}(E \cap \{f = t\}) \, d\mathcal{H}^k(t), \qquad \forall E \in \mathcal{B}(\mathbb{R}^n). \qquad (A.27)$$

When $k = n$, this is the area formula (A.18). When f is affine and $k \le n - 1$, (A.27) is equivalent to Fubini's theorem on $\mathbb{R}^n \equiv \mathbb{R}^{n-k} \times \mathbb{R}^k$. The proof is analogous to the case $k = 1$ (the only differences being found in the underlying linear algebra), see [EG92, Section 3.4] and [AFP00, Section 2.12].

Remark A.3   There is a sometime confusing case, which is met for example in the proof of Theorem 19.7, and that may require clarification. Consider a situation when $f : \mathbb{R}^n \to \mathbb{R}^n$ is a smooth map such that

$$f(\mathbb{R}^n) \subset M, \, M \text{ an } k\text{-dimensional surface with } 1 \le k \le n - 1. \qquad (A.28)$$

In this situation the area formula for f contains no information: the right-hand side of (A.18) is zero (since $\nabla f(x) \in \mathbb{R}^n \otimes \mathbb{R}^n$ has rank at most $k < n$, and thus $\det \nabla f(x) = 0$), as well as its left-hand side (the integration in $d\mathcal{H}^0(x)$ happens on $\{f = y\}$, which is empty unless $y \in f(\mathbb{R}^n) \subset M$, so that the integration in dy happens on M, which is Lebesgue negligible). One expects, however, that nontrivial information may be contained in a variant of the coarea formula where the target lower dimensional Euclidean space $\mathbb{R}^k$ is replaced by (the non-flat, but still k-dimensional) M. This is indeed the case if one identifies $\nabla f(x)$, more properly, as an element of $(T_{f(x)}M) \otimes \mathbb{R}^n$ (rather than of $\mathbb{R}^n \otimes \mathbb{R}^n$), and considers the coarea factor of f to be

$$\mathbf{C}f(x) = \det \left(A(x) \right)^{1/2}, \qquad A(x) = \nabla f(x) \, \nabla f(x)^* \in (T_{f(x)}M) \otimes (T_{f(x)}M).$$
$$(A.29)$$

With this choice of $\mathbf{C}f$ and under (A.28) we indeed have the coarea formula

$$\int_E \mathbf{C}f = \int_M \mathcal{H}^{n-k}(E \cap \{f = t\}) \, d\mathcal{H}^k(t), \qquad \forall E \in \mathcal{B}(\mathbb{R}^n). \qquad (A.30)$$

A.15 Rectifiable Sets

Given $k \in \{1, \ldots, n-1\}$, a Borel set $M \subset \mathbb{R}^n$ is **countably $\mathcal{H}^k$-rectifiable** if it can be covered, modulo a set of $\mathcal{H}^k$-null measure, by countably many sets of the form $f(E)$, where $E \subset \mathbb{R}^k$ is a Borel set and $f : \mathbb{R}^k \to \mathbb{R}^n$ a Lipschitz map. Countably $\mathcal{H}^k$-rectifiable sets M such that $\mathcal{H}^k \llcorner M$ defines a Radon measure on $\mathbb{R}^n$, i.e., such that $\mathcal{H}^k(M \cap B_R) < \infty$, for every $R > 0$, are called **locally $\mathcal{H}^k$-rectifiable sets**; or, simply, $\mathcal{H}^k$-**rectifiable sets**, when $\mathcal{H}^k(M) < \infty$. A Borel set $M \subset \mathbb{R}^n$ is locally $\mathcal{H}^k$-rectifiable in $\mathbb{R}^n$ if and only if for $\mathcal{H}^k$-a.e. $x \in M$ there is a k-dimensional plane P_x in $\mathbb{R}^n$ such that $\mathcal{H}^k \llcorner [(M - x)/r] \xrightarrow{*} \mathcal{H}^k \llcorner P_x$ as $r \to 0^+$; since this property uniquely identifies P_x, we set $P_x = T_x M$ and call $T_x M$ the **approximate tangent plane to M at x**. Given M a locally $\mathcal{H}^k$-rectifiable set in $\mathbb{R}^n$ and a Lipschitz map $f : \mathbb{R}^n \to \mathbb{R}^m$, the restriction of f along $x + T_x M$ is differentiable at x, and the corresponding differential is a linear map $d^M f_x : T_x M \to \mathbb{R}^m$, to which we can associate a tensor $\nabla^M f(x) \in \mathbb{R}^n \otimes (T_x M)$, called the **tangential gradient of f along M**. An useful fact to keep in mind is that when f is both differentiable and tangentially differentiable along M at some $x \in M$, then

$$\nabla^M f(x) = \nabla f(x) \circ \mathbf{p}_x, \qquad \mathbf{p}_x \text{ the projection of } \mathbb{R}^n \text{ onto } T_x M. \qquad (A.31)$$

Coming to area and coarea formulas: **if $k \le m$**, then the formula

$$J^M f(x) = \det(\nabla^M f(x)^* \nabla^M f(x))^{1/2},$$

defines the **tangential Jacobian of f at x along M**, and the **area formula** holds

$$\int_M (g \circ f) \, J^M f \, d\mathcal{H}^k = \int_{\mathbb{R}^m} g(z) \, \mathcal{H}^0(\{f = z\}) \, d\mathcal{H}^k(z).$$

In the situation when $f(M)$ is a locally $\mathcal{H}^\ell$-rectifiable set in $\mathbb{R}^m$ **with $k \ge \ell$**, then, for $\mathcal{H}^k$-a.e. $x \in M$, $T_{f(x)}(f(M))$ exists and the differential $d^M f_x$ takes values in $T_{f(x)}[f(M)]$, i.e., $d^M f_x : T_x M \to T_{f(x)}[f(M)]$; correspondingly, we can identify the tangential gradient as a tensor $\nabla^M f(x) \in (T_{f(x)}[f(M)]) \otimes (T_x M)$, define the **tangential coarea factor of f along M** as

$$\mathbf{C}^M f(x) = \det(\nabla^M f(x) \nabla^M f(x)^*)^{1/2},$$

and have the **coarea formula**

$$\int_E \mathbf{C}^M f = \int_M \mathcal{H}^{n-k}(E \cap \{f = t\}) \, d\mathcal{H}^k(t), \qquad \forall E \in \mathcal{B}(\mathbb{R}^n). \quad \text{(A.32)}$$

A.16 The $C^{1,1}$-Version of the Whitney Extension Theorem

Given a C^1-function $f : \mathbb{R}^n \to \mathbb{R}$, we trivially notice that, on every compact set $K \subset \mathbb{R}^n$, the C^0-map $T = \nabla f$ is such that

$$\lim_{\delta \to 0^+} \sup_{x,y \in K \; 0 < |x-y| < \delta} \frac{|f(y) - f(x) - T(x) \cdot (y - x)|}{|x - y|} = 0. \quad \text{(A.33)}$$

Thus, the question raises if given any $f \in C^0(K)$ and $T \in C^0(K; \mathbb{R}^n)$ satisfying (A.33) on a compact set K, there exist $g \in C^1(\mathbb{R}^n)$ such that $g = f$ and $\nabla g = T$ on K. The affirmative answer to this question is the content of the celebrated Whitney extension theorem. We present here a $C^{1,1}$-version of Whitney's theorem, and refer to [Bie80] for a detailed proof.

Theorem A.4 (Whitney extension theorem, $C^{1,1}$-version) *If $K \subset \mathbb{R}^n$ is compact, and $f \in C^0(K)$ and $T \in C^0(K; \mathbb{R}^n)$ are such that T is Lipschitz continuous on K and (A.33) holds, then there is $g \in C^{1,1}(\mathbb{R}^n)$ such that $g = f$ and $\nabla g = T$ on K.*

Appendix B
Bibliographical Notes

The main goal of these bibliographical notes is to provide references to the material used in the preparation of this book. Occasionally, for some of the topics covered, some additional references are also included as examples of possible further readings. Both the choices of these examples, and of the topics for which they are offered, are very much biased by the personal interests of the author and are not intended, by any means, to provide an even remotely exhaustive review of such topics. In other words, of the hundreds of many interesting papers in OMT that may have been mentioned here, only a few have been, and many more have not! The reason is that citing more of these papers in a meaningful way, i.e., by providing some context for each them, would have required either peppering the main body of the text with citations (something that was avoided to keep the reader's focus entirely on the mathematical exposition) or writing a much longer set of bibliographical notes. The latter option seemed somehow unnecessary for two reasons. First, very complete bibliographical accounts and historical perspectives, covering both the more theoretical and the more applied aspects of OMT, are found in the monographs [Vil09, AGS08, San15]. Second, the best places to find updated, commented bibliographies on specific subtopics of OMT are the many survey papers and lecture notes that have been (and continue to be) written by several authors, and these can be easily found by a quick search on any mathematical database. With all these disclaimers:

Chapter 1: The construction of the Knothe map in Section 1.5 originates in [Kno57]. A very influential presentation of Knothe's construction is the one by Gromov found in [MS86].

Chapter 2: A standard reference for convex functions on $\mathbb{R}^n$ is Rockafellar's classic monograph [Roc97].

281

Chapter 3: Most of these results presented in this chapter continue to hold if $\mathbb{R}^n$ is replaced by a complete and separable metric space (X, d), with the linear cost defined by $c(x, y) = d(x, y)$, and the quadratic cost by $c(x, y) = d(x, y)^2$; see, e.g., [Vil03, Vil09, AGS08, ABS21, FG21, San15]. More structure is required to extend the part of Theorem 3.20, which establishes the relation between d^2-convexity (i.e., (3.19) with $c = d^2$) and standard convexity (i.e., (2.12)); see the notes to the next chapter.

Chapter 4: Our treatment of the Brenier theorem, as much as the ones found in other OMT books, follows Brenier's original paper [Bre91]. The Brenier theorem holds for more general transport costs and in more general ambient spaces than $\mathbb{R}^n$. For example, in $\mathbb{R}^n$, the existence of uniqueness of transport maps has been proved for transport costs of the general form $c(x, y) = h(y - x)$, h strictly convex, by Gangbo and McCann in [GM96]. Concerning more general ambient spaces, the two most direct and natural generalizations of the Brenier theorem are one to (infinite-dimensional) Hilbert spaces due to Ambrosio, Gigli, and Savaré [AGS08, Theorem 6.2.10] (see the contemporary [FU04] for the case of the Wiener space) and to Riemannian manifold by McCann [FM02b] (see [Fig07] for the case of non-compact manifolds). The extension of the Brenier theorem to infinite-dimensional "non-flat" spaces is the subject of current research, see e.g., [Ber08, Gig12, RS14, GRS16, CH15, CM17a].

Chapter 5: The elegant argument used to prove Theorem 5.1 comes from the theory of monotone maps developed by Alberti and Ambrosio in [AA99]. The implicit function theorem for convex functions (Theorem 5.3) is found, with a slightly different proof that avoids smoothing, in [McC95, Theorem 17].

Chapter 6: Theorem 6.1 is the main theorem in McCann's paper [McC95], and the original presentation has been followed closely.

Chapter 7: The second order distributional and a.e. differentiability of convex functions are presented following [EG92, Sections 6.3 and 6.4]. The discussion of the a.e. first order differentiability of the subdifferential is modeled after [BCP96].

Chapter 8: This chapter is based on [McC97, Section 4 and Appendix]. The validity of the Monge–Ampère equation for Brenier maps is the starting point for the regularity theory of optimal transport maps. The regularity theory for convex solutions to the Monge–Ampère equation is a very vast and layered topic. The very recent survey [DPF14] is a good starting point to enter the subject.

Chapter 9: Section 9.2: The proof of the Euclidean isoperimetric inequality given in Theorem 9.2 is due to Knothe [Kno57] (see also [MS86]). The argument still works using Knothe maps in place of Brenier maps (as done in Knothe's paper), since gradients of Knothe maps can be naturally represented as triangular matrices with nonnegative diagonal entries and thus satisfy a suitable variant of Proposition 9.1. Without additional effort the same argument actually proves the more general *anisotropic* isoperimetric inequality and thus solves Wulff's problem (since, e.g., [Mag12, Section 20.2], for an introduction to the Wulff's problem). A careful examination of Knothe's argument (which involves the use of more GMT than it would have been reasonable to use here) leads to a characterization of equality cases in the Wulff inequality (and thus also in the Euclidean isoperimetric inequality); see [BM94]. A difference between running Knothe's argument with Knothe maps rather than with Brenier maps is met when Knothe's argument is used as the starting point for discussing quantitative versions of these inequalities: in that problem, the use of Brenier maps is decisive for obtaining sharp stability rates; see [FMP10]. Knothe's argument also provides a starting point for studying isoperimetric inequalities on minimal surfaces; see [Cas10, BE22]. There is also an evident (and maybe not completely understood) relation between Knothe's argument and Cabré's proof of the isoperimetric inequality [Cab00, Cab08]: for sure, both arguments reduce the proof of isoperimetric theorems to the arithmetic-geometric mean inequality, which in turn can be interpreted as an "infinitesimal isoperimetric principle for volume elements." Cabré's argument is usually cited as the "ABP method," in reference to envelope/contact set argument it shares with the classical proof of the Alexandrov–Bakelman–Pucci estimate for elliptic PDEs. As much as Knothe's argument, Cabré's argument can be adapted to more general isoperimetric problems, for example to prove isoperimetry on minimal surfaces [Bre21] (which predates [BE22]) and isoperimetry with weights [CROS16]. Finally, there is a last interesting connection to mention with Almgren's isoperimetric theorem in arbitrary codimension [Alm86], which is also based on an envelope/contact set argument and infinitesimal isoperimetry of volume elements. **Section 9.3:** The Sobolev inequality was originally established in sharp form by Aubin [Aub76] and Talenti [Tal76]. The OMT proof of the sharp Sobolev inequality presented in Theorem 9.3 is due to Cordero-Erausquin, Nazaret, and Villani [CENV04] and has led to several developments; see [MV05, Naz06, MV08, MN17, MNT22]. We have commented in some detail on the geometric interest of establishing the Sobolev inequality in sharp form but have only mentioned the existence of strong physical motivations. Among them, the problem of the *stability of matter* in Quantum Mechanics; see, e.g., Lieb's survey [Lie90].

Chapter 10: This chapter follows closely McCann's seminal work [McC97].

Chapter 11: In this chapter we have touched briefly on the geometric properties of the Wasserstein space. The reader will find much more exhaustive discussions in references like [Vil09, AGS08].

Chapter 12: The interested reader can expand on this informal introduction to gradient flows and the minimizing movements scheme by reading some of the original papers by De Giorgi [DG93b, DG93a] as well as the introduction to [AGS08].

Chapter 13: This chapter follows very closely the seminal paper [JKO98] by Jordan, Kinderlehrer, and Otto. Otto's contemporary papers on the porous medium equation [Ott98, Ott01] are also very inspiring readings. For the most general treatment of minimizing movements schemes and OMT, see, of course, [AGS08].

Chapter 14: Our introduction to the Euler equations is strongly influenced by the beautiful book by Marchioro and Pulvirenti [MP94]. The connection between the Euler equations and Brenier maps is the central theme of the already mentioned paper by Brenier [Bre91], as well as of the paper by Benamou and Brenier [BB00], where the action of an Eulerian fluid is for the first time related to the Monge problem with quadratic cost.

Chapter 15: This chapter discusses the basic elements needed to formalize the ideas originating in [Ott01] and follows closely the treatment of the continuity equation in the study of curves of probability measures given in [AGS08]. The proof of the Benamou–Brenier formula using the Hopf–Lax semigroup is based on [GKO13].

Chapter 16: The solution of one-dimensional transport problems by the use of cumulative distribution functions is found in every account on OMT. Refined and more complete statements than the one provided here can be found, for example, in [Vil09, San15].

Chapter 17: The disintegration theorem is a basic tool in Probability Theory, and the starting point of the theory of Young's measures in the Calculus of Variations. The latter has several interesting connections with the study of minimal surfaces and geometric variational problems (the theory of varifolds) [All72, Sim83], as well as with the study of problems in Continuum Mechanics and Nonlinear Elasticity [Mul99, Ped97]. The approximation result in Remark 17.10 is found in [Gan99, Proposition A.3], although the sketch of proof presented here is taken from [Amb03, Theorem 9.3].

Chapter 18: Our presentation of the solution to the Monge problem with linear cost is based on the lecture notes by Ambrosio [Amb03], on the subsequent paper by Ambrosio and Pratelli [AP03], and on Pratelli's PhD thesis [Pra03]. Sudakov's argument was originally presented in [Sud79], where Sudakov wrongly claimed (using our notation and referring to the statement of Theorem 18.7) the validity of (18.32) (i.e., of $\mu_\sigma \ll \mathcal{H}^1 \llcorner P_G^{-1}(\sigma)$ for $(P_G)_\# \mu$-a.e. $\sigma \in \mathcal{S}^\circ$) for general partitions into segments, *without* assuming hypothesis (b) (i.e., that τ_G has countable Lipschitz property on E). This assertion is false in this generality, but, as we have seen, is correct in the special case when $G = G(f)$ for a Kantorovich potential f, which is all that matters for solving the Monge problem. In any case, this problem was not properly addressed in Sudakov's original paper. The first complete proof of Theorem 18.1 seems thus to be the one given by Evans and Gangbo in [EG99]. There the authors exploit fine regularity estimates for p-Laplacian equations to study, in the limit as $p \to \infty$, the Euler–Lagrange equation of the Kantorovich dual problem satisfied by any Kantorovich potential. In describing this limit procedure, they introduce various useful ideas concerning transport sets (see, e.g., [EG99, Lemma 5.1]), and they study the approximate $C^{1,1}$-regularity of Kantorovich potentials (see [EG99, Lemma 4.1]). The PDE character of the proof forces some additional assumptions on the origin and final measures (their supports are the closures of disjoint bounded open sets with smooth boundary). Shortly after [EG99], Caffarelli, Feldman, and McCann [CFM02] and Trudinger and Wang [TW01] independently proposed another strategy, which eliminates the smoothness and disjointness assumptions in [EG99] and, in the presentation given in [CFM02], also addresses the case when the distance in the Monge problem is not necessarily Euclidean but is rather induced by a general smooth and uniformly elliptic norm on $\mathbb{R}^n$. Their approach consists in studying a different approximation; precisely, for $p > 1$, they consider optimal pairs in the dual Kantorovich problem to $\mathbf{K}_p(\mu, \nu)$ and then prove their convergence, as $p \to 1^+$, to a pair $(f, -f)$, where f is a Kantorovich potential from μ to ν. With reference to [CFM02], these limit Kantorovich potentials are then proved to be approximately $C^{1,1}$-regular [CFM02, Lemma 16], an information that is then used to construct a change of variables that "straighten the transport rays of f" [CFM02, Lemma 22] and thus leads to the reduction to one-dimensional transport problems between *absolutely continuous* densities by a direct application of Fubini's theorem. Interestingly, as implied by our presentation, these arguments can be used to provide a full justification of Sudakov's argument, although this was definitely not the point of view of these authors.[1] The existence of a gap in Sudakov's presentation and the possibility of fixing it in

---

[1] For example, in [CFM02], the authors introduce their work as a third possible approach to the solution of the Monge problem, distinct from Sudakov's and the Evans–Gangbo ones.

Sudakov's original framework of disintegration theory are both points explicitly made in Ambrosio's lecture notes [Amb03]. As already noticed, [Amb03] (which acknowledges [EG99, CFM02]) has been, together with [AP03, Pra03], the main source for preparing our presentation. An interesting result contained in [AP03] (but omitted in our presentation) is the convergence of the optimal transport maps T_p in $\mathbf{M}_p(\mu, \nu)$ to an optimal transport map T in $\mathbf{M}_1(\mu, \nu)$ as $p \to 1^+$. Another paper using the same technical setting is [AKP04], by Ambrosio, Kirchheim, and Pratelli, where the Monge problem with respect to a *crystalline* norm on $\mathbb{R}^n$ is solved and which also contains an example, based on previous unpublished work by Alberti, Kirchheim, and Preiss, proving the gap in Sudakov's argument.

The Monge problem with linear cost can be considered in more general situations. The approach of De Pascale and Champion, which combines careful density estimates on transport sets with finely tuned approximations of the linear cost, solves the Monge problem on $\mathbb{R}^n$ with a general norm [CDP11] (see also the earlier [CDP10] addressing strictly convex norms, and the later [DPR11] addressing the Monge problem in the Heisenberg group, and pointing to the applicability of the method in more general ambient spaces). Sudakov's original approach has also been extensively developed in a long series of papers [BC11, Car11a, Car11b, Cav12b, Cav12a, BC13, Cav14b, BD18]. In particular, in [Car11a], appeared concurrently to [CDP11], Caravenna also solves the Monge problem in $\mathbb{R}^n$ with respect to a generic norm. The most general available results seem to be the ones contained in [BC13] and [Cav14b], with slightly different assumptions in terms on the non-branching structure of geodesics in the ambient metric space and of the specific form of the curvature/dimension condition under consideration. For more comments on these papers, see also the notes to the next chapter.

Chapter 19: The proof of Theorem 19.1 follows the original argument in [PW60] with two important differences: first, the one-dimensional Poincaré inequality is proved under the weaker and actually sharp condition that the weight ℓ is log-concave (rather than concave as in [PW60], or power-concave as in [Beb03]); second, following [FNT12], the splitting algorithm uses areas of two-dimensional projections rather than n-dimensional volumes as in [PW60], which allows for an immediate control on waist sizes. The rest of the chapter is modeled after the work of Klartag [Kla17], where the localization theorem is proved on (weighted) Riemannian manifolds. Once the arguments in this chapter are understood, a reader with a basic familiarity in Riemannian geometry should be able to comfortably access [Kla17]. It should also be noted that, from the technical viewpoint, [Kla17], and thus our presentation of it, draws on previous works on the Monge problem: to make two examples, the proof of Theorem 19.6 is [FM02a, Proposition 15], while step one of the proof of Theorem 19.4 is [EG99, Lemma 5.1]. Extending the localization theorem from the smooth context of weighted Riemannian manifolds to the non-smooth setting of metric-measure spaces requires an in-depth revision of the arguments presented in Chapters 18 and 19. An illuminating example is the impossibility of using the coarea formula to prove the absolute continuity (and discuss the concavity properties – or, more precisely, the curvature-dimension bounds) of the one-dimensional measures resulting from disintegration, which requires entirely new ideas (compare with [Cav14a, Lemma 4.6], [Cav14b, Section 6], and [CM17b, Lemma 4.1, Theorem 4.2]). Readers can get a first idea of the broad spectrum of applications unlocked by the validity of the localization theorem in metric-measure spaces by looking, for example, at [CM17b, CM17c, CMM19].

References

[AA99]   G. Alberti and L. Ambrosio. A geometrical approach to monotone functions in $\mathbf{R}^n$. *Math. Z.*, 230(2):259–316, 1999.

[ABS21] L. Ambrosio, E. Brué, and D. Semola. *Lectures on optimal transport*, volume 130 of *Unitext*. Springer, Cham, 2021. ix+250 pp. La Matematica per il 3+2.

[AFP00] L. Ambrosio, N. Fusco, and D. Pallara. *Functions of bounded variation and free discontinuity problems*. Oxford Mathematical Monographs. The Clarendon Press, Oxford University Press, New York, 2000. xviii+434 pp.

[AGS08] L. Ambrosio, N. Gigli, and G. Savaré. *Gradient flows in metric spaces and in the space of probability measures*. Lectures in Mathematics ETH Zürich. Birkhäuser Verlag, Basel, second edition, 2008. x+334 pp.

[AKP04] L. Ambrosio, B. Kirchheim, and A. Pratelli. Existence of optimal transport maps for crystalline norms. *Duke Math. J.*, 125(2):207–241, 2004.

[All72] W. K. Allard. On the first variation of a varifold. *Ann. Math.*, 95:417–491, 1972.

[Alm86] F. Almgren. Optimal isoperimetric inequalities. *Indiana Univ. Math. J.*, 35 (3):451–547, 1986.

[Amb03] L. Ambrosio. Lecture notes on optimal transport problems. In *Mathematical aspects of evolving interfaces (Funchal, 2000)*, volume 1812 of *Lecture Notes in Mathematics*, pages 1–52. Springer, Berlin, 2003.

[AP03]   L. Ambrosio and A. Pratelli. Existence and stability results in the L^1 theory of optimal transportation. In *Optimal transportation and applications (Martina Franca, 2001)*, volume 1813 of *Lecture Notes in Mathematics*, pages 123–160. Springer, Berlin, 2003.

[Aub76] T. Aubin. Problèmes isopérimétriques et espaces de Sobolev. *J. Differential Geom.*, 11(4):573–598, 1976.

[BB00]   J.-D. Benamou and Y. Brenier. A computational fluid mechanics solution to the Monge–Kantorovich mass transfer problem. *Numer. Math.*, 84(3):375–393, 2000.

[BC11]   S. Bianchini and F. Cavalletti. The Monge problem in geodesic spaces. In *Nonlinear conservation laws and applications*, volume 153 of *IMA Volumes in Mathematics and Its Applications*, pages 217–233. Springer, New York, 2011.

[BC13] S. Bianchini and F. Cavalletti. The Monge problem for distance cost in geodesic spaces. *Comm. Math. Phys.*, 318(3):615–673, 2013.

[BCP96] G. Bianchi, A. Colesanti, and C. Pucci. On the second differentiability of convex surfaces. *Geom. Dedicata*, 60(1):39–48, 1996.

[BD18] S. Bianchini and S. Daneri. On Sudakov's type decomposition of transference plans with norm costs. *Mem. Amer. Math. Soc.*, 251(1197):vi+112, 2018.

[BE22] S. Brendle and M. Eichmair. Proof of the Michael–Simon–Sobolev inequality using optimal transport. 2022.

[Beb03] M. Bebendorf. A note on the Poincaré inequality for convex domains. *Z. Anal. Anwendungen*, 22(4):751–756, 2003.

[Ber08] J. Bertrand. Existence and uniqueness of optimal maps on Alexandrov spaces. *Adv. Math.*, 219(3):838–851, 2008.

[BG03] Y. Brenier and W. Gangbo. L^p approximation of maps by diffeomorphisms. *Calc. Var. Partial Differ. Equ.*, 16(2):147–164, 2003.

[Bie80] E. Bierstone. Differentiable functions. *Bol. Soc. Brasil. Mat.*, 11(2):139–189, 1980.

[BM94] J. E. Brothers and F. Morgan. The isoperimetric theorem for general integrands. *Mich. Math. J.*, 41(3):419–431, 1994.

[Bre91] Y. Brenier. Polar factorization and monotone rearrangement of vector-valued functions. *Comm. Pure Appl. Math.*, 44(4):375–417, 1991.

[Bre21] S. Brendle. The isoperimetric inequality for a minimal submanifold in Euclidean space. *J. Amer. Math. Soc.*, 34(2):595–603, 2021.

[Cab00] X. Cabré. Partial differential equations, geometry and stochastic control. *Butl. Soc. Catalana Mat.*, 15(1):7–27, 2000.

[Cab08] X. Cabré. Elliptic PDE's in probability and geometry: symmetry and regularity of solutions. *Discrete Contin. Dyn. Syst.*, 20(3):425–457, 2008.

[Car11a] L. Caravenna. A proof of Monge problem in $\mathbb{R}^n$ by stability. *Rend. Istit. Mat. Univ. Trieste*, 43:31–51, 2011.

[Car11b] L. Caravenna. A proof of Sudakov theorem with strictly convex norms. *Math. Z.*, 268(1–2):371–407, 2011.

[Cas10] P. Castillon. Submanifolds, isoperimetric inequalities and optimal transportation. *J. Funct. Anal.*, 259(1):79–103, 2010.

[Cav12a] F. Cavalletti. The Monge problem in Wiener space. *Calc. Var. Partial Differ. Equ.*, 45(1–2):101–124, 2012.

[Cav12b] F. Cavalletti. Optimal transport with branching distance costs and the obstacle problem. *SIAM J. Math. Anal.*, 44(1):454–482, 2012.

[Cav14a] F. Cavalletti. Decomposition of geodesics in the Wasserstein space and the globalization problem. *Geom. Funct. Anal.*, 24(2):493–551, 2014.

[Cav14b] F. Cavalletti. Monge problem in metric measure spaces with Riemannian curvature-dimension condition. *Nonlinear Anal.*, 99:136–151, 2014.

[CDP10] T. Champion and L. De Pascale. The Monge problem for strictly convex norms in $\mathbb{R}^d$. *J. Eur. Math. Soc. (JEMS)*, 12(6):1355–1369, 2010.

[CDP11] T. Champion and L. De Pascale. The Monge problem in $\mathbb{R}^d$. *Duke Math. J.*, 157(3):551–572, 2011.

[CENV04] D. Cordero-Erausquin, B. Nazaret, and C. Villani. A mass-transportation approach to sharp Sobolev and Gagliardo-Nirenberg inequalities. *Adv. Math.*, 182(2):307–332, 2004.

[CFM02] L. A. Caffarelli, M. Feldman, and R. J. McCann. Constructing optimal maps for Monge's transport problem as a limit of strictly convex costs. *J. Amer. Math. Soc.*, 15(1):1–26, 2002.

[CH15] F. Cavalletti and M. Huesmann. Existence and uniqueness of optimal transport maps. *Ann. Inst. H. Poincaré C Anal. Non Linéaire*, 32(6):1367–1377, 2015.

[CM17a] F. Cavalletti and A. Mondino. Optimal maps in essentially non-branching spaces. *Commun. Contemp. Math.*, 19(6):1750007, 27, 2017.

[CM17b] F. Cavalletti and A. Mondino. Sharp and rigid isoperimetric inequalities in metric-measure spaces with lower Ricci curvature bounds. *Invent. Math.*, 208(3):803–849, 2017.

[CM17c] F. Cavalletti and A. Mondino. Sharp geometric and functional inequalities in metric measure spaces with lower Ricci curvature bounds. *Geom. Topol.*, 21(1):603–645, 2017.

[CMM19] F. Cavalletti, F. Maggi, and A. Mondino. Quantitative isoperimetry à la Levy-Gromov. *Comm. Pure Appl. Math.*, 72(8):1631–1677, 2019.

[CROS16] X. Cabré, X. Ros-Oton, and J. Serra. Sharp isoperimetric inequalities via the ABP method. *J. Eur. Math. Soc. (JEMS)*, 18(12):2971–2998, 2016.

[CV77] C. Castaing and M. Valadier. *Convex analysis and measurable multifunctions*, volume 580 of *Lecture Notes in Mathematics*. Springer-Verlag, Berlin and New York, 1977. vii+278 pp.

[DG93a] E. De Giorgi. Motion by variation. In *Current problems of analysis and mathematical physics (Italian) (Taormina, 1992)*, pages 65–70. Univ. Roma "La Sapienza," Rome, 1993.

[DG93b] E. De Giorgi. New problems on minimizing movements. In *Boundary value problems for partial differential equations and applications*, volume 29 of *RMA Research Notes in Applied Mathematics*, pages 81–98. Masson, Paris, 1993.

[DPF14] G. De Philippis and Alessio Figalli. The Monge-Ampère equation and its link to optimal transportation. *Bull. Amer. Math. Soc. (N.S.)*, 51(4):527–580, 2014.

[DPR11] L. De Pascale and S. Rigot. Monge's transport problem in the Heisenberg group. *Adv. Calc. Var.*, 4(2):195–227, 2011.

[EG92] L. C. Evans and R. F. Gariepy. *Measure theory and fine properties of functions*. Studies in Advanced Mathematics. CRC Press, Boca Raton, FL, 1992. viii+268 pp.

[EG99] L. C. Evans and W. Gangbo. Differential equations methods for the Monge–Kantorovich mass transfer problem. *Mem. Amer. Math. Soc.*, 137(653): viii+66, 1999.

[Fed69] H. Federer. *Geometric measure theory*, volume 153 of *Die Grundlehren der mathematischen Wissenschaften*. Springer-Verlag New York Inc., New York, 1969. xiv+676 pp.

[FG21] A. Figalli and F. Glaudo. *An invitation to optimal transport, Wasserstein distances, and gradient flows*. EMS Textbooks in Mathematics. EMS Press, Berlin, 2021. vi+136 pp.

[Fig07] A. Figalli. The Monge problem on non-compact manifolds. *Rend. Semin. Mat. Univ. Padova*, 117:147–166, 2007.

[FM02a] M. Feldman and R. J. McCann. Monge's transport problem on a Riemannian manifold. *Trans. Amer. Math. Soc.*, 354(4):1667–1697, 2002.

[FM02b] R. McCann. Polar factorization of maps on Riemannian manifolds. *Geom. Funct. Anal.*, 11(4):589–608, 2001.

[FMP10] A. Figalli, F. Maggi, and A. Pratelli. A mass transportation approach to quantitative isoperimetric inequalities. *Inv. Math.*, 182(1):167–211, 2010.

[FNT12] V. Ferone, C. Nitsch, and C. Trombetti. A remark on optimal weighted Poincaré inequalities for convex domains. *Atti Accad. Naz. Lincei Rend. Lincei Mat. Appl.*, 23(4):467–475, 2012.

[FU04] D. Feyel and A. S. Üstünel. Monge–Kantorovitch measure transportation and Monge-Ampère equation on Wiener space. *Probab. Theory Relat. Fields*, 128(3):347–385, 2004.

[Gan99] W. Gangbo. The Monge mass transfer problem and its applications. In *Monge Ampère equation: applications to geometry and optimization (Deerfield Beach, FL, 1997)*, volume 226 of *Contemporary Mathematics*, pages 79–104. American Mathematical Society, Providence, RI, 1999.

[Gig12] N. Gigli. Optimal maps in non branching spaces with Ricci curvature bounded from below. *Geom. Funct. Anal.*, 22(4):990–999, 2012.

[GKO13] N. Gigli, K. Kuwada, and S.-Ichi Ohta. Heat flow on Alexandrov spaces. *Comm. Pure Appl. Math.*, 66(3):307–331, 2013.

[GM87] M. Gromov and V. D. Milman. Generalization of the spherical isoperimetric inequality to uniformly convex Banach spaces. *Compositio Math.*, 62(3): 263–282, 1987.

[GM96] W. Gangbo and R. J. McCann. The geometry of optimal transportation. *Acta Math.*, 177(2):113–161, 1996.

[GRS16] N. Gigli, T. Rajala, and K.-T. Sturm. Optimal maps and exponentiation on finite-dimensional spaces with Ricci curvature bounded from below. *J. Geom. Anal.*, 26(4):2914–2929, 2016.

[JKO98] R. Jordan, D. Kinderlehrer, and F. Otto. The variational formulation of the Fokker-Planck equation. *SIAM J. Math. Anal.*, 29(1):1–17, 1998.

[Kla17] B. Klartag. Needle decompositions in Riemannian geometry. *Mem. Amer. Math. Soc.*, 249(1180):v+77, 2017.

[KLS95] R. Kannan, L. Lovász, and M. Simonovits. Isoperimetric problems for convex bodies and a localization lemma. *Discrete Comput. Geom.*, 13(3–4): 541–559, 1995.

[Kno57] H. Knothe. Contributions to the theory of convex bodies. *Mich. Math. J.*, 4: 39–52, 1957.

[Lie90] E. H. Lieb. The stability of matter: from atoms to stars. *Bull. Amer. Math. Soc. (N.S.)*, 22(1):1–49, 1990.

[LS93] L. Lovász and M. Simonovits. Random walks in a convex body and an improved volume algorithm. *Random Struct. Algorithms*, 4(4):359–412, 1993.

[Mag12] F. Maggi. *Sets of finite perimeter and geometric variational problems: an introduction to Geometric Measure Theory*, volume 135 of *Cambridge Studies in Advanced Mathematics*. Cambridge University Press, Cambridge, 2012.

[McC95] R. J. McCann. Existence and uniqueness of monotone measure-preserving maps. *Duke Math. J.*, 80(2):309–323, 1995.

[McC97] R. J. McCann. A convexity principle for interacting gases. *Adv. Math.*, 128 (1):153–179, 1997.

[MN17] F. Maggi and R. Neumayer. A bridge between Sobolev and Escobar inequalities and beyond. *J. Funct. Anal.*, 273(6):2070–2106, 2017.

[MNT22] F. Maggi, R. Neumayer, and I. Tomasetti. Rigidity theorems for best Sobolev inequalities. 2022.

[MP94] C. Marchioro and M. Pulvirenti. *Mathematical theory of incompressible nonviscous fluids*, volume 96 of *Applied Mathematical Sciences*. Springer-Verlag, New York, 1994. ISBN 0-387-94044-8. xii+283 pp.

[MS86] V. D. Milman and G. Schechtman. *Asymptotic theory of finite-dimensional normed spaces. With an appendix by M. Gromov*. Number 1200 in Lecture Notes in Mathematics. Springer-Verlag, Berlin, 1986. viii+156 pp.

[Mul99] S. Müller. Variational models for microstructure and phase transitions. In *Calculus of variations and geometric evolution problems (Cetraro, 1996)*, volume 1713 of *Lecture Notes in Mathematics*, pp. 85–210. Springer, Berlin, 1999.

[MV05] F. Maggi and C. Villani. Balls have the worst best Sobolev inequalities. *J. Geom. Anal.*, 15(1):83–121, 2005.

[MV08] F. Maggi and C. Villani. Balls have the worst best Sobolev inequalities. II. Variants and extensions. *Calc. Var. Partial Differ. Equ.*, 31(1):47–74, 2008.

[Naz06] B. Nazaret. Best constant in Sobolev trace inequalities on the half-space. *Nonlinear Anal.*, 65(10):1977–1985, 2006.

[Ott98] F. Otto. Dynamics of labyrinthine pattern formation in magnetic fluids: a mean-field theory. *Arch. Rational Mech. Anal.*, 141(1):63–103, 1998.

[Ott01] F. Otto. The geometry of dissipative evolution equations: the porous medium equation. *Comm. Partial Differ. Equ.*, 26(1–2):101–174, 2001.

[Ped97] P. Pedregal. *Parametrized measures and variational principles*, volume 30 of *Progress in Nonlinear Differential Equations and their Applications*. Birkhäuser Verlag, Basel, 1997. xii+212 pp.

[Pra03] A. Pratelli. Existence of optimal transport maps and regularity of the transport density in mass transportation problems. 2003. Available at https://cvgmt.sns.it/paper/758/.

[PW60] L. E. Payne and H. F. Weinberger. An optimal Poincaré inequality for convex domains. *Arch. Rational Mech. Anal.*, 5(1960):286–292, 1960.

[Roc97] R. Tyrrell Rockafellar. *Convex analysis*. Princeton Landmarks in Mathematics. Princeton University Press, Princeton, NJ, 1997. xviii+451 pp. Reprint of the 1970 original, Princeton Paperbacks.

[RS14] T. Rajala and K.-T. Sturm. Non-branching geodesics and optimal maps in strong $CD(K, \infty)$-spaces. *Calc. Var. Partial Differ. Equ.*, 50(3–4):831–846, 2014.

[San15] F. Santambrogio. *Optimal transport for applied mathematicians*, volume 87 of *Progress in Nonlinear Differential Equations and their Applications*. Birkhäuser/Springer, Cham, 2015. xxvii+353 pp. Calculus of variations, PDEs, and modeling.

[Sim83] L. Simon. *Lectures on geometric measure theory*, volume 3 of *Proceedings of the Centre for Mathematical Analysis*. Australian National University, Centre for Mathematical Analysis, Canberra, 1983. vii+272 pp.

[Sud79] V. N. Sudakov. Geometric problems in the theory of infinite-dimensional probability distributions. *Proc. Steklov Inst. Math.*, (2):i–v, 1–178, 1979. Cover to cover translation of Trudy Mat. Inst. Steklov 141 (1976).

[Tal76] G. Talenti. Elliptic equations and rearrangements. *Ann. Scuola Norm. Sup. Pisa Cl. Sci. (4)*, 3(4):697–718, 1976.

[TW01] N. S. Trudinger and X.-J. Wang. On the Monge mass transfer problem. *Calc. Var. Partial Differ. Equ.*, 13(1):19–31, 2001.

[Vil03] C. Villani. *Topics in optimal transportation*, volume 58 of *Graduate Studies in Mathematics*. American Mathematical Society, Providence, RI, 2003. xvi+370 pp.

[Vil09] C. Villani. *Optimal transport*, volume 338 of *Grundlehren der Mathematischen Wissenschaften [Fundamental Principles of Mathematical Sciences]*. Springer-Verlag, Berlin, 2009. xxii+973 pp. Old and new.

Index

Printed in the United States
by Baker & Taylor Publisher Services